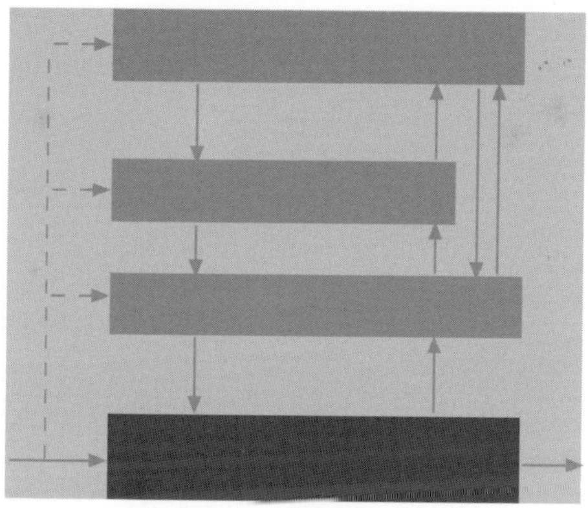

Iterative Algorithms
for Multilayer Optimizing Control

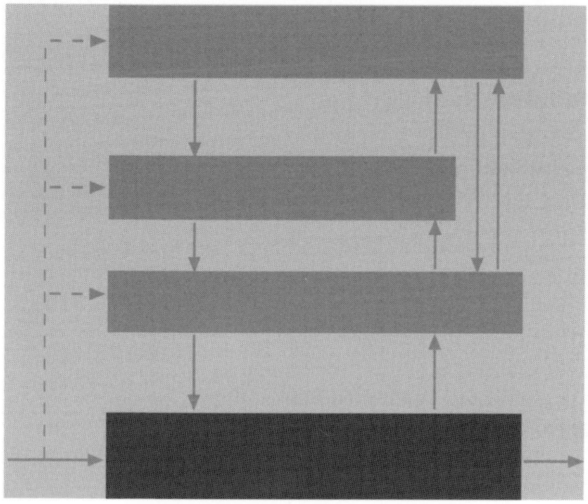

Iterative Algorithms for Multilayer Optimizing Control

Mietek A Brdys

The University of Birmingham, UK

Gdańsk University of Technology, Poland

Piotr Tatjewski

Warsaw University of Technology, Poland

ICP

Imperial College Press

Published by

Imperial College Press
57 Shelton Street
Covent Garden
London WC2H 9HE

Distributed by

World Scientific Publishing Co. Pte. Ltd.
5 Toh Tuck Link, Singapore 596224
USA office: 27 Warren Street, Suite 401-402, Hackensack, NJ 07601
UK office: 57 Shelton Street, Covent Garden, London WC2H 9HE

British Library Cataloguing-in-Publication Data
A catalogue record for this book is available from the British Library.

ITERATIVE ALGORITHMS FOR MULTILAYER OPTIMIZING CONTROL

ISBN 1-86094-514-7

Printed by FuIsland Offset Printing (S) Pte Ltd, Singapore

To our wives

Preface

In spite of limited knowledge about the process dynamics and under varying and unmeasured disturbance inputs affecting the process operational performance, design of a control system capable of maintaining an optimized operation of a complex dynamic process has attracted considerable attention of the industry and research community over last couple of decades. Although excellent progress has been made the problem is still challenging and of increasing importance. Design of a controller as a single entity with a homogeneous centralized decision mechanism is possible only in situations in which the process dynamics, uncertainty and the control objectives, are relatively simple, e.g., in the case of a stabilization of angular velocity of a d.c. motor. In many cases of complex multivariable processes, with seriously uncertain and nonlinear dynamics and under varying in time disturbance inputs, it is too difficult or simply impossible. Moreover, a general objective of the industrial complex process control in a market economy is to maximize economical efficiency over a long time horizon while maintaining its sustainable operation — and it is usually not easy and straightforward to translate these goals into trajectories or even steady-state values of the process controlled inputs. A well-established way to cope with a design of a controller in such complex situations is to apply a *hierarchical control structure*. The hierarchical structuring of the control system allows efficient handling of the process dynamics complexity and the uncertainty. Such an approach to the control system design is well-established in industrial practice and discussed in many papers and monographs. The main idea is to decompose the original control task into a sequence of different, simpler and hierarchically structured sub-tasks, handled by dedicated control layers. The reason is to make the control system design easier by creating technically well-defined and simpler subtasks requiring different theoretical and technical tools, as well as to improve the control performance and reliability.

In Chapter 1 and Chapter 2, the book comprehensively covers a multilayer control architecture for systems with a steady-state but varying in time optimal operation mode. The optimizing control algorithms are presented in Chapter 3 and Chapter 4. The modified model-based optimization and the model parameter estimation is integrated in the algorithms in such a manner that a true process optimum is reached after a number of control applications to the process have been made. The corresponding output responses are measured after the transients have died and the process has reached its new steady-state. The measurements are utilized in order to suitably modify the model based optimization problem and to update the model parameters. The resulting control technology is called *Integrated System Optimization and Parameter Estimation (ISOPE).* It is extremely important that uncomplicated models of the process are needed but simple point-parametric models are sufficient. The resulting optimizing control algorithms are of the *iterative* type. The controlled process-based iterations are needed in order to cope with the uncertainty. A case when the constraints on the process output are active is separately considered in Chapter 5. The control structure is further developed as well as the optimizing control algorithms in order to efficiently handle the output constraints.

The algorithms are tested by simulation and applications to several case study examples are also presented. These are: ethylene distillation column, vaporizer pilot scale plant and styrene distillation line consisting of three columns.

The multilayer control structure remains the same for dynamic processes operating in a batch mode. However, the algorithms then become more complex, they are covered in detail in Chapter 6. An industrial furnace case study illustrates an overall process of control design and the application results.

Control structures and iterative optimizing control algorithms for complex controlled processes with a spatial structure are considered in Chapter 7. The *spatial decomposition* is applied, in addition to the *functional decomposition,* in order to support the latter in coping with the complexity. It is based on a division of the control task (or a functionally partial task — within one layer of the described multilayer structure) into local subtasks of the same functional kind but related to individual spatially isolated parts of the entire complex control process — subtasks of smaller dimensionality, smaller amount of processed information. This procedure leads to so-called *multilevel structures* or *multilayer-multilevel structures*.

The most demanding is the dynamic continuous optimizing control, applicable to cases of systems operating in truly dynamic modes that cannot be split into the consecutive batches. A case study of an Integrated Wastewater System is given in Chapter 1 in order to illustrate the multilayer control concept applied to synthesize the dynamic optimizing control. In this case study most recent results on application of optimizing control to Integrated Wastewater Systems are presented, with temporal decomposition and repetitive Robust Model Predictive Control (RMPC) as key control technologies used, within the multilayer structure. The *temporal decomposition* is applied when, as in this case study, the task of control generation is formulated as a dynamic optimization problem and the controlled dynamic system (and/or disturbances) is multi-scale, i.e., there is significant difference between the rate of change of fast and slow state variables (and/or disturbances) of the system.

Strengthening the RMPC information feedback mechanism by combining it with the ISOPE feedback mechanism within the hierarchical structure that is presented in the book is seen as a difficult but very promising future research direction for further development of optimizing control systems.

The book presents basic structures, concepts and algorithms in the area of multilayer optimizing control of industrial systems, as well as the results of research that was carried out over the last two decades. It is addressed to several categories of readers, first of all to the research community and postgraduate students, but also to practitioners. In order to make the book more useful, longer or more difficult proofs of theorems were moved to the Appendices.

Acknowledgements

We are grateful to many people who have made contributions in various ways to this book. Our Ph.D. and D.Sc. degree research has been carried out in a group led by Professor W. Findeisen and located in the Institute of Control and Computation Engineering (formerly the Institute of Automatic Control), Warsaw University of Technology, where the second author is still employed. We are very grateful to Professor Findeisen and colleagues from the group, especially to Profesor K. Malinowski and Dr. A. Woźniak, and from the Institute for the fruitful scientific atmosphere supporting the research.

The first author spent a year in 1983 – 1984 in the Control Engineering Center at City University, London led by Professor P. D. Roberts, as Senior Visiting Research Fellow of SRC, UK. The second author visited Professor Robert's group at City University during 1986 and he has also been supported by SRC, UK. Our interests in ISOPE technology, a backbone control technology considered in this book, have been developed during those visits. Also, the vaporizer and furnace applications have been performed in the Control Engineering Center research laboratory. We have benefited immensely from these visits and we are very grateful to Professor Roberts.

We wish to thank our numerous Ph.D. students who have enthusiastically carried out their research and have successfully completed it achieving excellent results.

Special thanks go to Mrs. Mariola Lubińska and Mr. Tomasz Rutkowski from the Intelligent Decision Support and Control Group at Gdansk University of Technology, for help in preparing many figures for Chapter 1, Chapter 6 and Chapter 7. Discussions with all members of the Group were very useful in preparing the presentation of application of the hierarchical dynamic optimizing control to Integrated Wastewater System at Kartuzy, northern Poland.

Finally, we would like to thank our wives for their love, support and encouragement. Without their help and sacrifices, we could not have completed this book.

Mietek A. Brdys — *Birmingham, Gdańsk*

Piotr Tatjewski — *Warszawa*

Notation and Acronyms

Notation

$\mathbb{R}^n$	n-dimensional space of real numbers
n_x	dimension of the vector $x \in \mathbb{R}^n$, $n_x = \dim x$
x^T, A^T	transpose of the vector x, of the matrix A
$\|x\|_R^2$	$x^T R x$
$\text{diag}\{a_1, ..., a_n\}$	diagonal matrix with $a_1, ..., a_n$ on the diagonal
(x, y)	pair of elements x and y, also $[x^T y^T]^T$ if x, y are vectors
$g(\cdot)$, $f(\cdot)$, *etc.*	scalar or vector functions
$g^{'}(x)$	derivative of g at x, if $g : \mathbb{R}^n \to \mathbb{R}$ then $g^{'}(x) = [\frac{\partial g(x)}{\partial x_1} \cdots \frac{\partial g(x)}{\partial x_n}]$; if $g : \mathbb{R}^n \to \mathbb{R}^m$ then $g^{'}(x) = [\frac{\partial g_i(x)}{\partial x_j}] = \begin{bmatrix} \frac{\partial g_1(x)}{\partial x_1} & \cdots & \frac{\partial g_1(x)}{\partial x_n} \\ \vdots & \ddots & \vdots \\ \frac{\partial g_m(x)}{\partial x_1} & \cdots & \frac{\partial g_m(x)}{\partial x_n} \end{bmatrix}$
$\nabla g(x) = (g^{'}(x))^T$	gradient
$g_x^{'}(x, y)$	partial derivative of g with respect to x, at (x, y)
$\dot{x}(t)$	time derivative of $x(t)$
$\overline{1:N} = 1, ..., N$	consecutive natural numbers from 1 to N
$\min_x f(x)$	minimal value of $f(x)$ with respect to x
$\min_{x \in X} f(x)$	minimal value of $f(x)$ with respect to $x \in X$
$\text{minimize}_x f(x)$	task of minimizing $f(x)$ with respect to x

Acronyms

General

DCS	Distributed Control System
IWWS	Integrated Wastewater System
LMI	Linear Matrix Inequalities
MPC	Model Predictive Control
PID	Proportional-Integral-Derivative
RMPC	Robust MPC
SCADA	Supervisory Control and Data Acquisition
SQP	Sequential Quadratic Programming
WWTP	Wastewater Treatment Plant

Optimizing Control

AIBMF	Augmented IBMF
AISOPE	Augmented ISOPE
AMOP	Augmented MOP
CAMOP	Conditioned AMOP
CFC	Constraint Follow-up Controller
CMMOP	Conditioned MMOP
DISOPE	Dynamic ISOPE
DMMOP	Dynamic MMOP
DOCP	Dynamic OCP
DPEP	Dynamic PEP
IBM	Interaction Balance Method
IBMF	IBM with Feedback
ISOPE	Integrated System Optimization and Parameter Estimation
ISOPEB	ISOPE Basic
ISOPED	ISOPE Dual
ISOPEY	ISOPE with constraints on y (outputs)
ISOPEDY	ISOPED with constraints on y (outputs)
ITS	Iterative Two-Step
LOC	Local Optimizing Controller
MMOP	Modified MOP
MOP	Model Optimization Problem
OCP	Optimizing Control Problem
PEP	Parameter Estimation Problem
RMMOP	Relaxed MMOP
SOCP	System OCP

Contents

Chapter 1
Multilayer Control

1.1 Control System

Controlling a process is usually defined as influencing it in such a way as to force it to operate fulfilling certain assumed requirements, see e.g., (Findeisen et al., 1980, 1994). This definition applies to all situations and processes which undergo control. Control problems are considered and investigated in a vast body of literature as they arise in numerous applications and form essential components of industrial, transportation, communication, environmental and economic systems. In this book we shall be dealing with control problems arising in industrial systems like chemical reactors, distillation columns or wastewater treatment plants.

A process which undergoes control (a controlled process) is always a certain isolated part of the environment in which it exists, undergoing influences of the environment, uncontrolled or controlled by certain control unit (algorithms executed by an automatic controller, human operator, *etc.*). For example, a control computer or a human operator tries to enforce the desired parameters of technological processes in a chemical reactor, in a distillation column or in a wastewater treatment plant by enforcing appropriate values of certain process variables which influence its behaviour (levels, flows, temperatures, pressures, *etc.*) — in the presence of changes in raw materials, utilities and ambient conditions which disturb the desired course of the processes. A significant general feature is the fact that a process is not isolated from its environment, it undergoes controlled and uncontrolled external influences — we talk about *controlled and uncontrolled input variables* (*inputs*) of a process. Uncontrolled inputs are also called *free* or *disturbing* inputs (*disturbances*). Evaluation of the state of a controlled industrial process is done on the basis of measurements or estimations of values of appropriate variables characterizing the process behaviour. These variables are called the *process outputs*. In case of controlling a chemical reactor or a wastewater treatment plant examples of the output variables are parameters of a reacting mixture, such as temperature or composition, as well as parameters characterizing the state of technological apparatus (fills, levels, *etc.*). Knowing objectives of control and analyzing values of process outputs and those uncontrolled inputs which are known (measured, estimated), the control unit makes decisions about whether to maintain or change appropriate values of the controlled inputs. General structure of a control system is presented in Fig. 1.1.

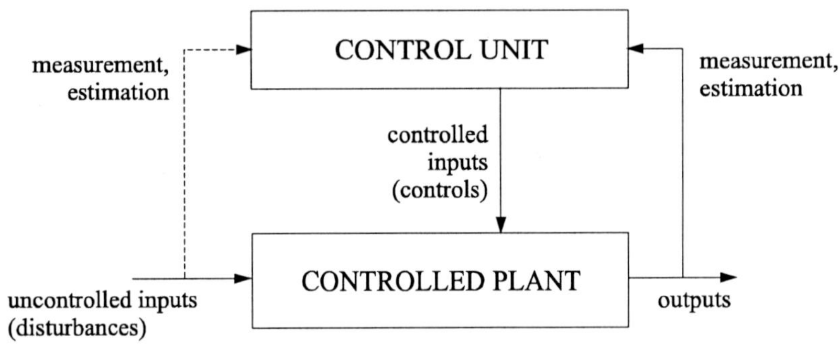

Fig. 1.1 The control system.

1.2 Hierarchical Control Structures

Design of a control unit (a controller) as a single entity with a homogeneous centralized decision mechanism is possible only in situations in which the process and the control objective are relatively simple, e.g., for the case of stabilization of angular velocity of an engine, *etc.* In more complex situations a straightforward task of designing and implementing a single centralized control unit is too difficult and in many cases of complex multivariable processes just impossible. Moreover, a general objective of the industrial complex process control in a market economy is to maximize economical efficiency over a long time horizon — and it is usually not easy and straightforward to translate this goal into trajectories or even steady-state values of the process controlled inputs. A well-established way to cope with a design of a controller in such complex situations is to apply a *hierarchical control structure*. The idea is well established in industrial practice and discussed in many papers and monographs, see e.g., (Lefkovitz, 1966; Findeisen *et al.*, 1980, 1994; Findeisen, 1974, 1997; Tatjewski, 2002). The main idea is to decompose the original control task into a sequence of different, simpler and hierarchically structured subtasks, handled by dedicated control layers. The reason is to make the control system design easier by creating technically well-defined and simpler subtasks requiring different theoretical and technical tools, as well as to improve the control performance and reliability.

There are two basic methods of decomposition of the overall control objective:

- *functional decomposition*
- *spatial decomposition*

The first one applies to a process treated as a whole, and is based on assigning a set of functionally different partial control objectives — in a structure of vertical, hierarchical dependence, called the *multilayer structure*. The decision unit connected with each layer makes decisions concerning the controlled process, but each of them makes decisions of a different kind. On the other hand, the spatial decomposition is connected with a spatial structure of a complex controlled process. It is based on a division of the control task (or a functionally partial task, e.g., within one layer of the described multilayer structure) into local subtasks of the same functional kind but related to individual spatially isolated parts of the entire complex control process — subtasks of smaller dimensionality, smaller amount of processed information. This procedure leads to so-called *multilevel structures*, or multilayer-multilevel structures, see e.g., (Mesarovic *et al.*, 1970; Findeisen *et al.*, 1980, Findeisen, 1974, 1997). In this book we shall be mainly interested in *control algorithms for multilayer structures of industrial processes*, although in Chapter 7 multilayer-multilevel control structures will be considered.

There is also another kind of decomposition — the *temporal decomposition*, see e.g., (Findeisen *et al.*, 1980; Findeisen, 1997). It is applied to cases when the task of control generation is formulated as a dynamic optimization problem and the controlled dynamic system (and/or disturbances) is multi-scale, i.e., there is significant difference between the rate of change of fast and slow state variables (and/or disturbances) of the system. This leads to a multilayer concept as well. But now at each layer functionally the same task is performed: dynamic optimization — only with longer horizons and less detailed models at higher layers.

In the following subsections the multilayer control structures, functional and temporal, will be presented. The multilevel structure will be presented and discussed in more detail at the beginning of Chapter 7, since only in this chapter structures of this kind are considered.

1.2.1 *Functional multilayer structure*

Let us consider the control task for an industrial plant. To maximize the economical objective of the process control the following partial objectives should be met:

(1) To maintain the plant in a safe operation mode, i.e., to constrain to an acceptable level the probability of undesirable, uncontrollable process behaviour (caused by faults, *etc.*),
(2) To meet demands on product quality and economical usage of technological apparatuses, i.e., keeping certain process input and output values within prescribed limits,
(3) To maximize the current production profit.

It is easy to notice that the first two mentioned partial objectives are also closely related to economics of the process control. Undesirable, uncontrollable process behaviour usually leads to serious losses connected with production breaks, losses which may be much larger than those caused by not optimal, but safe production running. On the other hand, failures with meeting demands on product quality parameters leads usually to decreased profits connected with the necessity to lower the product price or even to dispose the product as a waste.

Let us consider the three partial objectives of controlling an industrial plant, mentioned at the end of the previous section. The order of enlisting this partial objectives is not incidental. Safety of the control system is most important, next in importance is to care about the quality of products. Only after ensuring the realization of these two aims, can there be room for on-line economic optimization of values of variables defining the state of processes in the plant. It is in this order that the layers of the basic functional *multilayer control structure* presented in Fig. 1.2 are located, on top of the controlled process located at the bottom (Tatjewski, 2002).

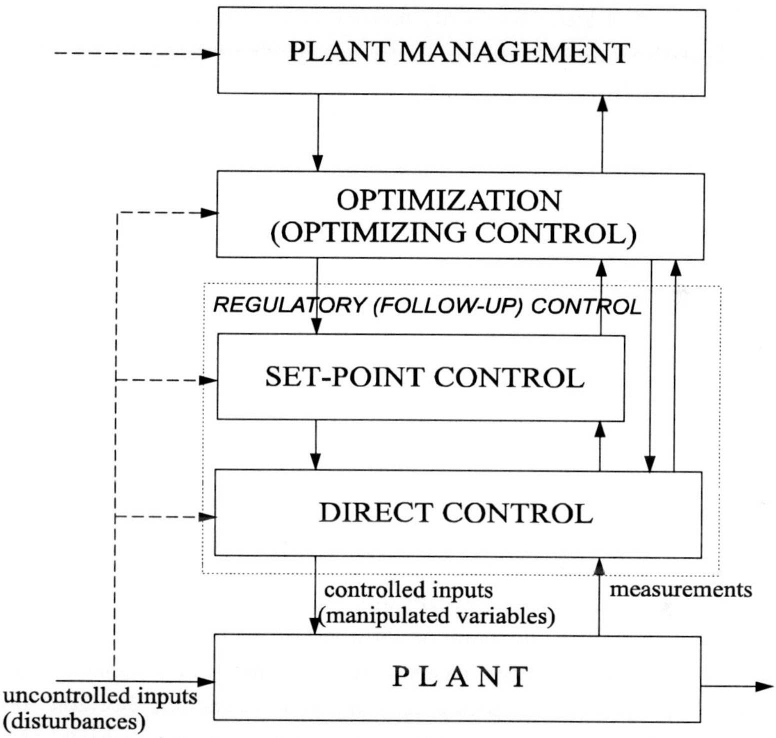

Fig. 1.2 Functional multilayer control structure.

The two lower control layers, the direct control layer and the set-point control layer, are both the *regulatory* control layers (with follow-up controllers, feedback or feedback-feedforward ones) — that is why they are additionally marked off with a thin dotted line in the figure.

The *direct control layer* is responsible for safety of dynamic processes in the plant. Only this layer has direct access to the plant, it can directly influence the controlled input variables (usually called manipulated variables), marked in the Figure with u. Technical realization of the tasks of the layer is nowadays ensured, for complex industrial processes, by distributed control systems (DCS). Algorithms of direct control should be robust and relatively easy, that is why classic PID algorithms are still dominant. However, the possibilities of DCS systems are much wider. Thus, in places where the classic PID control leads to unsatisfactory control quality, more advanced control algorithms can be employed, especially appropriate modifications of the PID algorithm and, recently, simple realization of predictive controllers. In the literature, especially in publications issued by companies delivering control equipment and software, one can find descriptions of *basic regulatory control*, also called *direct (regulatory) control*, as opposed to *advanced control*, located at higher layer. However, it should be strongly emphasized that the generic feature distinguishing all direct (basic) controllers is the direct access to the controlled process (process manipulated variables are outputs of the direct controllers) and high frequency of intervention (small sampling period), not the kind of algorithm (advanced or not) employed. Therefore, we shall be describing controllers (control algorithms) acting as higher layer controllers, with outputs being *set-point values* for direct controllers located below, as *set-point controllers* constituting the *set-point control layer*, see also (Tatjewski, 2002).

The objective of the *set-point (regulatory) control layer* is to control slower process variables which decide mainly on the *product quality* parameters, such as concentrations in reactors or distillation columns. Therefore, it is required that set-point control algorithms should keep high quality of operation, most frequently in cases of multivariable, non-linear processes. The most typical, modern solutions applied are receding horizon model-based predictive control algorithms, commonly described as MPC (Model Predictive Control) algorithms. Most popular (at least in petrochemical industries) are applications based on the DMC (Dynamic Matrix Control) algorithm, developed in the 1970s. It should also be mentioned that the set-point control layer can not always occur. It is not distinguished in cases when there is no need for the set-point control in the sense described above. Moreover, this layer usually does not fully separate the direct control layer from the optimization layer — some of the set-points for basic controllers can be assigned and directly transmitted from the optimization layer, as can be seen in Fig. 1.2. One should also not be too rigorous when it is reasonable, e.g., usually primary SISO controllers of standard cascade control loops are included into direct control layer, although these controllers also act as set-point controllers for secondary (inner loop) controllers (but sampling periods of both are usually the same).

The *optimization layer,* or *optimizing control layer* is next, situated directly above the regulatory control layers, see Fig. 1.2. The objective of its operation is to calculate the process *optimal operating point* or *optimal operating trajectory*, i.e., optimal set-point values for controllers of its directly subordinate layers: set-point control layer and direct control layer, see Fig. 1.2. These values result generally from optimization of certain economic objective function which defines usually profit or running costs of the process operation, under constraints on process variables and a process model describing relations between process inputs and outputs. We shall clearly distinguish between set-point *optimization* and set-point *optimizing control* in this book. The difference is as explained in what follows.

The technological process operation is always *under uncertainty.* The source of the uncertainty is the behaviour of disturbances (uncontrolled process inputs), like variable properties of raw materials and utilities, changes in ambient conditions. Usually, part of these variables is measured or estimated, certain are not measured or are even not measurable. Optimal values of the set-points are dependent on disturbance values and vary when these values vary. The optimal operating point should be calculated for current values of disturbances, and recalculated after significant changes in these values. Certainly, when calculating optimal values of the set-points a *process model* must be used, describing dependence of process outputs on controls and disturbances. This is additional source of uncertainty, since modeling and identification of complex technological processes rarely leads to accurate models, due to complex nature od input-output dependencies and inevitable simplifications to obtain computationally tractable models, especially those for on-line applications.

Therefore, we define as the *set-point optimization* (at the *optimization layer*) a single process model optimization, to obtain *model-optimal set-points*, for current measurements or estimates of the disturbances taken into account in the model. Certainly, after sufficiently significant changes encountered in the disturbance values the optimization should be recalculated leading to corrections in the set-point values. Nevertheless, the described set-point optimization will result in set-points close to their true optimal values only when the model is sufficiently accurate and uncertainty in disturbance measurements or estimates is sufficiently low. Unfortunately, this happens rather very rarely in practical cases of complex industrial plant control. Therefore, a single optimization usually leads to solutions being only *suboptimal set-points* for the real process, with a degree of suboptimality dependent on the level of uncertainty.

On the other hand, one can try to improve the model-optimal set-points performing certain number of appropriately designed iterations (set-point changes) on the controlled process, exploiting additional measurement information gained in this way at each iteration. This is precisely what will be called the *steady-state optimizing control* (performed at the *optimizing control layer*), since the iterative process of set-point improvement uses feedback information at each iteration (from steady-state measurements) to cope with the uncertainty. Therefore, the iterative optimizing control algorithms can be treated as *open-loop-with-feedback* control algorithms, with set-points as *optimizing controls*. Iterative optimizing control algorithms are the main subject of this book, and they will be presented and analyzed in subsequent chapters.

Obviously, the multilayer control structure with set-point optimization or optimizing control is reasonable only for processes with disturbances slowly varying (with respect to controlled process dynamics) or rarely changing their values (but possibly in an abrupt way, like e.g., when switching to different delivery source of raw materials resulting in step changes in raw materials parameters), see e.g., (Findeisen *et al.*, 1978, 1980; Findeisen, 1997; Tatjewski, 2002). Only in this situation there is reasonable to keep constant optimal (or suboptimal) values of set-points over longer time intervals, there is time to perform iterations of optimizing control algorithms. Certainly, we are talking here only about slowly changing disturbances (uncontrolled inputs) which affect the process economical performance, not fast disturbances like those causing measurement noise or, e.g., parameter changes of heating media in the case of temperature stabilization. These much faster disturbances can be well filtered out by feedback controllers at the regulatory control layers.

In classic works devoted to multilayer control of industrial processes and complex systems, see e.g., (Findeisen *et al.*, 1980), there were three layers distinguished following functional decomposition of the control task: regulatory control, optimization and adaptation as the highest layer (with plant management often not mentioned). In what has been presented in this chapter, see Fig. 1.2, the task of adaptation was not introduced, especially as a task forming a separate layer direct at the top of the optimization layer. The reason is that development of microelectronics and computer technology has led to radical changes in the equipment of automatic control. Along with appearance of microprocessor controllers there came a new age in the range of regulatory control techniques and general industrial processes control techniques. In each of the functional and equipment layers of the control system it is now possible to implement tasks of diagnostics of signals and of controller operation, with automatic shift to redundant controllers or algorithms when needed. Therefore, tasks of supervision and diagnostics dispersed significantly. Along with the still existing central tasks of supervision, diagnostics and adaptation of certain structures, especially for the optimization layer, we also have similar tasks which are realized locally as local algorithms more or less integrated with local control algorithms. Such a control structure ensures greater speed of action and increased reliability, according to a general rule: make information-decision feedback loops as short as possible. A multilayer control structure considering the tasks of supervision, diagnostics and adaptation is presented in Fig. 1.3 (compare with Fig. 1.2), see also (Tatjewski, 2002).

1.2.2 *Temporal multilayer structure*

The *temporal decomposition* is applied to cases when the task of control generation is formulated as a dynamic optimization problem and the controlled dynamic system (and/or disturbances) is multi-scale, i.e., there is significant difference between the rate of change of fast and slow state variables (and/or disturbances) of the system. This leads to a multilayer concept as well, based on different time horizons at different layers. At each layer functionally the same task is performed: dynamic optimization. However, the higher the layer the longer the optimization horizon and the less detailed (more aggregated) the model of the controlled system and disturbances — only slower modes considered as dynamic. Main information passed from a higher layer to a direct subordinate one is the value of the system state — calculated at intermediate time instant at the higher layer and passed as a target state (for the end of the shorter horizon) for calculations at the lower layer. This idea is presented in Fig. 1.4.

Multilayer structures of the kind presented in Fig. 1.4 are often emerging as a reasonable realization of one functional task of the functional multilayer structure of an industrial system, i.e., as realization of the dynamic

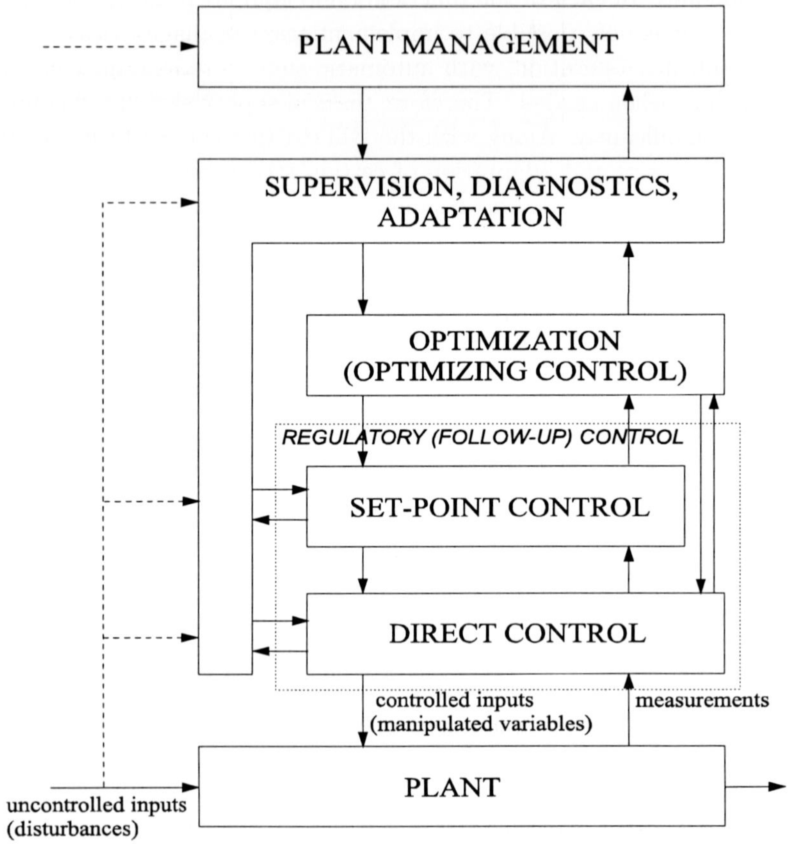

Fig. 1.3 Functional multilayer control structure with the tasks of supervision, diagnostics and adaptation.

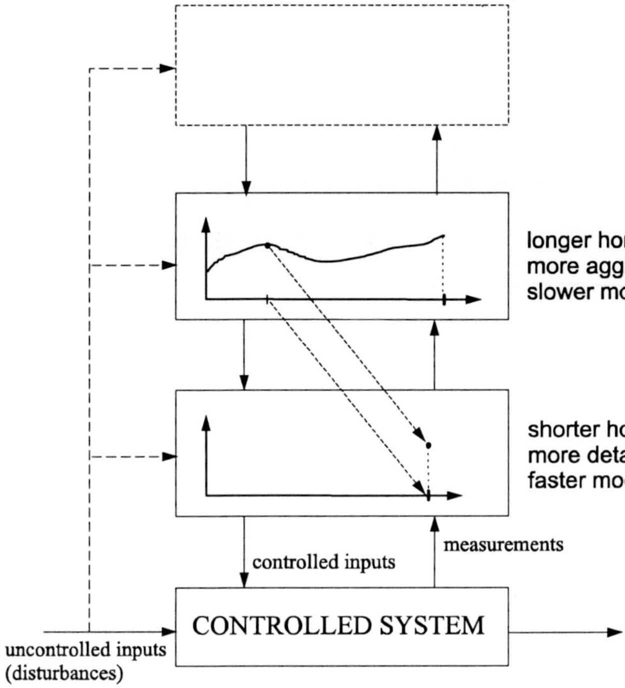

Fig. 1.4 Temporal multilayer control structure.

optimizing control task or dynamic plant management task — in cases of mentioned above complex multi-scale models of the controlled system. In this situation the layers of the temporal multilayer hierarchy can be called *sub-layers* of the overall functional multilayer structure.

1.3 Decomposition and Modeling in Functional Multilayer Structure

Let us now consider in more detail the multilayer structure resulting from the functional decomposition of the overall economic control objective, as presented in the Fig. 1.5, comp. Fig. 1.3, see also (Duda *et al.*, 1995). The novel feature of this structure is a decomposition of dynamic processes occurring in the plant into *fast* processes influenced by the input variables u and fast changing disturbances z and *slow* processes influences by in-

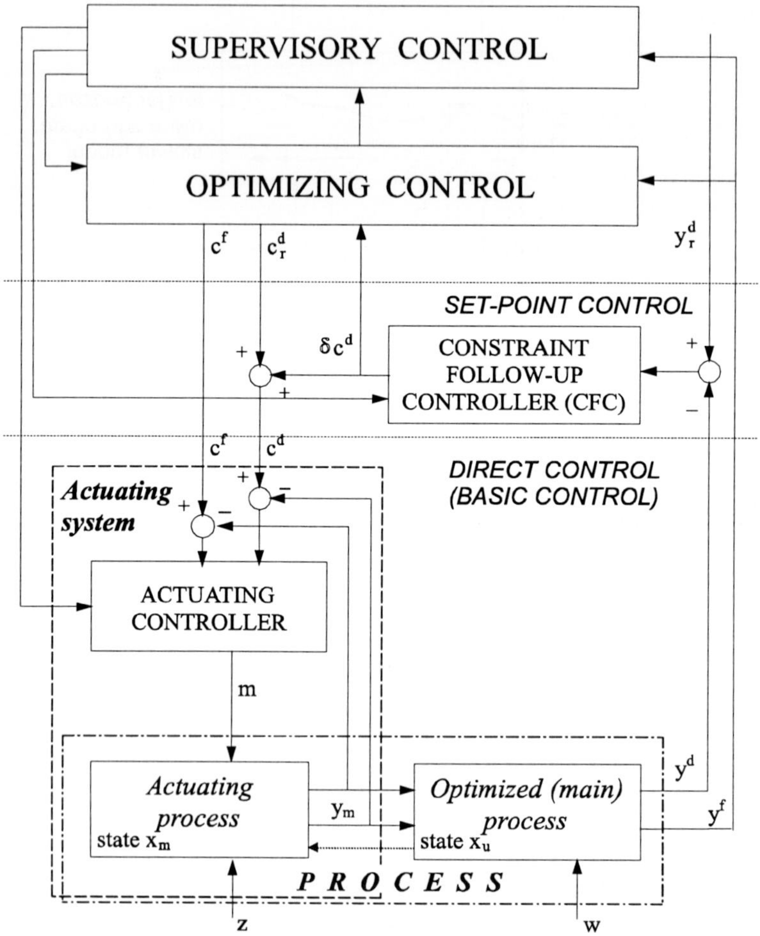

Fig. 1.5 Multilayer control structure with decomposition of the plant.

termediate plant outputs y_m and slow varying disturbances w. Outputs $y = (y^f, y^d)$ from the "slow" part of the plant are variables which are significant for the task of optimization (optimizing control) and supervision.

The objective of the set-point control presented in Fig. 1.5 is to enforce the equality constraint

$$y^d(t) = y_r^d \tag{1.1}$$

i.e., to enforce that certain components of process output vector y, marked as sub-vector y^d, are stabilized at prescribed values y_r^d. This corresponds to a regulatory control of important, in normal operating conditions always active, constraints on certain output variables. Such a situation is encountered in practical applications, e.g., when it is desired to maintain concentration of a pollution in a stream of a product on a fixed value which is appropriately (safely) close to maximum admissible value. Therefore, the set-point controller in Fig. 1.5 has been defined as *constraint follow-up controller (CFC)*. The remaining, free (uncontrolled) components of the output vector y has been marked as a sub-vector y^f.

Let us have a deeper look at the structure presented in Fig. 2.8. If direct control systems are operating properly, then, apart from periods directly following fast (step) changes of set-points c or disturbances z, we can assume

$$y_m(t) = c(t). \tag{1.2}$$

It can then be assumed that, from the point of view of set-point controllers and optimizing control algorithms, dynamically slow processes in the plant are only visible. An input-output relation defined by these processes can be described by an operator which will be denoted by F_c,

$$y(t) = F_c(c(t), w(t)). \tag{1.3}$$

Therefore, for set-point controllers and optimization algorithms, fast processes in the plant together with the direct control layer can be treated as an *actuating system* — which enforces set-point values $c(t)$ of controlled variables $y_c(t)$, that is enforces the equality $y_m(t) = c(t)$. The term *actuating* system has been introduced analogically to an *actuating* element which is, for example, a valve with a positioner (Duda *et al.*, 1995). That is why fast processes in Fig. 1.5 are named *actuating processes*. On the other hand, slow processes are called *optimized processes*, because the plant behaviour characterized by these processes is seen by higher layers, especially the optimizing control layer.

Analytical shape of the process model operator (1.3) is rarely available. It results from solving the following implicit model, which is typically assumed for continuous systems with lumped parameters

$$\begin{aligned} \frac{dx_c(t)}{dt} &= f_c(x_c(t), c(t), w(t)) \\ y(t) &= g_c(x_c(t), c(t)) \end{aligned} \tag{1.4}$$

where x_c is a the state vector of slow processes, see Fig. 1.5, and $y_m(t) = c(t)$ was assumed eliminating variables corresponding to the controlled outputs $y_m(t)$, i.e., assuming ideal operating of the actuating system.

A description of the entire dynamic plant can be assumed, analogically, in the following general form

$$\begin{aligned} \frac{dx_u(t)}{dt} &= f_1(x_u(t), x_c(t), u(t), z(t), w(t)) \\ \frac{dx_c(t)}{dt} &= f_2(x_u(t), x_c(t), u(t), w(t)) \\ y(t) &= g(x_u(t), x_c(t), u(t)) \end{aligned} \tag{1.5}$$

where the vector $x(t)$ was written in a divided form corresponding to fast and slow states, $x(t) = (x_u(t), x_c(t))$, and, consequently, no direct influence of fast varying disturbances $z(t)$ on the sub-vector of slow states $x_c(t)$ was assumed. Decomposition, namely a division of the whole state vector x into sub-vectors of fast and slow states, x_u and x_c, is in each case an individual question resulting from process characteristics and requirements concerning the controlled variables. Models (1.4) and (1.5) should of course be completed by a set of appropriate initial conditions, essential during formal analytical considerations or any numerical calculations.

Assuming equation defining controlled variables in the form, see (Findeisen *et al.*, 1980; Findeisen, 1974, 1997)

$$y_m(t) = h(x(t), u(t)) \tag{1.6}$$

one can consider relations between descriptions (1.5) and (1.4). Assuming equality (1.2) the following results

$$\begin{aligned} \frac{dx_c(t)}{dt} &= f_c(x_c(t), h(x_u(t), x_c(t), u(t)), w(t)) \\ &= f_2(x_u(t), x_c(t), u(t), w(t)) \end{aligned} \tag{1.7}$$

$$\begin{aligned} y(t) &= g_c(x_c(t), h(x_u(t), x_c(t), u(t))) \\ &= g(x_u(t), x_c(t), u(t)). \end{aligned} \tag{1.8}$$

1.4 Optimizing Control

Although Chapter 2 is entirely devoted to general aspects of optimizing control structures and algorithms, we shall briefly introduce now the three most general cases encountered, in order to be able to comment properly on an involved case study presented in the next section of this chapter.

Let us remind the reader that optimizing control algorithms are the open-loop-with-feedback algorithms, with set-points of lower layer controllers as decision variables — optimizing controls (Subsection 1.2.1). Three most general cases can be distinguished here:

(1) Steady-state optimizing control
(2) Dynamic batch optimizing control
(3) Dynamic continuous optimizing control

The simplest and relatively best understood is the first case of *steady-state optimizing control*. This case applies to situations when the uncontrolled (and mostly unknown) process inputs (disturbances) are slowly varying, with respect to controlled process dynamics, or rarely changing their values, but possibly even in an abrupt way (e.g., when switching to different delivery source of a raw material resulting in step changes in their parameters). In this situation there is reasonable to keep constant optimal (or suboptimal) values of set-points over longer time intervals, there is time to perform iterations of optimizing control algorithms. The optimizing control algorithms are the open-loop-with-feedback, a steady-state (static) optimization is performed in open loop (at each iteration), whereas feedback information consists of process outputs measurements in consecutive steady-states arising in the controlled plant after the transients have died, after each change in the set-points. The core of the optimizing control algorithms lies in appropriate use of this feedback information to modify the consecutive steady-state optimizations (performance function and/or constraints) in a way forcing convergence to improved set-points, in a sense of process economic performance. Most chapters of the book are devoted to steady-state optimizing control.

The *dynamic batch optimizing control* applies to cases of several consecutive batches with the same or very similar initial conditions (and uncontrolled inputs trajectories, when applicable), with significant uncertainty in the process model and, possibly, initial conditions. The optimizing control algorithm is, as always, open-loop-with-feedback. But this time the open loop optimization is a dynamic one over the one batch horizon. Optimizing controls are now controlled inputs trajectories (set-point trajectories) over the batch, and feedback information consists of measured process output trajectories over the batch. Thus, an optimizing control algorithm modifies iteratively, after every batch, the dynamic optimization problem forcing convergence to improved set-point trajectories. Dynamic batch optimizing control algorithms, although much more computationally involved than those for steady-state control, are also sufficiently mature to be successfully applied. Chapter 6 is devoted to dynamic batch optimizing control.

The *dynamic continuous optimizing control* should apply to cases of systems operating in truly dynamic modes that cannot be split into the consecutive batches. Unfortunately, it is far from being so mature as the optimizing control in two preceding cases, especially when considering iterative algorithms. The main difficulty is how to define an iterative feedback information coping with uncertainty here, if attempting to design iterative open-loop-with-feedback algorithms. In the opinion of the authors, the most promising approach seems to be to further develop the receding horizon (repetitive) model based predictive optimization to cope better with the uncertainty. However, this is still rather an open research area, except cases with uncertainty treated repetitively by output (state) measurement and correction only, as in predictive control algorithms. This is a *repetitive*, not *iterative* approach — therefore, the dynamic continuous optimizing control will not be tackled in the book. However, in the subsequent section a case study of a complex system will be given, where the dynamic continuous optimizing control is successfully applied. It illustrates well the multilayer control concepts presented in this chapter (including predictive optimizing control just mentioned), whereas case studies with iterative steady-state optimizing control and batch optimizing control will be given in other chapters of the book, illustrating theory and algorithms presented there.

1.5 Multilayer Control of Integrated Wastewater System: A Case Study

We shall now introduce a process (system) that is composed of the sewer and wastewater treatment plant (WWTP) to give an integrated wastewater system (IWWS). An operation of this technological plant aims at cleaning a water that is polluted due to the municipal and industrial usage. The wastewater is released to the sewer system and it is delivered to the WWTP to be cleaned before is released to the receiving waters such as lakes or rivers. In order to protect the environment the cleaned water must obey requirements with regard to concentrations of the pollutants. The pollutant concentration limits are set up by legislation and they have now been unified across the European Community. A biological treatment that is based on an activated sludge technology is able to treat the wastewater so that the desired requirements are met. Hence, it is not that difficult to keep the plant running and to meet the discharge quality standards. However, the associated operational cost could be too high. Hence, an optimizing control is wanted. There are variety of reasons which make the optimizing control synthesis and implementation a very challenging task.

In this section we shall present a hierarchical control structure that has been recently developed within the EU funded project Smart Control of Integrated Wastewater Systems (SMAC) and successfully applied to four pilot sites across Europe.

1.5.1 *Presentation of integrated wastewater system at Kartuzy in Poland*

Kartuzy is a town located in northern Poland, with a population of about thirty thousands. The WWTP there serves the town needs and also the needs of the surrounding villages. It was one of the pilot sites in the SMAC project. A structure of the IWWS in Kartuzy and its layout are given in Fig. 1.6 and Fig. 1.7, respectively. A view on the WWTP in Kartuzy is shown in Fig. 1.8 and Fig. 1.9.

The biological treatment is based on an activated sludge with the microorganisms involved in order to remove pollutants from the wastewater, and it is carried out in the reactor zones. A phosphorus removal is supported by a chemical precipitation that consists in dosing the iron sulphate (PIX) to the aerobic zone of the biological reactor in order to ensure a

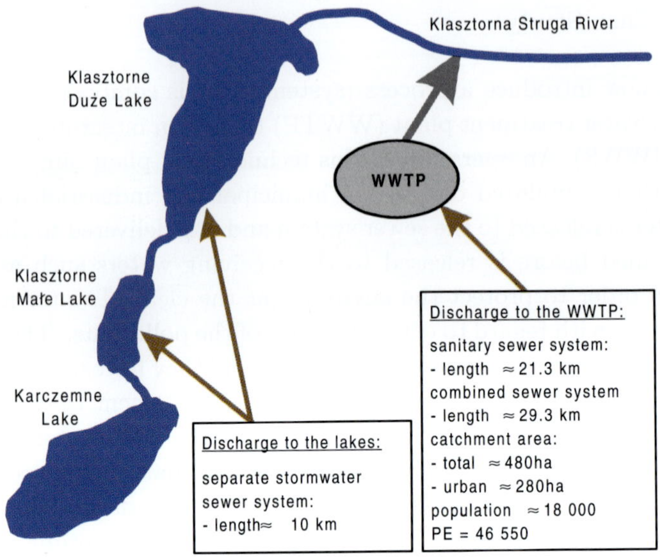

Fig. 1.6 Structure of the integrated wastewater system in Kartuzy (northern Poland).

desired level of the removal. This is illustrated in Fig. 1.10.

The system operational diagram is illustrated in Fig. 1.12 while the details are shown in Fig. 1.13 showing also the control handles available at the plant. The control handles or manipulated variables are:

$$\begin{array}{l} S_{oi},\ i \in \overline{1:4} \text{ - dissolved oxygen concentrations at aerobic aerated zones} \\ Q_{rec\_internal,i},\ i \in \overline{1:2} \text{ - internal recirculation flow rate} \\ Q_{rec\_external} \text{ - external recirculation flow rate} \\ Q_{ret\_in},\ Q_{ret\_out} \text{ - pumping in and out of the equalization tank} \\ Q_{sanit\_out} \text{ - pumping from the septic tank} \\ m_{PIX,\mathrm{inf}\,low},\ m_{PIX,aerobic} \text{ - metal salt (PIX) dosing} \\ Q_{excessive\_sl\,\mathrm{dge}} \text{ - wasted sludge flow rate} \end{array} \tag{1.9}$$

The first biological reactor zone is anaerobic (see Fig. 1.12 and Fig. 1.13), where the release of phosphorus should occur. The internal recirculation of mixed liquor originates from the anoxic zone. The second zone is anoxic where denitrification occurs. The returned activated sludge from the bottom of the clarifiers and the internal recirculation from the end of the

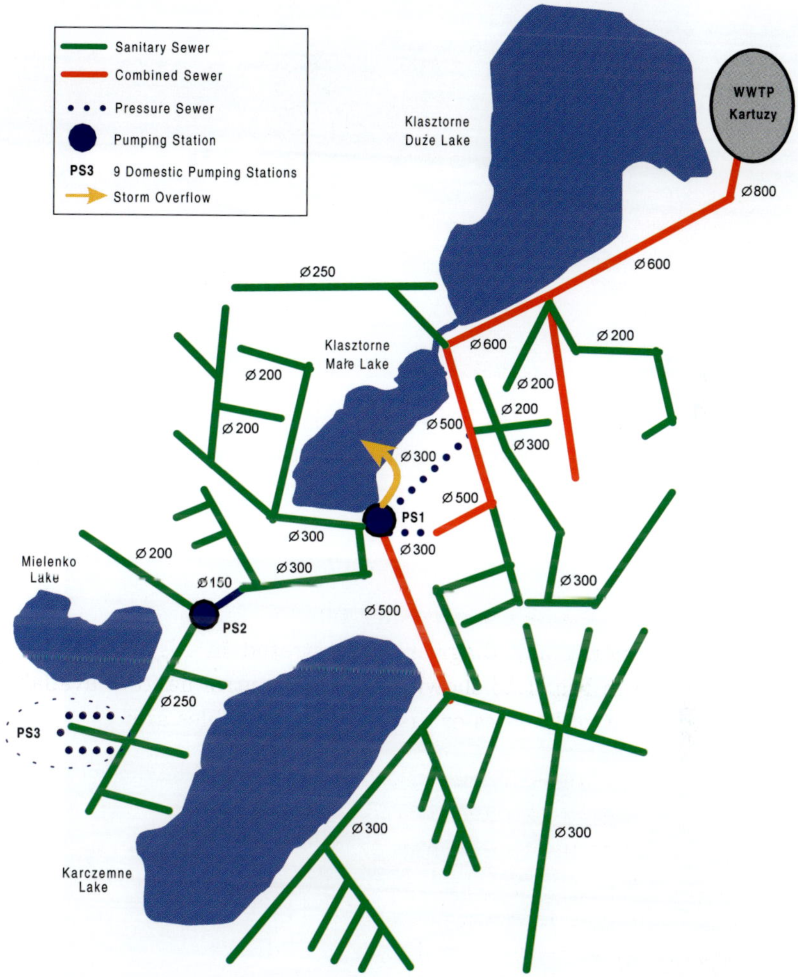

Fig. 1.7 Layout of the integrated wastewater system in Kartuzy (northern Poland).

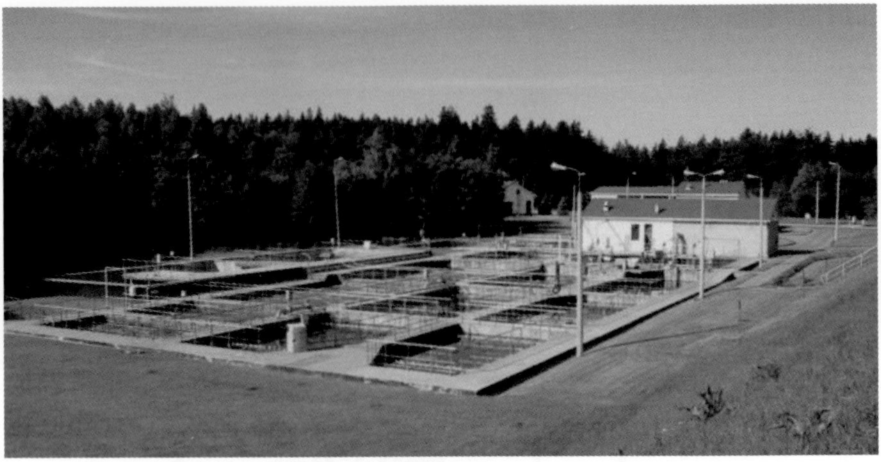

Fig. 1.8 The WWTP in Kartuzy (northern Poland) - the plant overview.

Fig. 1.9 The WWTP in Kartuzy (northern Poland) - the biological reactor.

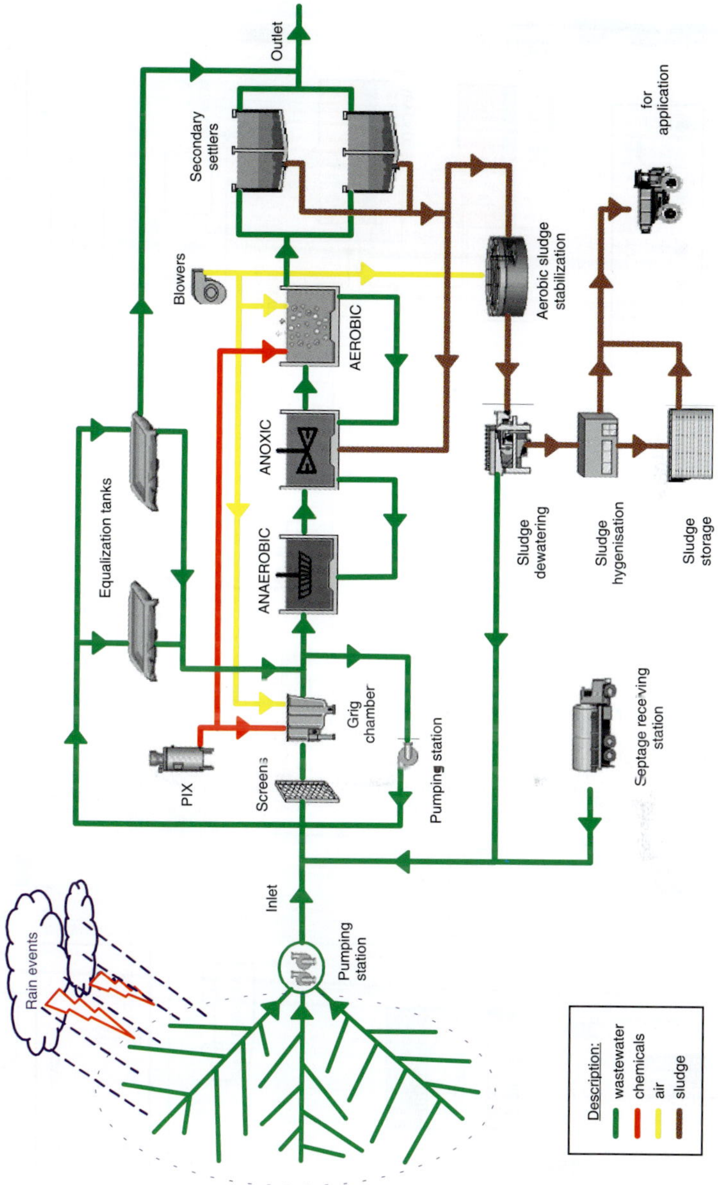

Fig. 1.10 Integrated wastewater system in Kartuzy (northern Poland) - schematic view.

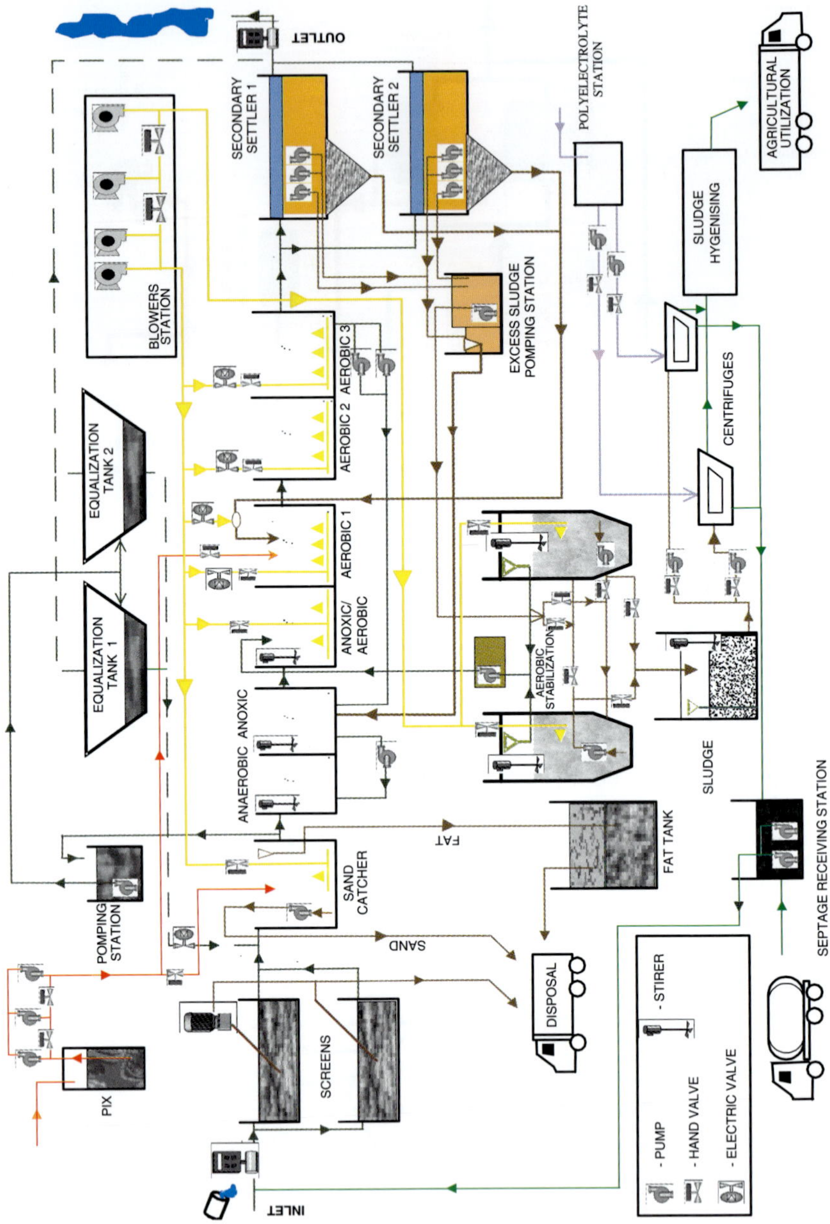

Fig. 1.11 Integrated wastewater system in Kartuzy (northern Poland) - detailed view.

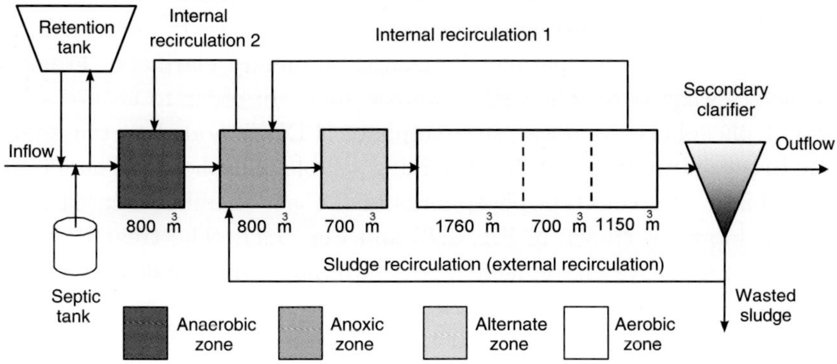

Fig. 1.12 Integrated wastewater system in Kartuzy (northern Poland) - the operational diagram.

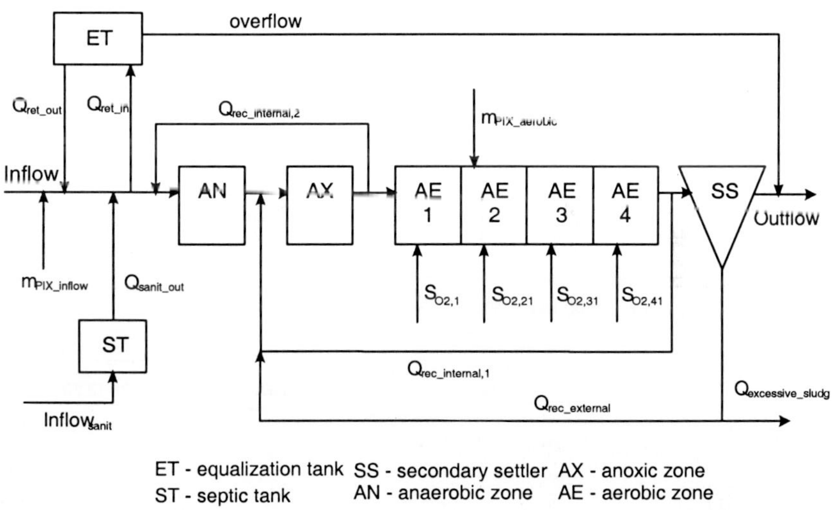

Fig. 1.13 Integrated wastewater system in Kartuzy (northern Poland) - the detailed operational diagram.

aerobic zone (containing nitrates) are directed to the anoxic zone. The last part of the reactor (aerobic) is aerated by a diffused aeration system. This zone is divided into three or four compartments of various intensity of aeration. The biologically treated wastewater and biomass (activated sludge) are separated into two parallel horizontal secondary clarifiers. From the clarifiers sewage is recirculated to anoxic zone. In order to ensure a high level of phosphorus removal, iron sulphate (PIX) is added to the aerobic zone to precipitate most of the remaining soluble phosphorus. There is also a possibility to precipitate phosphorus in the grit chamber located at the WWTP input, as shown in Fig. 1.10 and Fig. 1.11. The treated sewage, outflow from the system (effluent), goes into the receiver that flows into the Klasztorna Struga River (see Fig. 1.6).

Practice at WWTPs show that the flow rate and pollutant composition at WWTP inlet are almost never constant in time (Olsson and Newell, 1999). This is illustrated in Fig. 1.14.

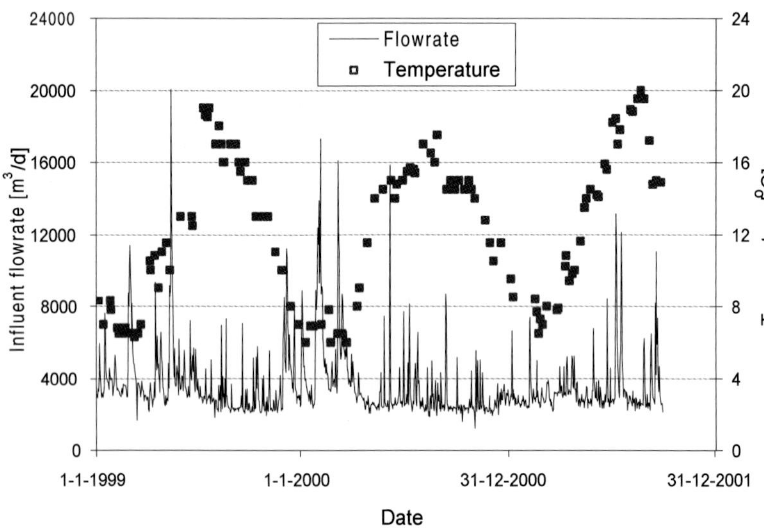

Fig. 1.14 Integrated wastewater system in Kartuzy (northern Poland) - variability of the inflow into to the WWTP.

1.5.2 *Control objectives*

In most of the WWTPs there are typically cyclic influent variations. In the case of unexpected events caused by: rainstorms, snowmelts, industrial discharges, toxic releases, *etc.*, the influent variations (flow rate and compositions) may differ several times from a normal average, which makes a control design the challenging task. In order to handle this, sewers system and retention tanks should be included into a control problem formulation. At high hydraulic loads, the control system will aim at maximizing utilization of the hydraulic capacity of the sewer (storage) and at maximizing the hydraulic WWTP capacity. The important thing is to deal with hydraulic control to avoid and/or minimize discharge and/or bypassing of untreated wastewater to the receiving waters. There is a hard constraint on the maximum flow rate throughout the WWTP because of a fear of biological sludge wash-out from the settler. The interaction between the sewer and treatment plant will firstly be hydraulic, since the hydraulic load produced by the sewer system will subsequently be passed through the treatment plant with its limited hydraulic capacity. The sewer system can be used to retain water upstream, to avoid treatment plant overloading. In that case there is also a hard constraint on the maximum storage volume in the sewer. Regarding the pollutant load there is a limited possibility of controlling the pollutant concentration at the WWTP inlet by mixing the sewer outflow in the equalization tanks and other mixing tanks located at the WWTP, especially the septic tank. Proper utilization of information from the sewer network gives the operator and/or the control system, an opportunity to prepare the plant for increased loading and/or toxic releases which enables more efficient control, and in particular, more efficient handling an overflow or/and a toxic pollution.

The system operation aims at meeting the following objectives:

- *to meet effluent discharge requirements,*
- *to keep biological sustainability of activated sludge in the treatment plant,*
- *to minimize operational costs (energy cost due to aeration and cost of chemicals),*
- *to minimize the load of untreated sewage discharged to the receiving waters.*

1.5.3 *Control structure*

A synthesis of the control structure for integrated wastewater system is very complicated, because of its specific following features:

- Multiple time scales in dynamics of the biological processes and tanks.
- Varying influent flow rate and its quality parameter concentrations.
- Heavily varying hydraulic and quality disturbances into the system due to different intensity of the rainfalls.
- Highly non-linear and mutually interacting dynamics of large dimension.
- Biological sustainability requires considering the system over a long time horizon and accurate predictions of disturbances over a such horizon are not available.
- Appearance of short-term and unpredictable heavy rainfall disturbance events having long-term consequences for the system's behaviour.
- Lack of models that are suitable for control design and implementation.
- There are a lot of state variables but only small parts of them is practically measurable.

In order to handle these difficulties a multilayer hierarchical control structure was recently proposed (Brdyś *et al.*, 2002; Grochowski, 2004; Grochowski *et al.*, 2004a). A hierarchical structuring of the control generation process enables us to properly and comprehensively utilize all available quantitative and qualitative information about plant structure and dynamics, its interactions with the environment and up-to-date operational experience. The control structure is illustrated in Fig. 1.15.

A *functional decomposition* of the control task generates the control *layers,* while *sub-layers* emerge within one of the functional layers from a *temporal decomposition* of the dynamics of the controlled processes (see subsections 1.2.1 and 1.2.2). Certain main parts in this structure can be distinguished: monitoring system, supervisory control layer (SuCL), optimizing control layer (OCL) and follow-up control layer (FuCL), together with intelligent units of SuCL: SAU (including KDU and RAU) and PAU. A placement of these units within the control structure is shown in the Fig. 1.15. The solid lines symbolize information flows of control objectives,

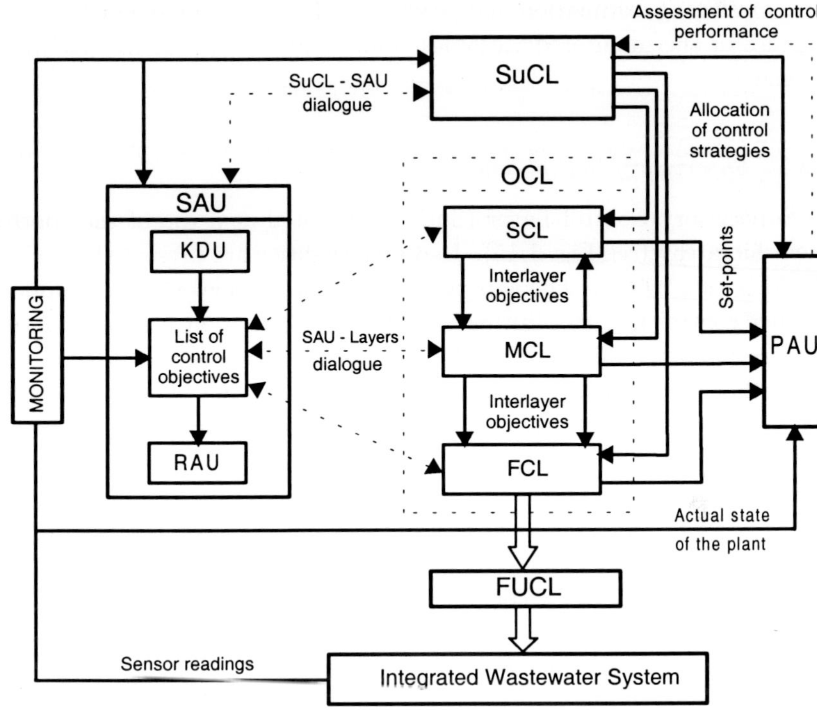

Fig. 1.15 Hierarchical structure for optimizing control of integrated wastewater systems.

set-points, constraints, measurements, risk and performance indicators *etc.*, whilst the dotted lines represent dialogue processes between the control and supervisory units.

Clearly, the most important objective of an entire control structure is to produce efficient and quality control actions. A receding horizon model predictive control (MPC) technology (Morari and Lee, 1999; Mayne *et al.*, 2000; Maciejowski, 2002) is a core control technology that is chosen for generating the optimized control actions and the OCL is the place where this process is carried out. Detailed information regarding OCL can be found in (Gmiński, *et al.*, 2004). The OCL must have the following information, in order to work properly: state of the system, prediction of the disturbances, control objectives and constraints. The first two are provided by a monitoring system. The monitoring in general, consists of three main parts: measuring, estimation and prediction. Its resolution depends on the dynamics of the monitored variables time scale and on the measurement equipment.

1.5.3.1 *Supervisory control layer*

The Supervisory Control Layer (SuCL) is located at a top of the control system hierarchy (see Fig. 1.15). Its functionalities are presented in (Grochowski, *et al.*, 2004a). The supervisory control provides the best support for other control and monitoring activities. Information from all control structure units is available at SuCL at every time step with a time resolution that is adequate to a time unit of a time scale being considered. Hence, SuCL has global knowledge about current activity of entire system and it is able to coordinate and schedule operations of other parts of the system, to ensure that most efficient control actions are produced and to suitably react to unwelcome events within the plant. Based on information delivered by the monitoring system, optimizing control layer and dedicated agents of the SuCL select a control strategy to be currently applied to the system.

The Situation Assessment Unit (SAU) and Performance Assessment Unit (PAU) agents play a supporting role in this structure. The SAU carries out routine activities but it also needs to be prepared to get involved in a dialogue with SuCL and to quickly answer questions stated by the SuCL. Based on an assessment of an *operational situation* of IWWS the SuCL allocates suitable control strategies (Brdyś *et al.*, 2002). Regardless of the quality of the SuCL mechanisms employed to make the final selection of a control strategy to be applied, the strategy is selected based on a prediction of the system's operational performance. However, due to the uncertainty the predicted performance is not the same as the performance that is achieved in the real system, after the control strategy has been applied to the system. The achieved performance is on-line monitored and assessed by PAU.

The MPC mechanisms at the control layers check the achieved performance indirectly by comparing the model responses with the IWWS states at discrete time instants ending the MPC's time steps. PAU generates more accurate information about achieved performance and based on this information SuCL may halt applying current control strategy and switch to another one.

The switching between control strategies is another important function of SuCL and it is described in (Grochowski, *et al.*, 2004b).

1.5.3.2 *Optimizing control layer*

In Fig. 1.15, the Optimizing Control Layer (OCL) is responsible for generating the optimized and *robustly feasible* trajectories of the manipulated variables (control trajectories). The control objectives at the OCL can be split into the *long term* (biological sustainability and operational cost), *medium term* (effluent quality, actuator constraints, technological constraints, operational cost) and *short term* (effluent quality during heavy and of short duration events, actuator constraints, meeting demand on desired carbon, PIX and dissolved oxygen and operational cost). The different time horizons of the objectives are as a matter of fact mainly implied by a multiple time scale feature of an internal dynamics of the biological treatment process and variability of the disturbance inputs. There can be various ways of controlling at each of the time scales. It depends on assumed/chosen control strategy and the associated objective function and constraints. The core control method at OCL is MPC, and temporal decomposition is used internally at the layer. In order to fulfil the desired objectives, OCL generates control trajectories over each of the control horizons.

1.5.3.3 *Follow-up control layer*

In Fig. 1.15, the Follow-up Control Layer (FuCL) is responsible for forcing the plant to follow the set-points prescribed by the optimizing part of the structure, it performs *direct control* actions (see Subsection 1.2.1). This is done by direct hardware maneuvering, namely simple PID controllers, sensors, servos, valves controlled via Programmable Logic Controllers.

1.5.3.4 *Temporal decomposition*

It is very difficult to efficiently handle the multiple time scale dynamics in the optimizing control problem by a centralized optimizing controller as the needed long prediction horizon and short time steps lead to an optimization problem of high dimension and under large uncertainty radius. In order to alleviate these two fundamental difficulties a *temporal decomposition* (see Subsection 1.2.2) of the optimizing controller time scale into the the *slow*, *medium* and *fast* time scales is proposed. As a result the optimizing con-

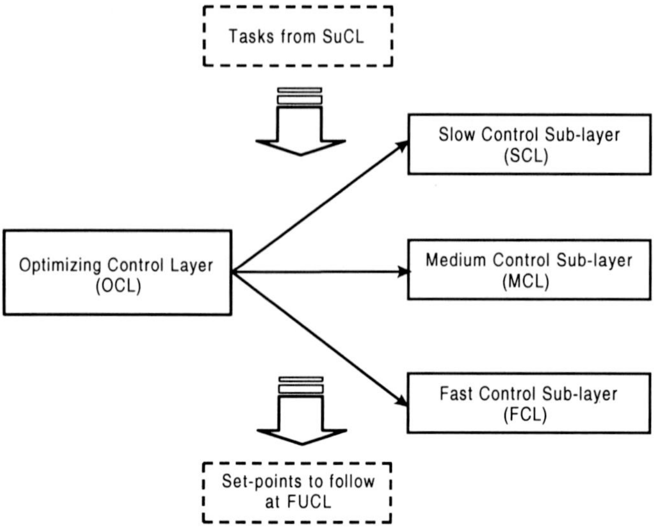

Fig. 1.16 Temporal decomposition of the optimizing control layer into three sub-layers operating at different time scales.

troller gets structured in a form of the three-layer hierarchy (see Fig. 1.15 and Fig. 1.16).

Each sub-layer controller operates at different time scale and handles objectives relevant to this time scale.

The *Slow Control sub-Layer* (SCL) operates at a slow time scale with one day control step and handles long-term objectives over a horizon of a week up to several months. This control layer is responsible for biomass biological sustainability, volume control of equalization and stormwater tanks and long-term economic objectives under as wide as possible range of disturbance inputs.

The *Medium Control sub-Layer* (MCL) operates at a medium time scale with one hour control step and handles medium-term objectives over a horizon of one day. This control layer is responsible for maintaining the effluent quality within required limits and optimizing the operating cost subject to technological and actuator constraints prescribed by the SCL. Also the target values of the manipulated variables prescribed by the SCL must be reached by the MCL at the end of its time horizon. Hence, this sub-layer constitutes an *actuating system* for the SCL (see Section 1.3). The targets are: sludge mass, sludge retention time, equalization tank level and septic tank level. Dynamics of these manipulated variables is slow and the MCL is no able to determine proper values of the variables by considering them over its own control time horizon only. The sludge mass and sludge retention time are key quantities for achieving sludge parameters that are desired for the biological sustainability of the plant operation. This is a long-term objective. Similarly, due to time constants of the equalization tank and septic tank their volume trajectories need to be seen over a long time horizon. This is illustrated for the equalization tank in Fig. 1.17.

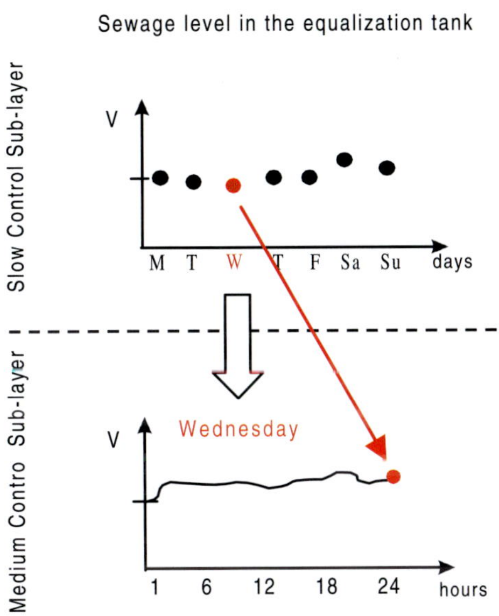

Fig. 1.17 Illustration of the interlayer control objectives at the medium control layer for the equalization tank volume.

On Monday morning the SCL determines the tank volume trajectory one week ahead. Due to the one day control time step at this control layer the trajectory is composed of the tank volumes at the end of each day over the week. The value at the end of Wednesday or on Thursday morning is the final volume in the tank for the dynamical optimizing control problem at MCL. The MCL starts to calculate on Wednesday morning the control actions to be applied to the system over a control horizon at MCL that is equal to one day. Hence, the controls from Wednesday morning till Thursday morning will be determined that are constant over one hour — the control step at the MCL. Now, with the MCL ability to change the manipulated variables every one hour, the optimized tank trajectory can be calculated at this layer without the worry that, for example, on Thursday morning the tank will be almost full and it would not be able to receive a rainfall that would occur just after the Thursday time has started. Thus, the MCL does not have to worry that an overflow of sewage to receiving waters would occur on Thursday. It is so, because the final state in the MCL optimizing control problem has been prescribed by the SCL, which was in a position, due to its longer (one week) control horizon, to properly predict the rainfall on Thursday — calculating such volume to be reached at the tank on Thursday morning that enables handling the rainfall without overflowing. The manipulated variables at MCL are: dissolved oxygen concentrations at the aerobic zones, recirculation flow rates and tank pump in/out flow rates, and chemical precipitation (PIX).

The *Fast Control sub-Layer* (FCL) operates at the fastest time scale with a one minute control step and handles short term objectives over a horizon of one hour. Generating the set-points for the Follow-Up Control Layer so that the process is forced to follow the manipulated variable trajectories that are prescribed by MCL, is the main functionality of the FCL. The process actuators are mostly simple Programmable Logic Control (PLC) controlled devices, except for the air flow rates that are provided by the aeration system in order to achieve required set-points for the dissolved oxygen concentrations. These airflow rates are the manipulated variables at the FCL. Meeting the prescribed set-points with least cost of energy consumed by the blower and valves in the aeration system is a complicated task, see (Piotrowski, 2004).

The targets passed from an upper control layer (sub-layer) to the adjacent control lower layer (sub-layer) are called interlayer targets that constitute the *interlayer control objectives.* The interlayer control objectives emerge as a result of the temporal decomposition of the optimizing control level. Manipulated variables at the lower sub-layer are also used in the upper sub-layer in the mathematical model utilized by the MPC there. However, optimized values of these variables at the upper sub-layer are rather crude due an excessive control time step used there. However, they can be used as a sort of average values to be followed at the lower layer.

This is illustrated in Fig. 1.18. These average values accommodate precious information about predicted long term behaviour of the manipulated variables and they are treated as additional control constraints, the *interlayer constraints*, by the MPC at the lower sub-layer.

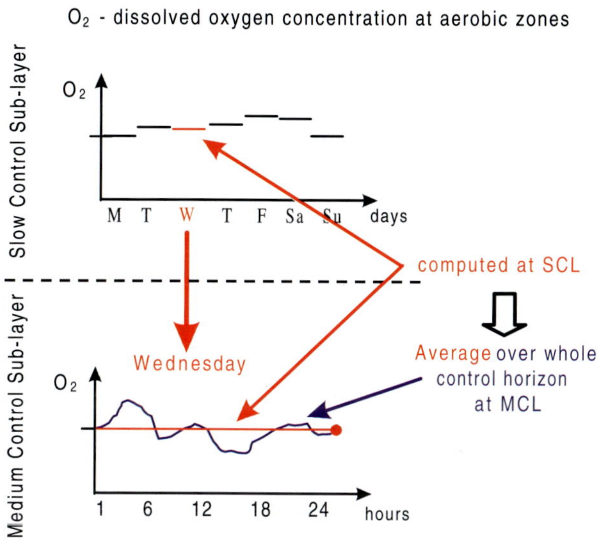

Fig. 1.18 Illustration of the interlayer control constraints.

An overall information structure of the controller at the Medium Control Sub-layer is illustrated in Fig. 1.19.

The structure consists of three main modules: Robust MPC (RMPC) module, Grey Box parameter estimation module and plant state estimation module. The RMPC (Brdyś and Chang, 2001) at MCL repetitively solves on-line a complex, model-based dynamic optimization problem and it needs the plant model. Available physical models of biological processes (Henze *et al.*, 2000) are too complex to meet constraints imposed in real time applications, within the allowed computing time. Hence, a simplified model of the plant called Grey-Box model was developed (Rutkowski, 2004). The RMPC at MCL uses the Grey Box model (Gmiński *et al.*, 2004). The model parameters must be estimated on-line and this is done by Weighted Least Squares with Moving Measurement Window algorithm (Konarczak

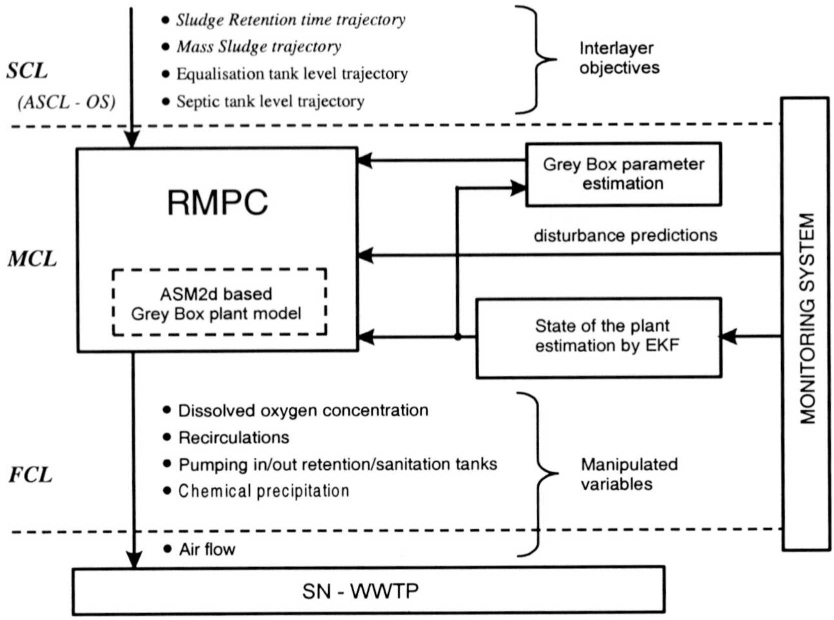

Fig. 1.19 An information structure of the optimizing controller operating at a medium time scale within the Medium Control Layer.

et al., 2004). Finally, the plant states needed by the RMPC are estimated from the plant measurements and are also based on an accurate physical plant model ASM2d (Henze *et al.*, 2000), that is periodically calibrated. An Extended Kalman Filter is used to carry out the state estimation in real time (Brdyś *et al.*, 2004). It is fast enough so that the more complex plant model ASM2d is allowed. It is necessary to estimate the states using different models because a joint state and parameter estimation entirely based on the Grey-Box model is not feasible. With the limited number of measurements available the parameters are not identifiable and the states are not observable. Thus, the state estimates produced by the filter are used as pseudomeasurements in the parameter estimation and this is emphasized in Fig. 1.19 by information exchange links between the estimation and the parameter estimation modules. The SCL in the structure shown in the Fig. 1.19 combines activities of plant operators and computers. It is still not feasible to automatically generate the sludge retention time and sludge mass trajectories that, if reached in the controlled plant, would guarantee desired biological sustainability of the activated sludge and hence, sustainable operation of the plant. These two key parameters are prescribed by the plant operators. It can be done and there is time to do it at the SCL, as there is always substantial operational experience available at the plant site and the control time step is typically equal to one day. The equalization tank trajectory and the septic tank trajectory are determined by computers. Hence, in this implementation of the SCL the optimizing computer control is supported by the plant operators. Consequently, the name *ASCL-OS* stands for *Auxiliary Slow Control Layer with Operator Support.*

Chapter 2
Optimizing Control Layer

2.1 Process Descriptions and Control Task Formulation

2.1.1 *Process description*

The task of the *optimizing control* in a multilayer control structure is to provide the controllers of the lower, follow-up (regulatory) control layers with the best possible values for their set points. The *best possible* means here dynamic trajectories or constant (steady-state) values of the set-points leading to maximum achievable values of the prescribed economic criteria of the process operation, while keeping the process variables within safe operation limits and satisfying certain additional constraints of a technological nature. The multilayer control structure simplified to regulatory (follow-up) control and optimizing control only is presented in Fig. 2.1, compare with Chapter 1 (Section 1.3).

A crucial factor for the design of the optimizing control layer is an adequate description of the process from the point of view of the task of this layer. The key factor in evaluation of this adequate description is the extraction of outputs and inputs of the process essential from the point of view of the layer task. Mappings relating these inputs and outputs define the process as seen by the layer. It is called the *optimized (sub)process* or *main (sub)process* in the decomposition structure of the process into two linked subprocesses introduced in Chapter 1, see Fig. 1.5 there and Fig. 2.2 in what follows. All more detailed, underlying process behaviour with faster dynamics and associated variables, not essential from the optimizing control point of view, constitute the *actuating (sub)process*.

The (fast) actuating subprocess is driven by manipulated inputs m and

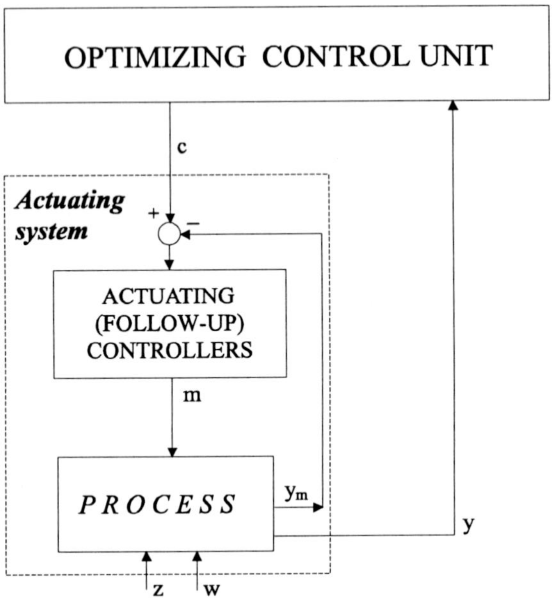

Fig. 2.1 Multilayer control structure with regulatory (follow-up) and optimizing control layers only.

fast disturbances z. Its dynamics depends on two state components: its own fast states x_m and slow states x_c coming from the main subprocess. The output vector of the actuating subprocess, y_m, is composed of technological decision variables essential for the process optimization and constitutes an input signal to the main, optimized subprocess of slower dynamics. This subprocess is also under influence of disturbances, namely slow disturbances w having a significant impact on the overall process performance. The manipulated variables m are driven by *direct (follow-up) controllers*, called *actuating controllers (AC)*. The task of these controllers is to force the actuating subprocess to generate the desired values of the outputs y_m regardless of the process fast disturbances z. These desired values are set-points for the AC and are denoted by c. The described structure of the decomposed process together with the actuating controllers is depicted in Fig. 2.2 (a). In Fig. 2.2 (b) an *equivalent slow dynamic process* is shown, with its (slow) dynamics modeled by the following state equations:

$$\dot{x}_c(t) = f_c(x_c(t), c(t), w(t)) \tag{2.1}$$

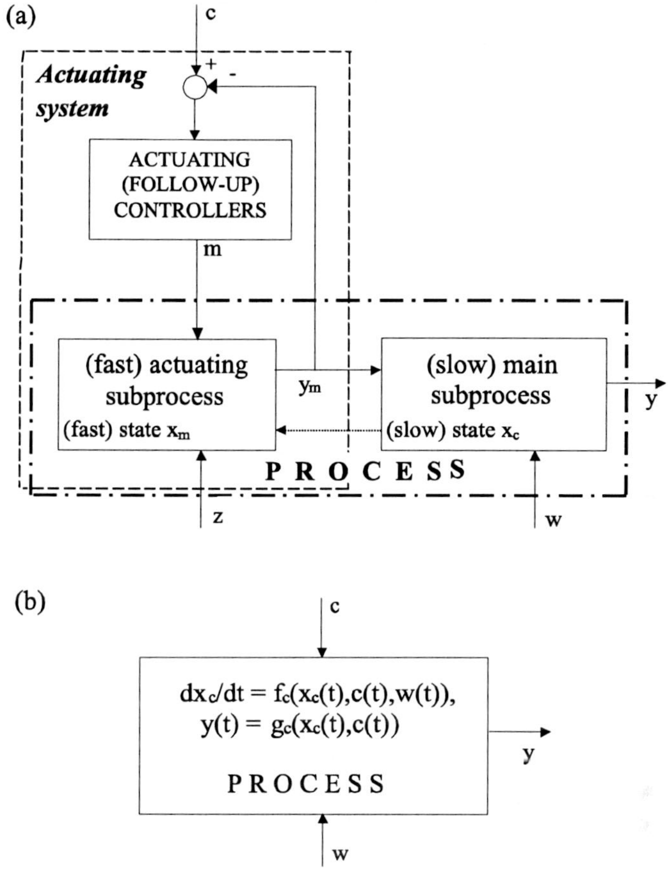

Fig. 2.2 (a) Original process with the actuating system, (b) Equivalent slow dynamic process (provided perfect actuating controller action).

It is the process representation which is seen by the optimizing control layer, under the assumption of perfect action of the actuating controllers, i.e.,

$$c(t) = y_m(t) \tag{2.2}$$

It is *the assumption underlying the multilayer decomposition* of the control task, making it possible to separate the design of the regulatory control and optimizing control. Thus, for the optimizing control layer the set-points c are the (optimized) controlled process inputs, while y are the (optimized) process outputs, see Fig. 2.2 (b). Throughout the book we will use *set-points* (desired values for set-points) or *optimizing controls* (optimizing control inputs) as equivalent descriptions for the variables c, as defined above.

2.1.2 *Optimizing control and optimal control problems*

The task of the optimizing control layer is to evaluate *optimal trajectories* of the set-points c for the lower layer regulatory controllers. We will assume that the dimensionality of c is n, $c \in \mathbb{R}^n$. The optimal trajectories of the optimizing controls c are those resulting in best possible performance of the optimized process, which is measured in *economical terms* and over *long time horizon.* Such formulation of the control task implies, obviously, that the operation of the control system under normal conditions is being considered — excluding such situations as process start-up and shut-down or emergency events, which are handled by special purpose emergency routines triggered by the operators or supervisory control unit of the multilayer control structure.

We shall assume that a scalar performance measure (performance function) of the process operation is used and that all other objectives (requirements) stemming from the plant management layer can be formulated as certain technological constraints (e.g., on product quality). The performance function for the normal operating conditions can be formulated as

$$J_d = \frac{1}{T} \int_0^T Q(c(\tau), y(\tau))d\tau \tag{2.3}$$

where Q is an instantaneous measure of the process performance and T stands for the control and optimization horizon, set according to the management layer requirements and technical possibilities following judgments of the future behaviour of the process and its environment (disturbance predictions or characteristics). The performance function is formulated, generally, in terms of set-points c and measured process variables directly influencing optimality of the process operation, called the process outputs y.

The decision variables c at the optimizing control layer are physical process variables, see y_m in Fig. 2.2 (a), and are therefore always constrained physically (e.g., non-negative values of flows) or technologically (e.g., maximal allowable temperatures). These constraints have to be taken into account at the optimizing control layer, where evaluation of optimal trajectories or points within the whole admissible region is the control task, not an operation around a well defined trajectory or point. Moreover, constraints on the process outputs y are often present and of primary importance, like constraints on final product composition. These constraints can be also dependent on control inputs c and disturbance inputs w.

Generally, the *objective of the optimizing control* can be formulated as follows:

> *To evaluate, on-line, values of the set-points $c(t)$ such that the performance function (2.3) is optimized on the horizon $[0, T]$, subject to all constraints imposed on c and y, where y are the process outputs dependent on control inputs c and on unknown or imperfectly measured disturbances w.*

This formulation states, certainly, an ideal and not realistic goal. The reasons are twofold:

- a lack of current and precise measurements of all disturbances and, more importantly, a lack of perfect knowledge of future behaviour of the process disturbances on the control horizon $[0, T]$,
- an imperfect knowledge of the input-output process mapping relating optimizing controls c and outputs y, i.e., imperfect process model available only.

Without the discussed unknown information it is impossible to determine in advance, i.e., at an initial time $t = 0$, the real optimal trajectories $c(t)$ of the optimizing controls, $t \in [0, T]$. Therefore, the task and real challenge of the optimizing control layer is to determine on-line "best possible" values of $c(t)$ in the presence of uncertainty in disturbance measurements and forecasting and in process modeling. A key element of all strategies of such dynamic optimizing control are solutions of properly defined optimal control problems. Assuming that the process model (process input-output mapping) is described by a set of process state and output equations

$$\dot{x}_c(t) = f_c(x_c(t), c(t), w(t)) \tag{2.4}$$

$$y(t) = g_c(x_c(t), c(t)) \tag{2.5}$$

and that disturbances are represented in a deterministic way by disturbance estimates $w_e(t)$, $t \in [0, T]$, we can state such an *optimal control problem* in the following form:

$$\text{minimize } \frac{1}{T}\int_0^T Q(c(\tau), y(\tau))d\tau \tag{2.6a}$$

subject to constraints :

$$\dot{x}_c(t) = f_c(x_c(t), c(t), w_e(t)) \tag{2.6b}$$

$$y(t) = g_c(x_c(t), c(t)) \tag{2.6c}$$

$$c(t) \in C \tag{2.6d}$$

$$\dot{c}(t) \in \dot{C} \tag{2.6e}$$

$$y(t) \in Y \tag{2.6f}$$

where (2.6d) are constraints on values of set-points and (2.6e) constraints on rate of change of set-points (optimizing controls) $c(t)$, while (2.6f) are constraints on process outputs. The latter may also happen to be dependent on c and w_e, although it is not a most common case. Solution of the defined complex optimal control problem may be difficult, requiring excessive computing time. It may even be impossible if the control horizon is long and if dimensionality of the decision vector c and state vector x_c is not small. Moreover, even if the solutions are available in an acceptable computation time, their application may be not practical on the whole control horizon $[0, T]$ — the less practical the more distant the current time t is from the computation time moment 0. Certainly, due to imprecise process model and different evolution of disturbances $w(t)$ than predicted in $w_e(t)$. The essence of dynamic optimizing control strategies is to cope with these uncertainties using feedback information from the process in the form of available measurements. A general and *non-iterative* approach would be here an application of *predictive control schemes* with receding horizon, with single step open-loop optimization utilizing the formulated optimal control problem (2.6a) – (2.6d), see e.g., (Findeisen *et al.*, 1980). However, in this book we are interested in *iterative* optimizing control algorithms. In the case of dynamic control such algorithms apply to batch processes. They will be discussed in Chapter 6. However, the case of *steady-state control* is important from a practical point of view. It is when the set-points for the follow-up controllers (actuating controllers) are assumed to be kept constant for relatively long periods of time and are adjusted on-line, periodically or on-request triggered by changes in disturbances or process management requirements. Iterative optimizing control algorithms for the steady-state process control is the main area of interest of the book.

2.1.3 *The steady-state case*

Certainly, the steady-state optimizing control can only be optimal when changes in disturbances or process characteristics are slow-varying or rare when abrupt, when compared to the process dynamics. This case is important from a practical viewpoint, since most of continuous industrial processes are designed to operate in steady-state conditions.

We introduced the set-points c and outputs y as functions of time, $c(t)$ and $y(t)$, because the optimizing control problem is generally a *dynamic problem*. When running the process *in the steady-state mode* it is assumed that for each normal process operating conditions *fixed optimal values* for the set-points (optimizing controls) $c(t)$ are to be evaluated by the optimizing control layer.

The process is set to operate in the steady-state mode in two main situations:

- Preliminary dynamical considerations indicate that steady-state operation is optimal.
- Steady-state process operation is decided in spite of the fact that dynamical considerations indicate dynamical (periodically changing) set-points would lead to better results. This usually follows from additional theoretical and practical considerations taking into account such factors as the value of the performance gain due to periodic control versus technical difficulties and disadvantages connected with periodic control implementation (frequent excitation of actuators, overall control structure reliability and maintenance, *etc.*).

However, even when the process is decided to operate in a dynamic periodic regime, the calculation of its optimal steady-states is usually required. The aim of the steady-state optimization here is to determine *mean values of the periodic inputs*. A steady-state model representing algebraic relations between average inputs and average outputs is used for this purpose. The task of the periodic control is then only to generate appropriate deviations from optimal mean values.

Following the dynamic process modeling as described by (2.4) and (2.5), the steady-state process model will be described by the following set of algebraic equations, generally nonlinear

$$0 = f_c(x_c, c, w) \tag{2.7}$$

$$y = g_c(x_c, c). \tag{2.8}$$

The above set of equations is, in fact, an implicit description of an explicit relation between the set-point values c and the states x_c,

$$x_c = \zeta(c, w). \tag{2.9}$$

Assuming that there exists a unique relation (2.9) we could eliminate explicit dependence on states converting the steady-state process model (2.7), (2.8) to the form

$$y = F(c, w, \alpha) \tag{2.10}$$

where

$$F(c, w, \alpha) = g_c(\zeta(c, w), c) \tag{2.11}$$

and α has been introduced to denote (adjustable) model parameters. However, one should realize that the explicit relation (2.9) is usually not easy or even not possible to be evaluated in real-life applications. Nevertheless, for each given value of c and w the resulting value y of the mapping F can be calculated numerically: first by solving the set of nonlinear equations (2.7) with respect to x_c, and then substituting the result to the output equations (2.8). A numerical procedure adequate to solve a preliminarily given set of nonlinear equations should always be carefully selected and preliminary tested, since for on-line applications it must be robust and sufficiently fast. A discussion of such procedures is out of the scope of this book, but it can be found in numerous monographs on numerical algorithms.

The steady-state description of the process has been introduced as a simplification of the dynamic model by setting time derivatives of the state variables x_c to zero. Although nicely introduced after the discussion of the dynamic case, it is not the way the steady-state (static) models are usually developed in practical industrial applications. Reliable, detailed steady-state models of technological processes are usually developed from general principles involving sets of complex nonlinear equations, often of a complexity not possible to deal with in dynamic modeling. Some or even all of the desired output variables y are then usually implicitly hidden in the state and output equations having, instead of (2.7), (2.8), the implicit form

$$0 = f_c(x_c, y, c, w) \tag{2.12}$$

$$0 = g_c(x_c, y, c, w). \tag{2.13}$$

In that case the full set of nonlinear algebraic equations (2.12), (2.13) of general dimensionality $\dim x_c + \dim y$ must be simultaneously solved by suitable numerical procedures, involving usually time-consuming and not very robust iterative resolution of algebraic loops. Such complex steady-state models are developed mainly for simulation purposes, with the development process supported by specialized, commercial software packages. These models are usually too complex, not adequate for *on-line* optimizing control applications. However, they constitute a good basis for development of much simpler models of the form (2.10), suitable for optimizing control purposes. These simplified models are often constructed as certain functional approximations of the input-output data sets generated by the full simulation models (or real process data, if available), e.g., by using the least squares technique to adopt parameters of *a priori* assumed polynomial functions.

It should be emphasized that steady-state operating mode of the process does not mean that optimal values of the set-points must be kept constant all the time, during the whole process operation horizon regardless of changes in disturbance values or process characteristics. The task of the optimizing control layer operating in a steady-state is to evaluate optimal constant set-point values *for current operating conditions.* This means that these values may and usually should be updated after each significant change in the disturbances (e.g., in feed inflow rate), management requirements or process characteristics (e.g., catalyst activity) affecting the overall process performance. The above comments also indicate again that the multilayer control structure with optimizing control layer operating in steady-state mode applies to the cases where process dynamics is significantly faster than dynamics of disturbances affecting the steady-state process characteristics and performance measure. By the process dynamics we understand here the dynamics of the main (optimized) process.

Assuming the case of the steady-state optimizing control the performance function is formulated as an instantaneous, not integral one as in (2.3). This function can be expressed as depending directly on values of the steady-states $c \in \mathbb{R}^n$ and the output variables $y \in \mathbb{R}^m$,

$$Q = Q(c, y). \tag{2.14}$$

Usually (2.14) represents instantaneous net production profit. Assuming linear dependence of costs of individual materials on their prices and control (stabilization) of the raw material streams and energy streams, the performance function (to be *minimized*) can be formulated in the following simple linear form

$$Q(c, y) = \sum_{j=1}^{n^J} p_j^c c_j - \sum_{j=1}^{m^J} p_j^y y_j \tag{2.15}$$

where p_j^c denote prices of the mentioned input streams, whereas p_j^y prices of output products and n^J is the number of input streams. Usually $n^J < n$ since the follow-up controllers usually stabilize more process variables than only input streams. The output variables y may represent production rates of both products and waste materials. In the latter case the prices will have negative sign. There may also exist components of the output vector y not entering the performance function, but important for the process constraint formulation, like product concentration in the output stream. Therefore, the situation when $m^J < m$ is possible and often encountered.

We assumed that there are always *constraints* on current values of the optimizing controls c, of physical and also technological nature. The constraints define the admissible set C (2.6d) — recall that the rate of change constraints (2.6e) cannot be taken into account when only steady states are considered. They should be dealt with by a special transition unit or directly by the follow-up control algorithms or structures, if possible, see e.g., (Goodwin *et al.*, 2001; Tatjewski, 2002) — we will address this point later on in this chapter.

The constraint set C will be assumed to have the following compact form

$$c \in C = \{c \in \mathbb{R}^n : \; g(c) \leq 0\} \tag{2.16}$$

where $g : \mathbb{R}^n \mapsto \mathbb{R}^r$ is a vector of constraint functions. Components of g represent usually simple limit constraints, $c_j \geq c_{j\min}$ or $c_j \leq c_{j\max}$, but may also represent more complicated constraints defined by functions of several components of the vector c. For example, bound on energy delivered by one power unit to several parts of the plant.

Any model is always valid for ranges of its variables for which it has been constructed, defining its validity set. We will not introduce the constraints describing this set, or distinguish these constraints formally from those defined above, in order not to complicate the presentation more than necessary. It means that the assumption is made that the process model is well defined on the set C (some optimization solvers require also it is defined on a certain neighborhood of C as well, it should be taken into account when choosing a solver).

Constraints on steady-state outputs (2.6f), i.e., constraints on certain elements of the output vector y are also often present, like requirements on minimal or maximal concentration of certain components in the product stream. These constraints define the admissible set Y assumed to have the form

$$y \in Y = \{y \in \mathbb{R}^m : \; \psi(y) \leq 0\} \tag{2.17}$$

where $\psi : \mathbb{R}^m \mapsto \mathbb{R}^{r_y}$ represents vector of constraint functions.

Having defined mathematical descriptions of the performance function and the constraints we are at the position to formulate the steady-state *optimizing control problem* (OCP):

$$\begin{aligned} &\text{minimize } Q(c, y) \\ &\text{subject to: } y = F_*(c, w) \\ &\qquad\qquad\quad g(c) \leq 0 \\ &\qquad\qquad\quad \psi(y) \leq 0 \end{aligned} \tag{2.18}$$

where

$$y = F_*(c, w) \tag{2.19}$$

denotes the true mapping between the process outputs and inputs in steady-states, unknown precisely for obvious reasons. The formerly introduced mapping F (2.10) is the available model of this mapping.

As in the dynamic case, a formulation of the (off-line) model optimization problem, i.e., the optimization problem with the process represented by its steady-state model equations as equality constraints, heavily depends on a knowledge of disturbances and the resulting disturbance model. We will address this problem in more detail in the next section assuming here that a deterministic estimate (e.g., last measurement) w_e is available —

the simplest and usually most practical disturbance model in process industries. In this way we have defined all major parts of an optimization problem, which can be treated as a basic version of the off-line model of the steady-state optimizing control task. Thus, we arrived at the following basic *model optimization problem* (MOP)

$$\begin{aligned} &\text{minimize } Q(c, y) \\ &\text{subject to: } \ y = F(c, w_e, \alpha) \\ &\qquad\qquad\quad\ g(c) \leq 0 \\ &\qquad\qquad\quad\ \psi(y) \leq 0. \end{aligned} \tag{2.20}$$

2.1.4 *Implementation of the set-point step changes*

For each new operating conditions modifications of the set-points of the follow-up controllers can result from a single solution of an optimization problem like the MOP problem introduced above, or from a more elaborate iterative on-line feedback optimizing control strategy when a single solution of a model-based optimization is only an element of every iteration and a convergent series of steady-states is generated. Regardless of the steady-state optimizing control strategy applied, it produces new values at certain distant instants of time, which have to be implemented as current set-points of the process follow-up controllers (actuating controllers). These modifications, if applied directly as pure step changes of the set-points may lead to excessive and thus unacceptable process state variations during transients in the plant. Moreover, the rate of change constraints should also be taken into account, as it cannot follow from steady-state considerations, for obvious reasons. Therefore, a dynamic shaping of set-point trajectories between consecutive optimized steady-state values c^i and c^{i+1} may be necessary, except in cases when these constraints can be taken into account directly at the follow-up control layer.

This dynamic shaping can be performed by the *transition unit* specially designed to this end. The transition unit takes into account the rate of change constraints for $c(t)$ and simplified slow process dynamics and translates the possibly too large step-like changes into acceptable smooth continuous trajectories between c^i and c^{i+1}, or a series of smaller steps approximating such trajectories. Certainly, these constraints are those for the original controlled process variables $y_m(t)$, comp. Section 2.1.1. The natural strategy of the transition unit is the feedforward (open loop) control technique. The multilayer control structure incorporating the transition unit is presented in Fig. 2.3, where $c_{\mathrm{resh}}(t)$ denotes set-point trajectories with dynamically reshaped step changes.

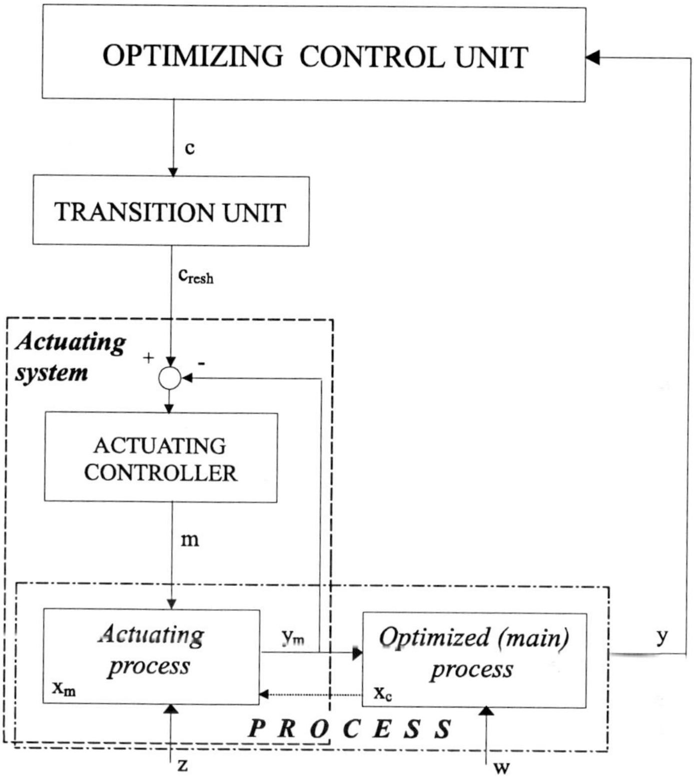

Fig. 2.3 Multilayer control structure with a transition unit.

A separate transition unit may not be needed if lower-layer follow-up controllers implementing the step changes of the set-points c are capable of taking into account rate of change constraints. It is the case when modern advanced control algorithms capable to handle amplitude and rate of change constraints are applied, see e.g., (Goodwin *et al.*, 2001; Maciejowski, 2002; Tatjewski, 2002) or at least classical two-degrees-of-freedom controllers like standard PID controllers, but with suitably adjusted dynamic transfer function blocks at the set-point input. It is possible when programmable controllers are used, equipped usually with specific function blocks or programmable blocks preprocessing the original set-point values before they are fed to the feedback control algorithm.

It is not the aim of this book to focus attention on implementation issues of consecutive steady-states generated at the optimizing control layer, although technically it is an important task in the multilayer control structure. The reason is it does not influence the design of iterative optimizing control algorithms which we are studying. Therefore, the transition unit will not be distinguished within the control structure further in this book, in order not to complicate the presentation and notation unnecessarily. Consequently, the variables $c_{\text{resh}}(t)$ will not be further used.

2.2 Steady-State Model Optimization

The basic formulation (2.20) of the steady-state model optimization problem (MOP) was introduced in the previous section, assuming a deterministic estimate of the disturbance vector. However, different formulations of the optimization problem are also possible and useful depending on the approach to the description of the disturbance uncertainty. The three practically sound approaches are:

- stochastic,
- set-bounded,
- deterministic.

In this section our attention will be focused primarily on their technical features and suitable solution techniques.

Appropriately selected general purpose constrained optimization routines of nonlinear programming, see e.g., (Fletcher, 1987; Bertsekas, 1997) are directly applicable to deterministic steady-state optimization problems, like the MOP problem. But they are also basic elements of solvers for stochastic and set-bounded optimization problem formulations. These routines constitute an important element of iterative optimizing control algorithms as well, the algorithms to be presented in the following section of this chapter.

The three basic formulations of the steady-state optimization problem will now be discussed in more detail.

The stochastic formulation is as follows

$$\begin{aligned} &\text{minimize } E_w\{Q(c,y)\} \\ &\text{subject to}: \ y = F(c,w,\alpha) \\ &\qquad\qquad\quad c \in C \\ &\qquad\qquad\quad y \in Y \end{aligned} \tag{2.21}$$

where E stands for expected value. This formulation requires proper knowledge of the probability distributions of the uncertain variables w. Inadequate assumptions, e.g., about type of the probability distributions can lead to misleading results.

The stochastic nonlinear optimization problem itself is difficult to solve and very time consuming. The main source of difficulties lies in the fact that nonlinearities usually change probability characteristics. To explain it let us assume that probability distribution of the uncertain variables w is given. For a nonlinear mapping F probability distribution of the output variables y can be quite different. Moreover, it usually also depends on the value of the decision (control) vector c, what is not the case when F is affine. It means that for each value of c (generated by the optimization routine) the probability density function of y must be calculated, which is a time-consuming numerical task. If the function $Q(c, y)$ is also nonlinear with respect to y, then its probability density function must be also calculated, in the same way.

Another important and difficult problem in stochastic formulation is created by possible constraints on the output variables y. These variables are now random variables and the constraints cannot be formulated in a deterministic way. The condition $y \in Y$ cannot be understood in a deterministic way, as a deterministic inequality $\psi(y) \leq 0$. Different approaches are possible here. In the simplest but important case when $\psi(y) = \{y \in \mathbb{R}^m : y \leq y_0\}$, where y_0 is a prescribed threshold value, they can be formulated as follows :

- Using expected value of the random variable in the same way as the original value in the deterministic case,

$$E_w\{y\} \leq y_0 \tag{2.22}$$

However, this formulation assures fulfillment of the constraints only "in average" and violations, even frequent, are possible. If the constraints are required to be kept in a more rigorous way, then the following formulation is more adequate.

- Demanding a prescribed guarantee of the constraint fulfillment, in a probabilistic sense:

$$P(y \leq y_0) \geq \beta \tag{2.23}$$

where a value of β, $0 < \beta \leq 1$, prescribes the guarantee — the closer β to 1 the greater the guarantee, i.e., the less probable violation of the constraint.

It is easily seen that in each of the above approaches the probability distribution of y, where $y = F(c, w, \alpha)$, must be known to calculate the constraint fulfillment or violation, for each value of the decision variables c. Hence, this probability distribution is needed for calculation of the values of both performance function and constraints, for every given value of the decision variables c. Optimization routines for generating consecutive values of c (approaching the searched optimum), are standard, as for deterministic problems. It does not matter for such a routine if the values of performance function and constraints (being numbers) stem from a simple deterministic algebraic formulation, or from complex calculations involving probability distributions. However, because these calculations can be extremely time consuming, the fastest possible routines should be selected, i.e., routines requiring as few steps as possible to get close to the optimum.

It should be realized that even in cases of simple constraint functions on y, like those in (2.22) or (2.23), the constraints are nonlinear with respect to decision variables c, due to the nonlinear output mapping $y = F(c, w, \alpha)$. Therefore, treating these constraints by addition to the performance function via suitable penalty terms may lead to a reasonable problem reformulation. Application of a quadratic penalty term results in the following modified performance function, for a single constraint of the type (2.22)

$$E_w\{Q(c, y)\} + \rho(E_w\{y\} - y_0)_+^2 \tag{2.24}$$

where ρ, $\rho > 0$, stands for penalty coefficient and $(\cdot)_+$ is defined in a standard way,

$$(x)_+ = \begin{cases} x, & \text{if } x \geq 0 \\ 0, & \text{if } x < 0 \end{cases} \tag{2.25}$$

Using the penalty function approach additionally softens the requirements imposed by the constraint formulation, especially when the value of the penalty coefficient is not too large. This may be acceptable. As a result, the complicated constraints are removed from the constraint set of the optimization problem. The remained constraints are those that are imposed on the original decision variables c ($c \in C$). In most cases they are simple linear constraints. Such problems are easier to solve. In fact, many standard optimization routines treat nonlinear constraints in a similar way by adding suitable multiplier and penalty terms to the original performance function, see e.g., (Fletcher, 1987; Bertsekas, 1997).

The *set-bounded formulation* constitutes the second approach to the uncertainty when setting the steady-state optimization problem. It assumes that only bounds on disturbances are given, say, $w_{j\min} \leq w_j \leq w_{j\max}$, for each disturbance component w_j – this can be denoted in a compact way by $w \in W$. In this situation *a safe (feasible) worst-case approach* is a reasonable alternative and can be formulated in the form

$$\begin{aligned} &\text{minimize } \max_{w \in W} Q(c, y) \\ &\text{subject to: } \; y = F(c, w, \alpha) \\ &\qquad\qquad\qquad c \in C_{sf} \end{aligned} \tag{2.26}$$

where $C_{sf} \subseteq C$ is a safe-feasible set assuring set-point feasibility irrespective of actual disturbance realization,

$$C_{sf} = \{c \in \mathbb{R}^n : \; \forall w \in W \;\; F(c, w, \alpha) \in Y\} \tag{2.27}$$

Geometric interpretation of the feasible set C, mean-feasible set C_{mf}, i.e., model feasible for the mean-value w_m of the disturbances w,

$$C_{mf} = \{c \in \mathbb{R}^n : \; F(c, w_m, \alpha) \in Y\} \tag{2.28}$$

and safe-feasible set C_{sf} is shown in Fig. 2.4, for a most simple one-dimensional case and an additive disturbance with a set-bounded uncertainty

$$W = \{w \in \mathbb{R} : \; w_m - \Delta w \leq w \leq w_m + \Delta w\} \tag{2.29}$$

where w_m, Δw are given constants.

It can be easily seen that the safe-feasible set is a proper subset of the mean-feasible set $C_{mf} \subseteq C$ – the larger the uncertainty represented by the value of Δw, the smaller C_{sf}. Too large values of Δw can

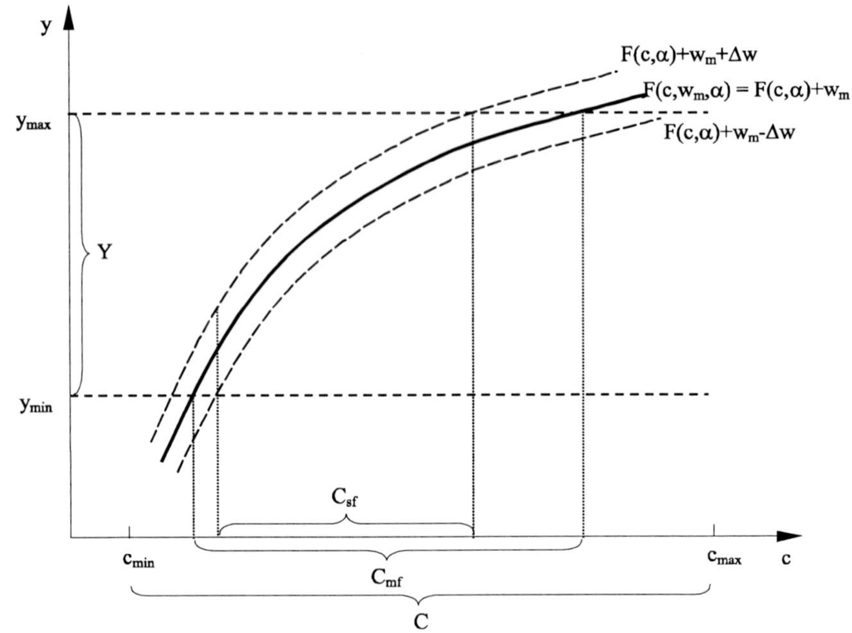

Fig. 2.4 Feasible, mean-feasible and safe-feasible sets for set-points in the case of set-bounded (additive) model uncertainty.

even result in an empty safe-feasible set, as would happen in Fig. 2.4 if $\Delta w > 0.5(y_{\max} - y_{\min})$. It should be pointed out that determination of safe-feasible sets is not usually an easy task for multivariable problems and more complex uncertainty structures. However, the described safe-feasible set-bounded approach to the uncertainty is the only method assuring strict output constraints satisfaction in the true process under uncertainty — the remaining two approaches (stochastic, deterministic with disturbance prediction) do not have this vital property.

The resulting optimization problem is a min-max problem. A possible algorithm for solving this problem is of an iterative double-loop (or two-level) type. It means that two optimization routines are used. An internal (or lower-level) routine solves, for each value of the original decision variables c, the following internal (or lower-level) optimization problem

$$\begin{aligned} &\text{maximize}_w \ Q(c, y) \\ &\text{subject to}: \ w \in W \\ &\qquad\qquad\quad y = F(c, w, \alpha). \end{aligned} \tag{2.30}$$

Let us denote solutions of this problem by $w^{mx}(c)$, then the external (or upper-level) optimization problem can be written as follows

$$\begin{aligned} &\text{minimize}_c \ Q(c, F(c, w^{mx}(c), \alpha)) \\ &\text{subject to}: c \in C_{sf}. \end{aligned} \tag{2.31}$$

From a practical point of view a knowledge of features of the performance function $Q^{mx}(\cdot)$ of the external problem (2.31),

$$Q^{mx}(c) = Q(c, F(c, w^{mx}(c), \alpha)) \tag{2.32}$$

is important in order to be able to choose an appropriate numerical optimization routine. Detailed mathematical analysis of properties of upper-level functions can be found, e.g., in (Findeisen *et al.*, 1980). Basic results concerning our problem are:

- If the function $Q(\cdot, \cdot)$ and the mapping $F(\cdot, \cdot, \alpha)$ are continuous and the set W is compact (i.e., constrained and closed) then the function $Q^{mx}(\cdot)$ is continuous.
- If additionally the mapping $w^{mx}(\cdot)$ is point-to-point, i.e., for each $c \in C$ there is a unique maximizing point $w^{mx}(c)$ for the problem (2.30), then the mapping $w^{mx}(c)$ is also continuous – and hence the mapping $F(c, w^{mx}(c), \alpha)$.
- If, additionally, the function $Q(\cdot, \cdot)$ and the mapping $F(\cdot, w, \alpha)$ have continuous first order derivatives (for each $w \in W$) then the upper-level function $Q^{mx}(\cdot)$ is differentiable as well. The formula for its derivative is

$$Q^{mx\,\prime}(c)^T = Q_c^{'}(c, y^{mx}(c))^T + Q_y^{'}(c, y^{mx}(c))\, F_c^{'}(c, w^{mx}(c), \alpha) \tag{2.33}$$

where

$$y^{mx}(c) = F(c, w^{mx}(c), \alpha) \tag{2.34}$$

and $Q_c^{'}(c, y)$ denotes partial derivative with respect to c, $Q_c^{'}(c, y) = \frac{\partial}{\partial c} Q(c, y)$, *etc.*

It can be easily seen that the external (upper-level) problem function Q^{mx} is continuous under rather natural assumptions. However, its differentiability does not follow only from differentiability of the functions involved in the original initial optimization problem. Uniqueness of solutions of the internal optimization problem (2.30) is the crucial requirement here. If the solution $w^{mx}(c)$ of this problem is not unique for every $c \in C$ then $Q^{mx}(\cdot)$ may be not differentiable. Therefore, when using a set-bounded approach to uncertainty, it is advisable to exploit possible freedom when constructing the process model in order to assure uniqueness of internal optimizations. This is particularly important when other problem features would enable application of gradient optimization techniques. The possibility to choose a fast convergent external optimization routine is vital because for each consecutive value of its decision variables c the optimization of the whole internal problem is performed, and that may be a quite involved numerical task.

The *deterministic formulation* of the process model steady-state optimization is the remaining and simplest approach to the uncertainty. A predicted, deterministic value of the disturbances is used. This results in the model optimization problem MOP, see (2.20), written here in a slightly more general form

$$\begin{aligned} &\text{minimize } Q(c, y) \\ &\text{subject to: } \; y = F(c, w_e, \alpha) \\ &\qquad\qquad\quad\;\; c \in C \\ &\qquad\qquad\quad\;\; y \in Y \end{aligned} \tag{2.35}$$

where w_e stands for a disturbance estimate, i.e., predicted values of disturbances (usually last estimated or measured values). The problem (2.35) is a standard nonlinear constrained optimization problem. Therefore, appropriate optimization routines can be directly applied.

However, it should be realized that the nonlinear constrained optimization problems discussed in this section are usually *difficult to solve*, even in simpler cases of deterministic models of uncertainty. This is caused mainly by nonlinearity of the equality constraints $y = F(c, w, \alpha)$ (modeled static characteristics of the process), as these nonlinearities are often severe. The number of inequality constraints is also rather excessive. Moreover, the feasible set C should cover *the whole range of possible input values.* This, combined with the problem nonlinearity, can easily result in a *multimodal optimization problem*, i.e., a problem having different local minima besides the searched global one. Thus, a global optimization routine should be then applied. Its choice and implementation is not easy even for an experienced user.

The iterative steady-state optimizing control algorithms presented further in this book will be derived for the case of a deterministic estimate of the disturbance uncertainty. The reasons are twofold. First and most important, it is nowadays the most practically sound solution in industrial applications, where current measurements of slowly-varying disturbance values provide good estimates for the optimizing control purposes, and there is usually lack of a sufficiently tight stochastic or set-bounded description. Second, the iterative optimizing control algorithms were developed for the deterministic case and only then offer complete and effective solutions. However, the control and algorithm structures developed for the deterministic case could be often also generalized to cases when stochastic or set-bounded approaches are more appropriate.

2.3 Steady-State Optimizing Control

2.3.1 *Basic control structures and strategies*

The simplest steady-state optimizing control strategy is to perform *a single solution* of the model optimization problem (MOP) after each sufficiently significant change in process disturbance estimates w_e and/or after a change in operational demands sent from the plant management level. By *sufficiently significant* we understand the change which requires, due to the process operator or as judged by a supervisory algorithm, a modification of the operating point of the process (a change of the set-points for the lower layer controllers). While changes in disturbance estimates affect directly the process model, the new operational requirements influence the constraints (e.g., desired production intensity, quality demands) or the performance function of the MOP problem (e.g., prices for products or optimizing controls — feed, energy streams).

After calculating a single solution of the model optimization problem MOP, the obtained current model-optimal point, say $\widehat{c}_m$, is applied to the process and remains unchanged until new significant changes in the operating conditions occur. What has just been described is precisely the classical *open-loop* optimizing control strategy. Notice that before solving the optimization problem a calibration (an adaptation) of the process model should be performed, as precisely as possible. It is particularly important in the presented open-loop control structure because the quality of the obtained steady-state control relies entirely on the precision of the process model used by the optimization routine. Having applied the steady-state control $\widehat{c}_m$ to the process it remains unmodified possibly for a longer time.

What has just been discussed clearly shows that operating in open loop only is reasonable if the process model and disturbance estimates *are accurate enough.* If it is not the case performance deterioration will be inevitable and violation of technological constraints may occur, the last phenomenon being even more important. Therefore, more elaborate control structures and algorithms involving certain feedback information should be then applied.

A basic underlying idea about how to improve the open-loop control is as follows: after applying the model-optimal optimizing control $\widehat{c}_m$ to the process (and decaying the following transient responses) a new steady-state equilibrium is achieved and the corresponding steady-state process outputs y^1 can be measured,

$$y^1 = F_*(\widehat{c}_m(\alpha^0), w) \tag{2.36}$$

where explicit dependence of $\widehat{c}_m$ on the current process model parameters $\alpha = \alpha^0$, i.e., $\widehat{c}_m = \widehat{c}_m(\alpha^0)$, has been introduced. Certainly, if the process model is not perfectly adequate (ideal) and/or the disturbance estimates w_e not equal to the real disturbance values w, then the measured process outputs y^1 will be different from the model-predicted values y^1_m ,

$$y^1_m = F(\widehat{c}_m(\alpha^0), w_e, \alpha^0). \tag{2.37}$$

This brings new information on the process and can be utilized to improve the accuracy of the model by performing its additional adaptation using an improved data set, i.e., the data set with the new measurements included. Let us denote the new model parameters obtained in this way by α^1. Now, we can repeat the solution of the model optimization problem MOP with α^0 replaced by α^1 and then apply the resulting steady-state optimizing control $\widehat{c}_m(\alpha^1)$ to the process. And again, new measurements are available in the process after the new steady-state has been established, and the whole procedure of model adaptation and the MOP solution can be again repeated, *etc.* What has just been described is precisely the *It-*

erative Two-Step method (ITS method) consisting of successively repeated model optimizations and model parameter estimations. The structure of the method is shown schematically in Fig. 2.5. It can be easily seen that the described ITS optimizing control method operates in the *open-loop-with-feedback* structure.

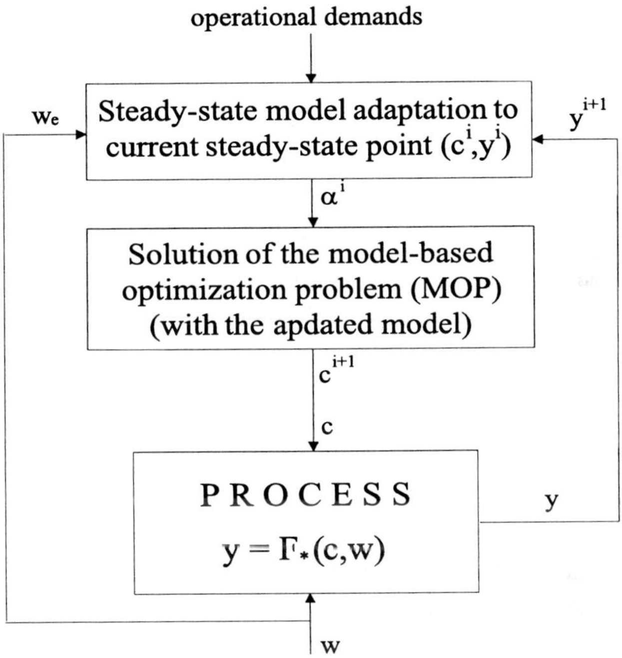

Fig. 2.5 Structure of the iterative two-step (ITS) method.

Fundamental questions arise concerning practical and theoretical properties of the ITS procedure:

- Is the original performance function improved at each iteration ?
- Is the whole procedure convergent, and if so, what can be said about the convergence limit in terms of its optimality and feasibility; is it possible at all to get the real optimal point in spite of model-reality differences?

Unfortunately, answers to the stated questions are generally not positive. Moreover, general theoretical analysis of the ITS method is not available, the results of its application do heavily depend on the actual situation, on the process nature and its model accuracy. We address these questions in a simple, illustrative way in the following example.

Example 2.1

The following simple MOP problem with a linear process model

$$\begin{aligned} &\text{minimize}_{c,y}\ Q(c, y) \\ &\text{subject to}:\ y = F(c, \alpha) = a \cdot c + \alpha \\ &\qquad\qquad\quad c \geq 0 \end{aligned} \tag{2.38}$$

will be considered, where $Q(\cdot, \cdot)$ is defined by its dashed contour lines depicted in Fig. 2.6. It is assumed for clarity of presentation that only the additive parameter α undergoes adaptation (a remaining constant), thus the parameter estimation consists of simple deterministic matching the model and plant responses, i.e., consists in finding $\alpha = \alpha(c)$ from the equation

$$F(c, \alpha) = F_*(c) \tag{2.39}$$

where the nonlinear plant output mapping $F_*(c)$ is shown graphically in the figure by a bold solid curve.

Starting from c^0 a sequence of points $\{c^n\}$ is generated by subsequent parameter estimations (2.39) and model optimizations (2.38):

$$c^0 \to \alpha^0 = \alpha(c^0) \to \hat{c}_m(\alpha^0) = c^1 \to \alpha^1 = \alpha(c^1) \to \hat{c}_m(\alpha^1) = c^2 \to \cdots \to c^\infty.$$

The sequence converges to the point c^∞, which is clearly not equal to the real optimal point $\hat{c}_*$. The point c^∞ is defined by the intersection of the plant output curve $F_*(c)$ with the curve of model-optimal outputs $F(\hat{c}_m(\alpha), \alpha)$, where $\hat{c}_m(\alpha)$ are minimizers of the model optimization problem (2.38). This simple example shows also that not every ITS iteration leads to an improvement in the true process performance function value, i.e., in the value $Q(c, F_*(c))$ — obviously a drawback extremely important from a practical viewpoint. Certainly, the results shown in the example could be improved if we used more elaborate model adaptation consisting in corrections of both a and α parameters. Nevertheless, the final result would generally still be suboptimal, with suboptimality difficult to be estimated in advance. However, it can be easily seen from Fig. 2.6 that $c^\infty = \hat{c}_*$

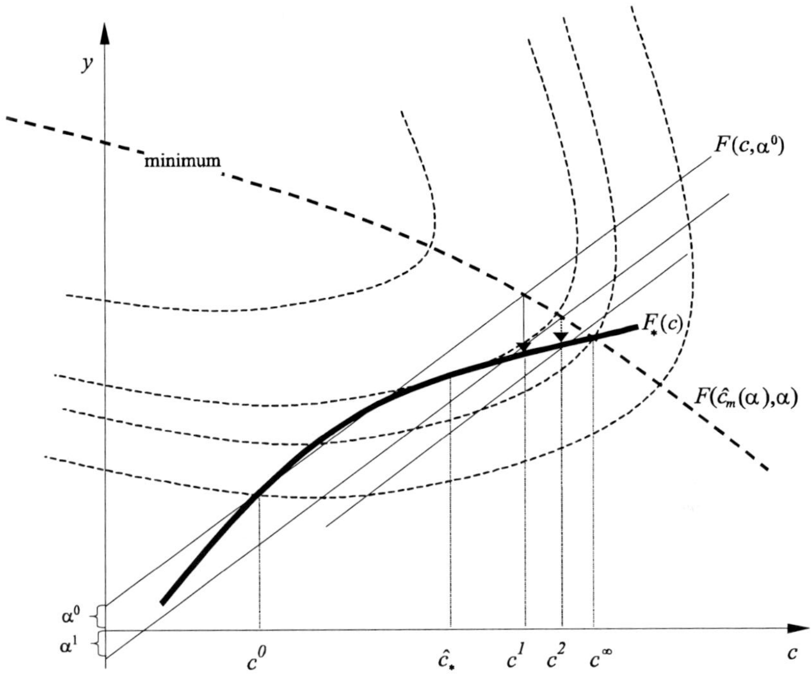

Fig. 2.6 Example iterations of the ITS algorithm.

only if $a = F_*'(\hat{c}_*)$, i.e., if the model mapping derivative matches the true process mapping derivative at the optimal point. □

This simple example reveals the important general condition which is necessary to achieve the true optimal point when using a plant model. Namely, derivatives of the plant mapping and its model must be equal at the optimal point, i.e.,

$$F_c'(\hat{c}_m(\alpha), w_e, \alpha) = (F_*)_c'(\hat{c}_*, w) \tag{2.40}$$

where $F_c'(c, w, \alpha)$ denotes partial derivative with respect to c, $F_c'(c, w, \alpha) = \frac{\partial}{\partial c}F(c, w, \alpha)$, *etc.* Many figures similar to Fig. 2.6 can be easily constructed showing that differences in derivatives can lead to quite significant differences in c^∞ and $\hat{c}_*$ and in the corresponding values of the true plant performance index, $Q(c^\infty, F_*(c^\infty, w))$ and $Q(\hat{c}_*, F_*(\hat{c}_*, w))$. The described procedure of interchanging steps of model parameter estimations and model optimizations can even happen to diverge — it is not difficult to construct such examples.

It follows from the argument presented here that special attention must be paid to the accuracy of the plant model and, especially, its derivatives when applying the single model optimization or the standard two-step procedure of interchanging parameter estimations and model optimizations. However, it is usually not possible to construct accurate plant models, especially when the range of admissible input signal variations is broad. The models are then sometimes quite crude. An alternative approach would be to try to acquire precise knowledge of the true plant mapping derivatives only locally based on suitable plant output measurements at current points. It turns out that this information can be used in the model-based optimizations in order to force the convergence of model-based solutions towards the true plant optimum. This is precisely the essence of the *modified iterative two-step method*, more commonly known as the *Integrated System Optimization and Parameter Estimation (ISOPE) method*. The ISOPE method will be presented in detail in Chapter 4, including derivation of several algorithms and their analysis. In this section, the aim is to present only the very basic idea of this method together with the corresponding control structure.

As explained briefly after the discussion of the properties of the standard ITS method, (2.40) is the necessary condition to achieve the real optimal point when applying an optimizing control algorithm which uses the plant model only. The condition states that at the optimal point values of the derivatives of the plant input-output mapping F_* and its model F must be equal. Let us assume that at a current operating steady-state point c^i the derivative $F_*'(c^i, w)$ of the plant mapping F_* can be approximated, certainly using current output measurements in an appropriate way. Then the idea about how to incorporate this knowledge into the optimization, in order to arrive at optimizing control algorithms with true optimality property, can now be explained in a very basic way.

Recall the basic model optimization problem MOP (2.20), but with inequality constraints on the process outputs omitted:

$$\begin{aligned} &\text{minimize } Q(c, y) \\ &\text{subject to: } \; y = F(c, w_e, \alpha) \\ &\qquad\qquad\quad\; g(c) \le 0. \end{aligned} \tag{2.41}$$

The output variables y can now be eliminated leading to the equivalent model optimization problem with independent decision variables c only

$$\begin{aligned} &\text{minimize}_c\ Q(c, F(c, w_e, \alpha)) \\ &\text{subject to}:\ g(c) \le 0. \end{aligned} \tag{2.42}$$

The corresponding optimizing control problem, compare with the OCP defined by (2.18), has the form

$$\begin{aligned} &\text{minimize}_c\ Q(c, F_*(c, w)) \\ &\text{subject to}:\ g(c) \le 0 \end{aligned} \tag{2.43}$$

with the solution (the true process optimal steady-state) denoted by $\hat{c}_*$. Assuming that values of the derivatives $F_*'(c^i, w)$ can be calculated at a current point c^i, let us consider the following modified model optimization problem (MMOP) at this point

$$\begin{aligned} &\text{minimize}_c\ \{Q(c, F(c, w_e, \alpha^i)) - \lambda(c^i, \alpha^i)^T c\} \\ &\text{subject to}:\ g(c) \le 0 \end{aligned} \tag{2.44}$$

where

$$\lambda(c^i, \alpha^i)^T = Q_y'(c^i, F(c^i, w_e, \alpha^i)) \cdot [F_c'(c^i, w_e, \alpha^i) - (F_*)_c'(c^i, w)]. \tag{2.45}$$

Now, it follows directly from the construction that *the performance function of the MMOP (2.44) has at the point c^i the derivative equal to the derivative of the performance function of the optimizing control problem (2.43),* provided the model and process outputs are equal after an appropriate model parameter estimation (yielding the parameter values α^i) at the point c^i,

$$F(c^i, w_e, \alpha^i) = F_*(c^i, w). \tag{2.46}$$

Imagine now the modified model optimization problem MMOP is used instead of the basic model optimization problem MOP in the iterative two-step method, i.e., the iterations of the set-points are done in such a way that a solution $\hat{c}(c^i)$ of the MMOP becomes the next process set-point c^{i+1}, $c^{i+1} = \hat{c}(c^i)$, *etc.* Then, if the sequence $\{c^i\}$ is convergent to a point, say $\tilde{c}$, this point satisfies $\tilde{c} = \hat{c}(\tilde{c})$ – being both the initial and the optimal point of the MMOP. Further, it can be then easily deduced that $\tilde{c}$ *satisfies also*

necessary optimality conditions for the optimizing control problem (OCP). The reasoning is as follows:
As $\tilde{c}$ is an optimal point for MMOP, then it satisfies its necessary optimality conditions. Provided the constraint set C is convex, these conditions are

$$\frac{d}{dc}\{Q(c, F(c, w_e, \alpha^i)) - \lambda(c^i, \alpha^i)^T c\} \cdot [c - \hat{c}(c^i)] \geq 0 \quad \text{for all } c \in C. \quad (2.47)$$

If the constraint set C is not active at $\hat{c}(c^i)$, these conditions simplify to the well known zero-derivative conditions

$$\frac{d}{dc}\{Q(c, F(c, w_e, \alpha^i)) - \lambda(c^i, \alpha^i)^T c\} = 0. \quad (2.48)$$

Now, because

$$\begin{aligned}\frac{d}{dc}\{Q(c, F(c, w_e, \alpha^i)) - \lambda(c^i, \alpha^i)^T c\} &= Q_y'(c^i, F(c^i, w_e, \alpha^i)) \cdot F_c'(c^i, w_e, \alpha^i) + \\ &-Q_y'(c^i, F(c^i, w_e, \alpha^i)) \cdot [F_c'(c^i, w_e, \alpha^i) - (F_*)_c'(c^i, w)] = \\ &= Q_y'(c^i, F(c^i, w_e, \alpha^i)) \cdot (F_*)_c'(c^i, w) \quad (2.49)\end{aligned}$$

the MMOP optimality conditions (2.47) can be rewritten in the form

$$[Q_y'(c^i, F(c^i, w_e, \alpha^i)) \cdot (F_*)_c'(c^i, w)] \cdot [c - \hat{c}(c^i)] \geq 0 \quad \text{for all } c \in C \quad (2.50)$$

yielding at the convergence point $\tilde{c}$

$$[Q_y'(\tilde{c}, F(\tilde{c}, w_e, \alpha^i)) \cdot (F_*)_c'(\tilde{c}, w)] \cdot [c - \tilde{c}] \geq 0 \quad \text{for all } c \in C. \quad (2.51)$$

From the other hand, necessary conditions for the point $\tilde{c}$ to be optimal for the OCP are

$$[Q_y'(\tilde{c}, F_*(\tilde{c}, w)) \cdot (F_*)_c'(\tilde{c}, w)] \cdot [c - \tilde{c}] \geq 0 \quad \text{for all } c \in C. \quad (2.52)$$

Because

$$F(\tilde{c}, w_e, \alpha^i) = F_*(\tilde{c}, w) \quad (2.53)$$

see (2.46), it follows from (2.51) and (2.52) that the point $\tilde{c}$ satisfies also necessary optimality conditions for the OCP. In other words, the iterations of the two-step method with the model optimization problem MOP replaced by the modified problem MMOP *converge to the point satisfying the necessary optimality conditions of the OCP (2.43)*. Therefore, a true optimal (precisely: satisfying the necessary optimality conditions) set-point for the process can be reached, not a suboptimal one as it was in the case of the standard ITS method.

What has just been explained is the underlying idea of the *modified iterative two-step method*, more commonly known as *the Integrated System Optimization and Parameter Estimation (ISOPE) method*. The structure of the method is depicted in Fig. 2.7.

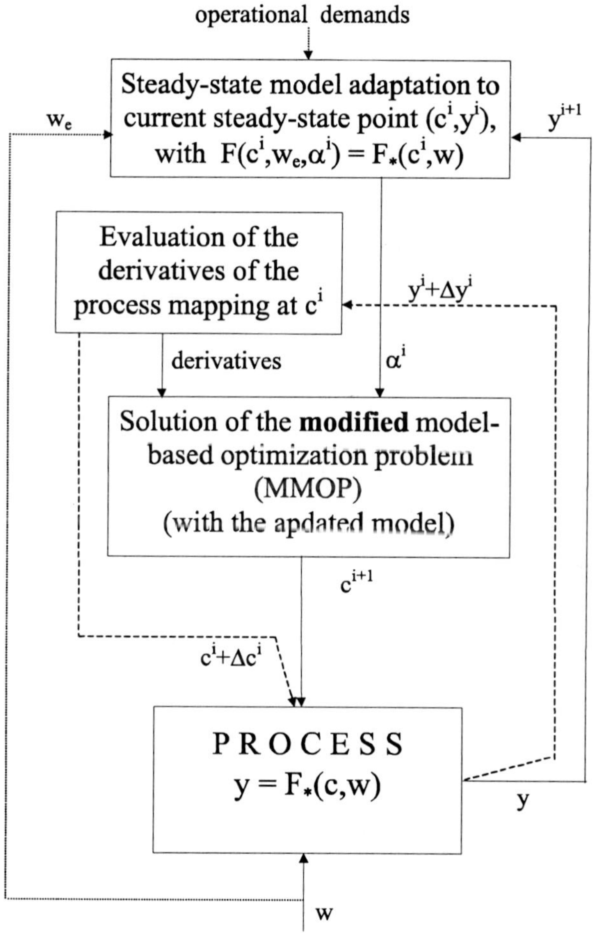

Fig. 2.7 Structure of the ISOPE method (modified two-step method).

When comparing the structures of the iterative two-step (ITS) and the modified iterative two-step (ISOPE) methods, compare Fig. 2.5 and Fig. 2.7, the main difference is in the introduction of the block evaluating process mapping derivatives in the ISOPE structure. This block performs the most difficult, from the practical viewpoint, part of each of the ISOPE algorithms. Moreover, differences between various ISOPE algorithms stem from different techniques of the derivative approximation. An inevitable part of each such technique are certain deviations of the current steady-state point c^i, $c^i + \Delta c^i$ (denoted in Fig. 2.7 by dashed lines), resulting in additional process outputs measurements, $y^i + \Delta y^i$. In more classical ISOPE algorithms additional deviations were necessary at each iteration point c^i (i.e., for all i), whereas in more recent algorithms the deviations are needed only during the so called initial phase and are also combined with the optimization task. The ISOPE algorithms, especially the more recent ones, can be quite involved. For the basic case without constraints on the process outputs, they will be described in detail in Chapter 4 both from theoretical and practical viewpoint. Let us recall here only that they differ from the ITS algorithms first of all in that *they converge to the process true optimal steady-state point.* Moreover, they result in the process performance function improvement after every or almost every iteration (depending on the algorithm applied).

2.3.2 *Problems with constraints on process outputs*

We shall now consider a more general and practically important case with constraints on the process outputs, $\psi(y) \leq 0$. However, as it was announced in the last subsection, a development of the ISOPE algorithms in Chapter 4 will be made for simpler, basic case without constraints on the outputs. The case with constrained outputs will then be treated separately in Chapter 5. The reason is that these two classes of constraints lead to on-line control structures *differing significantly,* in particular when iterative optimizing control algorithms are applied.

The constraints on decision variables c are relatively easy to be treated. Namely, $g(c^{i+1}) \leq 0$ means constraint satisfaction both in the model and in the real process at the control algorithm iteration i, since the values of the set-points c^{i+1} are directly applied to the process. However, the situation is quite different for constraints on outputs y. Satisfying the constraint $\psi(y^{i+1}) \leq 0$ in the model (the model-based algorithm can assure that) means that

$$\psi(F(c^{i+1}, w_e, \alpha^i)) \leq 0. \tag{2.54}$$

However, this *would not guarantee* that in the process

$$\psi(F_*(c^{i+1}, w)) \leq 0 \tag{2.55}$$

because generally,when model inaccuracy is assumed, we have

$$F(c^{i+1}, w_e, \alpha^i) \neq F_*(c^{i+1}, w). \tag{2.56}$$

As a result, an output constraint active in the model will usually be violated or inactive in the process, both situations are hardly acceptable from safety or optimality reasons. However, the constraint violation is usually too risky or unacceptable, leading either to dangerous operating conditions or unacceptable decrease in product quality. Therefore, a practical solution consisting in introduction of the constraints with *safety zones* has been applied for years in process industries, to assure output constraint satisfaction. In our notation, a constraint $\psi(y) + \delta\psi \leq 0$ instead of $\psi(y) \leq 0$ must be put into the optimization problem, where $\delta\psi > 0$ is a safety zone, the more severe the model uncertainty the larger the zone. Obviously, appropriate values of the zones $\delta\psi$ can be estimated only experimentally and, therefore, these zones are usually chosen in a rather conservative way.

Introduction of safety zones for active, critical constraints usually leads to significant losses in process efficiency and in productivity. Therefore, there is a common interest in control algorithms that are capable to operate with significantly smaller safety zones. In Chapter 3 and, first of all, in Chapter 5 iterative optimizing control algorithms will be presented that are capable *to operate with significantly decreased safety zones.* Two approaches will be developed:

- Iterative algorithms for the *control structure with feedback controlled output constraints,* keeping the process output constraints satisfied at each iteration with the accuracy of a feedback control error.
- Iterative ISOPE algorithms *with algorithmic implementation of output constraints*, i.e., incorporating the output constraints into the algorithm itself in such a way that they are satisfied at the end of iterations. However, during the iterations the constraints can be violated and therefore more significant safety zones may be needed.

For important constraints always active under normal operating conditions the *control structure with feedback controlled output constraints* has been found as a sound solution. The structure is shown in Fig. 2.8, it has been in fact introduced in Chapter 1 (see Section 1.3).

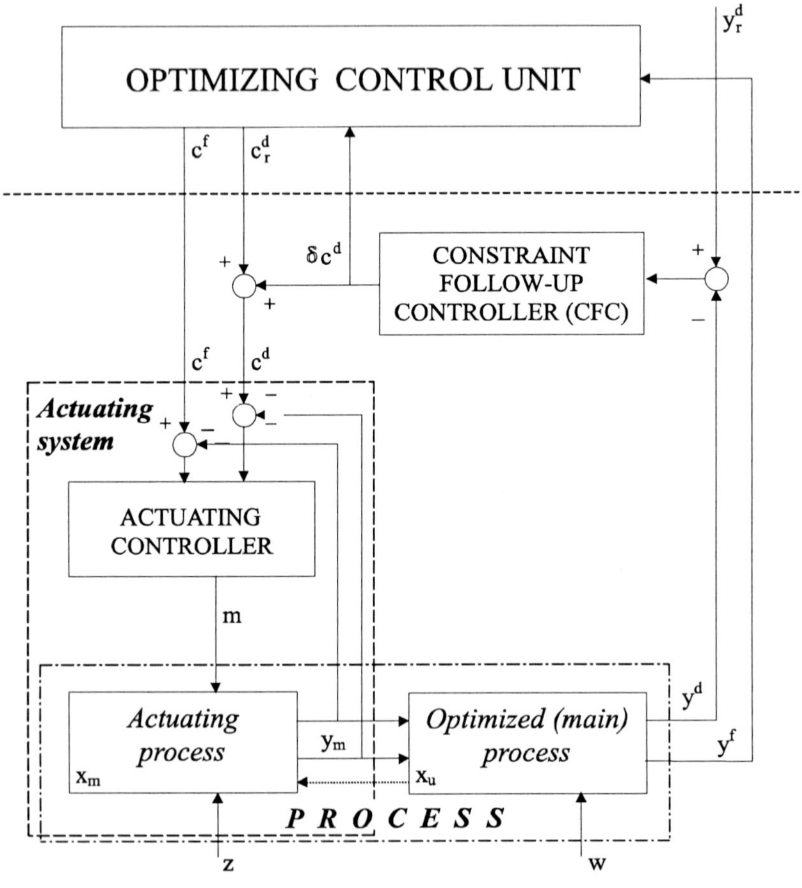

Fig. 2.8 Multilayer structure with feedback controlled output constraints.

It incorporates a dedicated set-point controller called a *constraint follow-up controller* (CFC) designed to keep the output constraints satisfied. It has been drawn for the case of constraints on process outputs simplified to the most encountered and simple form $y^d \le y_r^d$. Assuming only always active constraints are taken into account, the CFC provides for the equality $y^d - y_r^d = 0$.

Introduction of the CFC at the upper layer (set-point control layer) implies the situation where the output vector y is partitioned into two parts: the subvector of constrained outputs y^d and the subvector of free (unconstrained) outputs y^f, $y = (y^f, y^d)$, see Chapter 1 (Section 1.3). The optimizing control (set-point) vector is also appropriately partitioned, $c = (c^f, c^d)$, where $\dim c^d = \dim y^d$. The process model is obviously also partitioned in the way corresponding to the partitioning of y, i.e., $F = (F^f, F^d)$, to get an explicit description of the constrained outputs. As a result, the corresponding model-based optimization problem MOP (2.20) takes then the following form :

$$
\begin{aligned}
&\text{minimize} \quad Q(c, y) \\
&\text{subject to :} \quad y^f = F^f(c, w_e, \alpha) \\
&\qquad\qquad\qquad y^d = F^d(c, w_e, \alpha), \;\; y^d = y_r^d \\
&\qquad\qquad\qquad g(c^f) \leq 0
\end{aligned}
\tag{2.57}
$$

where the constraint on outputs $\psi(y^d) \leq 0$ has been simplified to the simpler form $y^d \leq y_r^d$ (and its activity is assumed).

Using the control structure of Fig. 2.8, the constraint $y^d = y_r^d$ is kept satisfied by the constraint follow-up controller (CFC), both in transient processes and in steady-states. Notice that the CFC is a correction controller, i.e., it corrects only the steady-state c_r^d evaluated (together with c^f) when solving the model-based or modified model-based optimization problem — the correction necessary to eliminate the impact of model uncertainty on the constraint fulfillment.

We will assume in most parts of the book that the set of active output constraints remains constant when searching for a new set-point under constant operating conditions. However, the activity of output constraints may change when these conditions change. This means that we assume that a single activation of an (iterative) optimizing control algorithm leading to optimal set-point adjustment operates *under the same active output constraints* (but active constraints on optimizing controls c may change during the search). The assumption is realistic and applies to many practical cases. First of all, the presented approach is always and easily applicable to cases of constantly active vital output constraints, such as composition constraints on product streams in distillation columns. Generally, the assumption could be relaxed allowing for more difficult cases when sets of active output constraints change. However, this would imply using variable structure constraint follow-up controllers (see Fig. 2.8) even during a single activation of an optimizing control algorithm under the constant operating conditions. It would therefore lead to much more complicated conditions for the optimizing controller and more complicated algorithms, still not well understood.

Applying the presented control structure with properly designed CFC controller should lead to significant decrease of safety zones associated with the constraints $y^d \leq y_r^d$, when compared to pure model-based treatment of the constraints. Notice that the CFC is in fact an element of the dynamic part of the control system. It is *dynamic* since it acts on c^d whenever a difference (an error) $e^c = y_r^d - y^d$ is encountered, with sampling time Δ_p much smaller than the time τ of the plant transient process. However, Δ_p is usually much greater than the basic sampling interval Δ_t of the follow-up actuating controllers acting directly on the plant inputs. The CFC should be designed in a way taking into account that it operates on the boundary of its process variable feasible region. It should avoid or minimize overshoots in the direction of this boundary violation. The set-point y_r^d itself is usually specified taking into account a safety zone around the true required output constraint value in order to assure the process operates without violating the true constraint.

The model-based optimization problem (2.57) for the optimizing control layer of the considered control structure of Fig. 2.8, which will be referred to as MOP2 in the sequel, can be rewritten as follows:

$$\begin{aligned} &\text{minimize } Q(c^f, c^d, y^f, y^d) \\ &\text{subject to: } \; y^f = F^f(c^f, c^d, w_e, \alpha) \\ &\qquad\qquad\quad F^d(c^f, c^d, w_e, \alpha) = y_r^d \\ &\qquad\qquad\quad g(c^f) \leq 0. \end{aligned} \tag{2.58}$$

The control structure and the model optimization problem (2.58) can form the basis of an iterative two-step (ITS) method, analogous to that described in the previous subsection. A special version of the ITS method can be derived for the structure, called the *iterative correction with disturbance estimation method*, suitable for cases with relatively accurate process model and a poor estimate of main disturbances. The method will be presented in the next chapter. While preserving the process output constraints with

CFC accuracy, methods belonging to the ITS class can only be suboptimal, the suboptimality depending on the level of model or disturbance estimate uncertainty.

The control structure of Fig. 2.8 and the corresponding *optimizing control problem*

$$\begin{aligned} &\text{minimize } Q(c^f, c^d, y^f, y^d) \\ &\text{subject to: } \; y^f = F_*^f(c^f, c^d, w) \\ &\qquad\qquad\quad F_*^d(c^f, c^d, w) = y_r^d \\ &\qquad\qquad\quad g(c^f) \leq 0 \end{aligned} \tag{2.59}$$

can be also the basis for development of dedicated algorithms of the class belonging to the modified two-step method (the ISOPE method). Such algorithms will be derived and presented in detail in the first section of Chapter 5.

The second possibility when considering optimizing control with constraints on process outputs is to *incorporate the output constraints into the ISOPE algorithm itself* in such a way that they are satisfied in the process at the end of the iterations. This can be achieved by suitably incorporating the constraints into the performance function of the appropriately formulated modified model optimization problem of the ISOPE algorithm. This obviously leads to more complicated ISOPE algorithms, which will be described in the second section of Chapter 5. These algorithms assure constraint fulfillment at convergence (at the end of iterations), but during the iterations this cannot be assured and therefore more significant safety zones than in the former case of feedback controlled constraints may be needed. Certainly, being more efficient makes the case with feedback controlled output constraints more expensive, and the CFC have to be designed and installed. Therefore, it is a matter of choice at the design stage which solution should be applied, depending on the desired accuracy of constraint fulfillment and related process productivity.

It should be added that in the latter case of constraints on outputs treated algorithmically only, the case with more general form of these constraints, namely dependent also on the set-point c,

$$\psi(c, y^d) \leq 0 \tag{2.60}$$

can be relatively easily incorporated into the algorithms.

Chapter 3

Iterative Correction with Disturbance Estimation

3.1 Correction Algorithm for Problems with Unconstrained Outputs

Let us consider cases when the following assumptions are met:

- The model of the controlled process (its structure and parameters) is sufficiently precise for optimization purposes, but
- modeling uncertainty is expressed mainly in inaccurate current estimation of the uncontrolled inputs (disturbances) w, i.e., in their inaccurate estimates w_e. These inputs may or may not be directly measured, but not sufficiently frequently for purposes of on-line repetitive optimization.

It is also assumed that the number of the uncontrolled inputs $n_w = \dim w$ is equal or greater than the number m of the outputs y, $m = \dim y$. If $n_w > m$, then m most important uncontrolled inputs will be selected, i.e., those most significantly influencing values of the outputs y, and the remaining ones will be treated as fixed parameters included into the model parameter vector α. In this way, it can be assumed that $n_w = m$, a technical assumption needed for the technique to be developed.

Suppose now that, for a currently available estimate $w_e = w_e^{(0)}$, the model optimization problem (MOP, see previous chapter)

$$\begin{aligned} &\text{minimize}_c \ Q(c, y) \\ &\text{subject to}: \ y = F(c, w_e^{(0)}, \alpha) \\ &\qquad\qquad\quad\ g(c) \le 0 \end{aligned} \tag{3.1}$$

has been solved, denoting the solution by $\hat{c}_m^{(0)}$. The set-point $\hat{c}_m^{(0)}$ was then applied to the controlled plant and, after transient processes have ended, the values of the outputs were measured yielding

$$y^{(0)} = F_*(\hat{c}_m^{(0)}, w). \tag{3.2}$$

On the other hand, calculating values of the outputs from the model only, at the current set-point $\hat{c}_m^{(0)}$ and disturbance estimate $w_e^{(0)}$, we get

$$y_m^{(0)} = F(\hat{c}_m^{(0)}, w_e^{(0)}, \alpha). \tag{3.3}$$

If the modeling had been perfect (including perfect estimates $w_e^{(0)} = w$ and perfect action of the regulatory controllers) then the equality

$$y_m^{(0)} = y^{(0)} \tag{3.4}$$

would obviously hold. However, it is usually not true and inaccurate estimation, i.e., $w_e^{(0)} \neq w$, results generally in the inequality $y_m^{(0)} \neq y^{(0)}$.

Generally, there may be different reasons implying this inequality:

(1) reasons connected with system failures,
(2) the use of approximate models of the process mapping F,
(3) errors in the estimate $w_e^{(0)}$ of the disturbances w.

The situation leading to the first case should be detected and handled by a supervisory control layer, which overrides then decision of the optimizing control layer. This case is out the scope of this book. Then, cases two and three should be taken into account. Static characteristics typical for the considered class of technological processes are rather regular. Therefore, for certain class of plants our assumption of sufficiently accurate modeling for optimization purposes is reasonable and we could assume that errors in the disturbance estimates dominate. This assumption, essential from technical reasons for further analysis, is also to a certain degree *conventional.* The reason is the errors in plant modeling can also be treated as certain disturbances. In particular, distinguishing between model parameters and external disturbances is often a matter of convenience and is not important for the algorithm which will now be presented. Additionally, imperfect action of the lower-layer (regulatory) controllers can also be treated as certain disturbances (therefore, we assume perfect action of these controllers throughout most part of this book, and in particular in this chapter).

It is possible, nevertheless, to correct the disturbance estimates $w_e^{(0)}$, and reasonable in the assumed class of systems. This possibility results from the application of the information contained in differences between values of the measured outputs $y^{(0)}$ and the model outputs $y_m^{(0)}$. The corrected disturbance estimate value $w_e^{(1)}$ can be evaluated solving the following set of equalities

$$F(\hat{c}_m^{(0)}, w_e^{(1)}, \alpha) = y^{(0)}. \tag{3.5}$$

In general, the set of m nonlinear equations (3.5) must be solved to get $w_e^{(1)}$. In practice, due to the uncertainty it may be sufficient to solve a set of linear equations

$$F_w^{'}(\hat{c}_m^{(0)}, w_e^{(0)}, \alpha)[w_e^{(1)} - w_e^{(0)}] = y^{(0)} - y_m^{(0)} \tag{3.6}$$

being a linear approximation of the original set (3.5) at the current value of the disturbance estimates. This set is simpler and quickly, robustly solvable. It is well defined if the matrix $F_w^{'}(\hat{c}_m^{(0)}, w_e^{(0)}, \alpha)$ is nonsingular, as can be usually assumed.

The solution of the model optimization problem (MOP) can now be repeated, with the corrected value of disturbance estimates $w_e = w_e^{(1)}$, *etc.*

In conclusion, the algorithm of *iterative correction with disturbance estimation* consists of two steps, iteratively repeated:

- Solution of the MOP with the current value of the disturbance estimates, application of the obtained set-point to the controlled process and measurement of the corresponding values of the plant outputs y, in the steady-state after transient processes have ended.
- Correction of the estimates of significant uncontrolled inputs (disturbances) w_e based on the obtained output measurements, performed as a solution of the set of linear equations (3.6), or the original set of nonlinear equations (3.5).

The structure of the presented correction algorithm is a special case of the structure of standard iterative two-step method (ITS), as described in the previous chapter (Section 2.3.1). The only difference is that the step of model parameter estimation is now performed as correction of disturbance estimates w_e, for appropriately chosen m most significant disturbances. Therefore, convergence and optimality properties of the correction algorithm are as for the ITS method. First, it is *suboptimal*, with suboptimality level dependent on the quality of the controlled plant model. In particular, structural accuracy of the model mapping $F(\cdot,\cdot,\cdot)$ is most important here, not precision of the evaluation of estimates of significant disturbances, as these estimates will be then corrected in the course of the correction algorithm. Second, convergence conditions of the algorithm are not known, although it happens to be usually convergent, with most improvement expected after the first iteration. It can be easily deduced, analogously as it has been done in Example 2.1 in Section 2.3.1, that the correction algorithm converges (provided it is convergent) to a point defined by intersection of the process output curve $F_*(c, w)$ with the curve of model-optimal outputs $F(\hat{c}_m(w_e), w_e)$, on the (c, y) plane.

Last but not least, it should be noted that the presented way of correction may lead to estimate values $w_d^{(i)}$ correcting, at least partially, not only estimation inaccuracies of the corresponding disturbances w. It may also correct both parametric and structural inaccuracies of the plant model being the source of differences between the measured output values and the output values calculated on the model only.

3.2 Correction Algorithm for Structures with Feedback Controlled Output Constraints

The optimizing control structure with feedback controlled output constraints will be considered, see Fig. 2.8 in the previous chapter (Subsection 2.3.2) and the discussion therein. Let us recall that this control structure is applicable, at normal operating conditions, for cases with active inequality constraints (or equality constraints) on some process outputs $y^d \in \mathbb{R}^{m_d}$. The constraints of a simple form $y^d = y_r^d$ will be assumed, i.e., the outputs y^d are to be kept constant at prescribed values y_r^d. In order to satisfy this requirement a set-point controller called *constraint follow-up controller* (CFC) is applied, acting on the set-points $c^d \in \mathbb{R}^{m_d}$.

Analogously as in the previous section, the cases will be addressed when:

- The steady-state model of the controlled process (its structure and parameters) is sufficiently precise for optimization purposes, but
- modeling uncertainty is expressed first of all in inaccurate on-line estimation of the uncontrolled inputs (disturbances) w, i.e., in their inaccurate estimates w_e. These inputs may be not directly mea-

sured or may be measured, but not sufficiently frequently for purposes of on-line repetitive optimization.

It is assumed the number of these inputs $n_w = \dim w$ is equal or greater than the number m_d of constrained and feedback controlled outputs y^d, $m_d = \dim y^d$. If $n_w > m_d$ then m_d most important uncontrolled inputs will be selected, i.e., those most significantly influencing values of the outputs y^d, and the remaining uncontrolled inputs will be treated as fixed parameters included into the model parameter vector α. In this way, it can be assumed that $n_w = m_d$.

In the correction algorithm considered in the previous section the correction was based on the information contained in differences between values of the measured plant outputs and outputs calculated on the model only. In this section feedback controlled outputs are considered, therefore both plant and model outputs are kept on the same values y_r^d. Thus, the difference is zero. Certainly, it is due to the feedback action of the CFC. To keep the prescribed outputs on the desired values in spite of changes of slow-varying disturbances w (and inaccuracies in the plant model), this controller generates appropriate non-zero set-point corrections δc^d, $c^d = c_r^d + \delta c^d$, see Fig. 2.8. These non-zero correcting optimizing controls δc^d are known and can be utilized to improve the disturbance estimates in the iterative correction algorithm, leading to improvement in the plant performance. The correction algorithm based on this idea was proposed by Duda (1991). It will be described in what follows.

It is assumed that there are significant slow-varying disturbances w influencing the process outputs y^d, i.e., the input-output relation F_*^d does explicitly depend on w,

$$y^p = F_*^d(c^f, c^d, w). \tag{3.7}$$

It is also assumed in this section (without loss of generality) that there are only constrained and CFC controlled outputs, i.e. $y = y^d$.

Let us assume now that for a given disturbance estimate $w_e^{(0)}$ the appropriate steady-state model optimization problem (MOP2, without free outputs)

$$\begin{aligned} &\text{minimize}_{c^f, c^d}\ Q(c^f, c^d, y^d) \\ &\text{subject to}: y^d = F^d(c^f, c^d, w_e^{(0)}, \alpha),\ y^d = y_r^d \\ &\qquad\qquad g(c^f) \le 0 \end{aligned} \tag{3.8}$$

has been solved, with the solution $\hat{c}^{(0)} = (\hat{c}^{f(0)}, \hat{c}_r^{d(0)})$. The set-point $\hat{c}^{(0)}$ is then applied to the controlled plant. However, only $\hat{c}^{f(0)}$ is directly applied (is to be forced by actuating controllers). The value $\hat{c}_r^{d(0)}$ is only a *recommended* value for c^d, which is to be corrected by the CFC, if needed, see Fig. 2.8. If the process model and disturbance estimates had been perfect then the steady-state correction δc^d introduced by the CFC would be zero, stabilizing c^d on the value $\hat{c}_r^{d(0)}$. As discussed previously, it is usually not true. Therefore, the CFC stabilizes c^d on the value $c^{d(0)} \neq \hat{c}_r^{d(0)}$, and the difference can be significant. Generally, there may be different reasons implying this inequality, as discussed in the previous section. It can be assumed, as discussed there, that errors in disturbance estimates dominate. To simplify the formal considerations it will also be assumed that the CFC controller action is perfect and the (unknown) value of the disturbance vector w is constant during the correction procedure. Let us remind the reader that $w_e^{(0)}$ is the estimate applied during last process steady-state optimization (3.8). Without action of the CFC, the outputs y^d would tend towards the value $y^{d(0)} \neq y_r^d$,

$$y^{d(0)} = F_*^d(\hat{c}^{f(0)}, \hat{c}_r^{d(0)}, w). \tag{3.9}$$

The CFC prevents this situation by changing its output to the value $c^{d(0)}$, $c^{d(0)} \neq \hat{c}_r^{d(0)}$ such that

$$F_*^d(\hat{c}^{f(0)}, c^{d(0)}, w) = y_r^d. \tag{3.10}$$

This automatic correction deteriorates the process performance in general. Therefore, a correction of the set-points is desirable to restore optimal operation. The main idea of the correction algorithm is to utilize information about $c^{d(0)}$ in order to evaluate a corrected value $w_e^{(1)}$ of the disturbance estimate. It is proposed to perform the correction in a way such that the following *nonlinear equation*

$$F^d(\hat{c}^{f(0)}, c^{d(0)}, w_e^{(1)}, \alpha) = y_r^d \tag{3.11}$$

is satisfied. If the model F^d had been perfect and action of the CFC ideal then $w_e^{(1)}$ would be equal to the "true" disturbance value w. Subsequent solution of the model optimization problem (3.8) with $w_e^{(0)}$ replaced by $w_e^{(1)}$ would then yield exact optimal set-point for the process. Of course, these assumptions are not realistic. Therefore, instead of solving the nonlinear set of equations (3.11) it is proposed to evaluate $w_e^{(1)}$ by solving the following set of *linear equations*

$$(F^d)'_{c^d}(\hat{c}^{f(0)}, \hat{c}_r^{d(0)}, w_e^{(0)}, \alpha)[c^{d(0)} - \hat{c}_r^{d(0)}] + \\ +(F^d)'_w(\hat{c}^{f(0)}, \hat{c}_r^{d(0)}, w_e^{(0)}, \alpha)[w_e^{(1)} - w_e^{(0)}] = 0. \quad (3.12)$$

This set of equations constitutes linearized version of (3.11) and is well defined if the matrix $(F^d)'_w(\hat{c}^{f(0)}, \hat{c}_r^{d(0)}, w_e^{(0)}, \alpha)$ is nonsingular, which can usually be assumed.

The steady-state optimization problem (3.8) can now be solved again, with $w_e^{(0)}$ replaced by $w_e^{(1)}$, yielding the solution $\hat{c}^{(1)} = (\hat{c}^{f(1)}, \hat{c}_r^{d(1)})$. Next correction of the disturbance estimate can then follow, yielding $w_e^{(2)}$ from a solution of the set of linear equations (3.12) with the iteration superscript of all variables incremented by 1, *etc.* The algorithm should be terminated when difference $w_e^{(i+1)} - w_e^{(i)}$ between two subsequent disturbance estimates is sufficiently small, i.e., does not cause essential change in the performance function value.

Convergence conditions and *optimality* of the presented algorithm are its two major important features. The general structure of the algorithm is analogous to the structure of the standard iterative two-step method, with the parameter estimation performed as a correction of the significant disturbance values. Therefore, the algorithm is generally only *suboptimal*, with suboptimality level and convergence conditions heavily problem dependent and generally not known. However, it can be easily seen that the generated sequence of points, if convergent, goes to a point $c^{(\infty)}$ defined by the intersection of the constrained process hyperplane $F_*^d(c^f, c^d, w) = y_r^d$ with the model curve $F^d(\hat{c}^f(w_e), \hat{c}_r^d(w_e), w_e, \alpha) = y_r^d$ (for varying disturbance estimates w_e), where $\hat{c}(w_e) = (\hat{c}^f(w_e), \hat{c}_r^d(w_e))$ is the solution of the model optimization problem (3.8). The point $c^{(\infty)}$ is generally different from the real optimal point $\hat{c}_*(w)$ due to differences between real process static characteristics F_* and its model F. A general question arises whether application of the algorithm is justified and can be recommended.

The assumed activity of the constraint $y^d = y_r^d$ implies that at each iteration the model-based optimized control $\hat{c}(w_e^{(i)}) \neq \hat{c}_*(w)$ would be either not feasible (i.e., corresponding $y^d > y_r^d$) or would yield $y^d < y_r^d$ and usually worse value of the performance function than the control resulting from an immediate action of the CFC. In practice, it can be assumed that application of the algorithm is recommended when during some prescribed time interval stabilization of the output constraint leads to a significant deterioration in the mean value of the performance function — as compared to the mean value in the preceding time interval. Moreover, the more accurate the mathematical model of the process statics the smaller the loss of optimality should be awaited. In conclusion, the algorithm can be recommended first of all for cases when mathematical models of the plant steady-state characteristics are relatively precise, the case met for a class of technological processes.

In (Duda, 1991) an attempt was made to find convergence conditions of the algorithm. However, only heavily problem dependent and not easily verifiable sufficient conditions have been formulated, of a rather limiting nature (a number of simple examples can be designed when these conditions are not satisfied but the algorithm converges).

We assumed that the feedback controllers ideally and sufficiently quickly stabilize the process outputs after each change of the set-points. In reality, these assumptions are not fulfilled since slow-varying disturbances are usually continuously changing (which does not exclude random step-type changes, of course). The controller action is also not ideal and elimination of certain disturbances may need longer actions. However, the algorithm can still be applied in these real conditions, because at each iteration it estimates the disturbance change. This change results either from actual external disturbance change or from other uncertain factors like model unstructured or structured uncertainty or stabilization errors. Therefore, the proposed algorithm of subsequent corrections of the process set-point should be treated, first of all, as an effective procedure of current estimation of too rarely or unmeasured disturbance inputs. However, its application weakens also in many cases the negative influence of other errors, like mathematical model inaccuracies and measurement errors. The reason is that the algorithm corrects the estimate of w from the process optimality viewpoint, choosing automatically an almost optimal bias of the estimate of w.

To simplify the analysis, the case with only feedback controlled (stabilized) outputs has been considered in this section. However, if there are both free and controlled plant outputs, $y = (y^f, y^d)$, then a simple combination of the correction algorithms presented in the previous and this section should be applied. At the first step, after solving the MOP, see (2.57), and application of the calculated set-point to the controlled plant, both free outputs and CFC corrections are measured. At the second step correction of significant disturbances in both free outputs and constrained outputs models is performed, solving the appropriate linear set of equations.

Chapter 4

Integrated System Optimization and Parameter Estimation (ISOPE)

The *Integrated System Optimization and Parameter Estimation* (ISOPE) method of steady-state iterative optimizing control is the subject of this chapter. The method was proposed by (Roberts, 1979), see also (Roberts and Williams, 1981), originally called a *modified two-step method.* However, it gained popularity under the name ISOPE, that is now usually used. As explained in Chapter 2 (Section 2.3) the feature that made the method so attractive is that it is able to generate a series of set-points *converging to the plant true optimal operating point* in spite of uncertainty, i.e., inaccuracy in process models and disturbance estimates. Structure and algorithms of the method will be derived and analyzed in detail in this chapter.

Optimality and convergence properties of the basic version of the ISOPE algorithm were first established by (Brdyś and Roberts, 1987). The improved, augmented version of the algorithm was published in (Brdyś *et al.*, 1987), together with thorough optimality and convergence analysis. The simple, regularizing augmentation of the performance function resulted in much less stringent, quite mild applicability conditions. The basic structure of the ISOPE algorithm will be derived in this augmented formulation in this chapter.

One of the basic assumptions underlying the multilayer control is that changes in values of uncontrolled inputs (disturbances) affecting the optimized process or changes of the process parameters itself are slow (or abrupt but rare) when compared to the process dynamics, as discussed in Chapters 1 and 2. Therefore, the disturbance values and process parameters can be assumed constant (although uncertain, not known precisely) during a single application of the ISOPE iterative algorithm, under given normal operating conditions. Taking it into account the symbols of slow disturbances w and their estimates w_e will be usually omitted in the input-output process mapping description F_* and its model F in this chapter, to simplify the notation. Thus, we will write $F_*(c)$ and $F(c,\alpha)$ instead of $F_*(c,w)$ and $F(c,w_e,\alpha)$, respectively. However, the model parameters α will remain and will be adjusted during the iterations, being among the factors responsible for coping with uncertainty.

We will also assume in this chapter that inequality constraints limiting the feasible set are on set-points (optimizing controls) c only. The ISOPE algorithms and control structures for problems with additional constraints on the process outputs will be presented in the next chapter. This way of presentation makes the description of basic ideas underlying the ISOPE technique simpler and more clear. Moreover, the introduction of additional (inequality) constraints on the process outputs leads to quite severe consequences in the considered case of on-line control under uncertainty, as first explained in Chapter 2 (Section 2.3). New control structures can than be needed and the algorithms completely reformulated.

4.1 Algorithm Structure

Taking into account the discussed assumptions and notation simplification, the basic *optimizing control problem* (OCP) can be formulated in the following form (compare with (2.18) in Chapter 2)

$$\begin{aligned}&\text{minimize } Q(c,y)\\&\text{subject to : } y = F_*(c)\\&\qquad\qquad\quad g(c) \le 0.\end{aligned} \tag{4.1}$$

To present the ISOPE methodology let us start with the equivalent formulation of this problem, with the output variables y eliminated

$$\begin{aligned}&\text{minimize}_{c,\alpha}\ q(c,\alpha)\\&\text{subject to: } F(c,\alpha) = F_*(c)\\&\qquad\qquad\quad g(c) \le 0\end{aligned} \tag{4.2}$$

where

$$q(c,\alpha) = Q(c, F(c,\alpha)). \tag{4.3}$$

The essential step in the derivation of the ISOPE technique is now a reformulation of the problem (4.2) in order to allow for separation of parameter estimation and modified optimization tasks. To achieve this, additional variables $v \in \mathbb{R}^n$ will be introduced, they will serve as decision variables of a modified model optimization. The problem (4.2) takes then the following equivalent form

$$\begin{aligned} &\text{minimize}_{c,\alpha} \; \{q(v,\alpha) + \rho\|c - v\|^2\} \\ &\text{subject to: } F(c,\alpha) = F_*(c) \\ &\qquad\qquad\quad g(v) \le 0 \\ &\qquad\qquad\quad v = c. \end{aligned} \tag{4.4}$$

The quadratic convexifying term (regularizing term) with a scalar parameter $\rho > 0$ has also been introduced in (4.4). This, obviously, does not destroy the problem equivalence and will contribute to nice applicability conditions, the method with this term is called the *augmented* ISOPE method — AISOPE (Brdyś *et al.*, 1987).

Let us write now the Lagrange function

$$L(c,v,\alpha,\lambda,\xi,\mu) = q(v,\alpha) + \rho\|c-v\|^2 + \lambda^T(c-v) + \xi^T(F(c,\alpha) - F_*(c)) + \mu^T g(v) \tag{4.5}$$

and consider the necessary optimality conditions for the problem (4.4):

$$q'_v(v,\alpha)^T - 2\rho(c-v) - \lambda + g'(v)^T\mu = 0 \tag{4.6a}$$

$$g(v) \le 0, \quad \mu \ge 0, \quad \mu^T g(v) = 0 \tag{4.6b}$$

$$2\rho(c-v) + \lambda + [F'_c(c,\alpha) - F'_*(c)]^T\xi = 0 \tag{4.6c}$$

$$q'_\alpha(v,\alpha)^T + F'_\alpha(c,\alpha)^T\xi = 0 \tag{4.6d}$$

$$F(c,\alpha) - F_*(c) = 0 \tag{4.6e}$$

$$c - v = 0. \tag{4.6f}$$

From (4.6d) the multipliers ξ can be explicitly derived (assuming the matrix in square brackets in the formula below is nonsingular):

$$\xi = [F'_\alpha(c,\alpha)F'_\alpha(c,\alpha)^T]^{-1}F'_\alpha(c,\alpha)\, q'_\alpha(v,\alpha)^T. \tag{4.7}$$

Using now the relation

$$q'_\alpha(c,\alpha) = Q'_y(c, F(c,\alpha))F'_\alpha(c,\alpha) \tag{4.8}$$

and the equality (4.6f) we get a simple formula for the multipliers

$$\xi = Q_y'(c, F(c,\alpha))^T. \tag{4.9}$$

Putting this into (4.6c) we obtain the formula for the multiplier λ,

$$\lambda(c,\alpha) = [F_c'(c,\alpha) - F_*'(c)]^T Q_y'(c, F(c,\alpha))^T \tag{4.10}$$

and the following reformulation of the necessary optimality conditions:

$$\begin{aligned} q_v'(v,\alpha)^T - 2\rho(c-v) - \lambda(c,\alpha) + g'(v)^T\mu &= 0 && \text{(4.11a)}\\ g(v) \le 0, \quad \mu \ge 0, \quad \mu^T g(v) &= 0 && \text{(4.11b)}\\ F(c,\alpha) - F_*(c) &= 0 && \text{(4.11c)}\\ c - v &= 0. && \text{(4.11d)} \end{aligned}$$

Let us now define the following *modified model-based optimization problem* (MMOP):

$$\begin{aligned} &\text{minimize}_v \ \{q(v,\alpha) - \lambda(c,\alpha)^T v + \rho\|c-v\|^2\}\\ &\text{subject to} : g(v) \le 0. \end{aligned} \tag{4.12}$$

Notice that (4.11a) and (4.11b) are precisely the necessary optimality conditions for the MMOP. Recall that the feasible set for the MMOP can be denoted by C,

$$C = \{c \in \mathbb{R}^n : \ g(c) \le 0 \ \}. \tag{4.13}$$

Let us also assume that the model $F(\cdot,\cdot)$ is *point parametric* on C, see (Brdyś, 1983), i.e.,

$$\textit{Assumption}: \text{for every } \bar{c} \in C \text{ there is } \bar{\alpha} \in \mathbb{R}^s \text{ such that } F_*(\bar{c}) = F(\bar{c},\bar{\alpha}). \tag{4.14}$$

The above assumption can be regarded as a necessary feature for the model to be well-defined. It states that it is possible to match exactly the model output and the plant output at any point of the feasible set associated with the original problem (4.1), by the appropriate choice of the parameters α. The assumption is, e.g., always satisfied when the model is additive with respect to a subset, say α_a, of the parameters, i.e.,

$$F(c,\alpha) = F(c,\alpha_n) + \alpha_a, \quad \alpha = (\alpha_n, \alpha_a), \quad \alpha_a \in \mathbb{R}^m. \tag{4.15}$$

The *parameter estimation problem* (PEP) will be defined as a problem of adapting the model parameters α at an operating point (set-point) c under the constraint (4.11c)

$$F(c, \alpha) - F_*(c) = 0. \tag{4.16}$$

PEP is well defined, for every $c \in C$, if the model is point-parametric on C. In particular, if the model has the structure (4.15), then condition (4.16) can always be satisfied by adapting only the additive parameters α_a, in fact by a simple substitution

$$\alpha_a = F_*(c) - F(c, \alpha_n). \tag{4.17}$$

Generally, there are also non-additive parameters α_n, and in such cases the PEP may have not a unique solution. However, as it will be clear from the presentation fo follow, it is not necessary to adjust the non-additive parameters α_n at every iteration of the ISOPE (at every PEP solution), these parameters may even be kept constant during a single run of the method, if reasonable (if there is not sufficient additional information gathered to perform new full adaptation during a single run).

We are now in a position to formulate *the ISOPE basic algorithm*:

Start. Given initial point c^0, relaxation coefficient k_c, $0 < k_c \leq 1$ and solution accuracy $\epsilon > 0$. Set $i := 0$.

Step 1. Apply c^i to the controlled plant and measure $y^i = F_*(c^i)$. Perform additional linearly independent perturbations around c^i and measure corresponding values of the plant outputs in steady-states (after the transient processes have died). Based on this measurements find a finite difference approximation of the process output mapping derivative $F_*^{'}(c^i)$.

Step 2. The *parameter estimation problem* PEP: using the obtained new measurements update parameters α under the restriction that the model outputs match the actual process outputs at c^i. This yields $\alpha^i = \hat{\alpha}(c^i)$ satisfying

$$y^i = F(c^i, \alpha^i) = F_*(c^i). \tag{4.18}$$

Step 3. For $c = c^i$ and $\alpha = \alpha^i$ and, therefore, $\lambda(c, \alpha) = \lambda(c^i, \alpha^i)$ solve the *modified model-based optimization problem* MMOP

$$\begin{aligned} &\text{minimize}_v \ \{q(v,\alpha) - \lambda(c,\alpha)^T v + \rho\|c-v\|^2\} \\ &\text{subject to}: \ g(v) \leq 0. \end{aligned} \tag{4.19}$$

Let $v^i = \hat{v}^i(c^i, \alpha^i)$ be the solution. If

$$\|c^i - v^i\| \leq \epsilon \tag{4.20}$$

then terminate (eq.(4.11d) satisfied – solution found).

Step 4. Set

$$c^{i+1} := c^i + k_c(v^i - c^i) \tag{4.21}$$

set $i := i + 1$ and continue from Step 1.

It can easily be seen that the algorithm is constructed in such a way that when it terminates, the necessary optimality conditions (4.11a)–(4.11d) for the OCP problem are satisfied. At each iteration equations (4.11a), (4.11b) and (4.11c) are fulfilled due to solving the PEP and MMOP problems in Steps 2 and 3. When the algorithm terminates the equation (4.11d) is fulfilled with the prescribed accuracy $\epsilon > 0$.

The whole ISOPE algorithm can be regarded as of fix-point type, since the set-points c are iterated in such a way as to fulfill the equation (4.11d), which in the algorithm realization takes the form

$$\hat{v}(c, \hat{\alpha}(c)) = c. \tag{4.22}$$

The iterative formula (4.21) is a simple adjustment rule for finding a fix point of (4.22), usually called iteration of a relaxation type and the parameter k_c is called the relaxation coefficient . Notice that if $k_c = 1$ then this formula becomes a direct substitution rule

$$c^{i+1} = v^i. \tag{4.23}$$

Being simple the formula (4.21) possesses a very important property, from a practical point of view: if the feasible set C is convex and the initial point $c^0 \in C$, then $c^1 \in C$ and, consequently, each point c^i of the generated sequence is feasible, provided $0 < k_c \leq 1$. Another, more complex iterative formula for solving the equation (4.22) was proposed by (Tatjewski and Roberts, 1987). Although very interesting, it will not be presented in this book because it is usually less practical for strongly constrained problems.

More importantly, the development of the ISOPE technique occurred to be more successful in a direction to improve significantly efficiency of the process derivative estimation than the set-point adjustment formula. This led to dual ISOPE algorithms presented further in this chapter.

It should be pointed out that at each iteration of the ISOPE algorithm the real process mapping derivative $F_*'(c^i)$ must be evaluated (approximated), in order to evaluate the modifier λ, see (4.10), necessary for the MMOP formulation. The necessity to know this derivative, at least locally, could have been awaited — it results from the argument in Chapter 2, see Subsection 2.3.1 therein. Nevertheless, the way this derivative is evaluated is one of the key-points of each ISOPE algorithm, important for its practical features and effectiveness. In the basic ISOPE algorithm structure given above this derivative is calculated in Step 2 using finite difference approximations based on output measurements from n, $n = \dim c$, additional set-point perturbations around a current value c^i. This is the simplest approach suggested by (Roberts, 1979) and utilized in many later papers, see e.g., (Roberts and Williams, 1981; Brdyś and Roberts, 1987; Brdyś *et al.*, 1987; Tatjewski and Roberts, 1987). However, it means n additional transient processes in the plant at each iteration of the algorithm, in addition to the single transient process associated with the set-point change from c^i to c^{i+1}. Therefore, this procedure is highly time and cost expensive. The ways to approximate the derivative $F_*'(c^i)$ more efficiently will be discussed later on in this chapter, leading to more effective and practical ISOPE algorithms.

Before passing to the outlined points, convergence and optimality of the basic algorithm will be investigated.

4.2 Convergence and Optimality

In order to perform convergence analysis, the ISOPE algorithm will be described as a mapping $\mathcal{A} : C \mapsto C$, being composition of other mappings representing various steps of the algorithm.

First, let us observe that the parameter estimation performed in Step 2 may not be unique, and define the sets

$$\hat{\alpha}_S(c) = \{\alpha \in \mathbb{R}^s : \; F(c, \alpha) = F_*(c) \} \tag{4.24}$$

$$A = \bigcup_{c \in C} \hat{\alpha}_S(c). \tag{4.25}$$

Therefore, $\hat{\alpha}_S$ is a point-to-set mapping and Step 2 of the algorithm can now be described in the following way: for a given c^i solve PEP finding $\alpha^i = \hat{\alpha}(c^i)$ such that $\alpha^i \in \hat{\alpha}_S(c^i)$. In the case of non-unique parameter estimation (possible when $s = \dim \alpha > m = \dim y$) the parameters α are chosen by the PEP not only to satisfy the point equality condition stated in (4.24), but also certain general model adaptation requirement like minimizing least squares modeling error. Obviously, $\hat{\alpha}(\cdot)$ reduces to $\hat{\alpha}_S(\cdot)$ and is unique when only additive parameters α_a are adjusted, see (4.15).

Similarly, assuming that the modified model-based optimization problem (MMOP) performed in Step 3 may be generally not unique, let us define the set of its solutions as

$$\hat{v}_S(c, \alpha) = \text{Argmin}_{v \in C}\{q(v, \alpha) - \lambda(c, \alpha)^T v + \rho\|c - v\|^2\}. \tag{4.26}$$

Step 3 of the algorithm can be, therefore, described as follows: for given c^i and α^i find v^i such that $v^i = \hat{v}(c^i, \alpha^i) \in \hat{v}_S(c^i, \alpha^i)$.

In Step 4 we allow the coefficient k_c, generally, to change during iterations, within the limits

$$\tau \le k_c \le B(c^i) \tag{4.27}$$

where $\tau > 0$ and $B : C \mapsto \mathbb{R}$ is a problem dependent function which will be defined in the sequel.

We can now describe the i-th iteration of the ISOPE algorithm as finding c^{i+1} such that

$$c^{i+1} \in \mathcal{A}(c^i) \tag{4.28}$$

where $\mathcal{A}(\cdot)$ is the algorithmic mapping (point-to-set mapping) defined as follows:

$$2^C \ni \mathcal{A}(c) = \{c + k_c(v - c) \in \mathbb{R}^n : \ v \in \hat{v}_S(c, \alpha), \ \alpha \in \hat{\alpha}_S(c), \ \tau \le k_c \le B(c)\} \tag{4.29}$$

Finally, let us define the ISOPE algorithm solution set Ω,

$$\Omega = \{c \in \mathbb{R}^n : \ \exists \hat{\alpha} \in \hat{\alpha}_S(c) \ \text{ such that } \ c \in \hat{v}_S(c, \hat{\alpha})\} \tag{4.30}$$

or, equivalently,

$$\Omega = \{c \in \mathbb{R}^n : \ c \in \mathcal{A}(c)\} \tag{4.31}$$

and let us denote

$$q_*(c) = Q(c, F_*(c)). \tag{4.32}$$

We are now in a position to state the convergence theorem. But, to make this statement shorter, let us first denote

$$b(\alpha) = \min_{c \in C} \lambda_{min}(q''_{cc}(c, \alpha)) \tag{4.33}$$

where $q''_{cc}(c, \alpha)$ is the second (Frechét) derivative of $q''_{cc}(\cdot, \alpha)$ (it is a symmetric square matrix of dimension $n = \dim c$) and $\lambda_{min}(q''_{cc}(c, \alpha))$ denotes its minimal eigenvalue.

Theorem 4.1 *Assume that*
(i) the set C is convex and compact,
(ii) the set A is compact,
(iii) mappings $F_*(\cdot)$ and $F(\cdot, \cdot)$ are continuous on C and $C \times A$, respectively,
(iv) the function $q_*(\cdot)$ is (Frechét) continuously differentiable on C,
(v) the function $q(\cdot, \cdot)$ is twice (Frechét) differentiable with respect to c on C, for every $\alpha \in A$, and $q(\cdot, \cdot)$, $q'_c(\cdot, \cdot)$, $q''_{cc}(\cdot, \cdot)$ are continuous on $C \times A$,
(vi) the point-to-set mapping $\hat{\alpha}_S(\cdot)$ is open on A.
Then there exist values ρ_1, τ, function $B(c)$ and some scalar $\epsilon > 0$ satisfying

$$\rho_1 > -0.5 \inf_{\alpha \in A} b(\alpha) \tag{4.34}$$

$$0 < \tau < \min\{1, \frac{2\inf_{\alpha \in A} b(\alpha) + 4\rho_1}{\delta + \epsilon}\} \tag{4.35}$$

$$B(c) = \min\{1, \frac{2\inf_{\alpha \in \hat{\alpha}_S(c)} b(\alpha) + 4\rho_1}{\delta + \epsilon}\} \tag{4.36}$$

where

$$\delta = \max_{c \in C} \|q'_*(c)\| \tag{4.37}$$

such that for every $\rho \geq \max\{0, \rho_1\}$:
(a) the algorithmic mapping $\mathcal{A}(\cdot)$ is well defined on C and $\mathcal{A}(c) \in C$ for every $c \in C$,
(b) each point generated by the algorithm satisfies the plant constraints and for every $i = 0, 1, \ldots$

$$q_*(c^{i+1}) < q_*(c^i) \qquad \text{if} \quad c^i \notin \Omega \tag{4.38}$$

(c) there is at least one cluster point of the sequence $\{c^i\}$ generated by the algorithm and each cluster point belongs to the solution set Ω.

Proof. The proof is given in Appendix A and utilizes Zangwill convergence theorem (Zangwill, 1969).

Let us discuss the theorem assumptions. The feasible set C is bounded in all realistic cases. Therefore, assumption (i) is satisfied if, additionally, the constraining functions $g_j(\cdot)$ are convex. Provided that for every $c \in C$ the set of model matching parameters $\hat{\alpha}_S(c)$ is constrained and closed for a reasonably constructed plant model, the assumptions (ii) follows from compactness of C. The smoothness assumptions (iii) and (iv) are satisfied in the vast majority of continuous industrial processes, differentiability of $q_*(\cdot)$ required in (iv) follows from differentiability of $Q(\cdot,\cdot)$ and $F_*(\cdot)$ (which must be actually assumed to assure the formula (4.10) is well-defined — in the proof of the Theorem equivalent formula (4.40) is used, defining the multipliers λ in terms of q_*). The smoothness assumption (v) concerns mathematical model of the performance function and output mapping only, and thus can usually be guaranteed by the model construction.

The last assumption, (vi), requires the point-to-set mapping $\hat{\alpha}_S(\cdot)$ to be open. Unfortunately, examples can be designed showing that there is no unique, elegant and general enough set of sufficient conditions for $\hat{\alpha}_S(\cdot)$ to be open in the general case of non-unique parameter estimation performed at each iteration. However, in most cases the parameter estimation is or can be made unique due to the used parameter adaptation criterion leading to a unique, i.e., point-to-point mapping $\hat{\alpha}(\cdot)$. This implies the assumption (vi) is satisfied. Theoretically most pleasant and practically often sound is the case when the parameters α are defined uniquely only by the set of equations stated in (4.24), i.e. only the additive parameters are adjusted during a single algorithm run or, more generally, when $\dim \alpha = \dim y$, the model is point-parametric and the set of equations stated in (4.24) is uniquely solved. In this case previous smoothness assumptions assure that the now point-to-point mapping $\hat{\alpha}(\cdot)$ is continuous on A.

In conclusion, one should admit that the sufficient conditions for the ISOPE algorithm to converge formulated in the Theorem 4.1 are reasonable. On the other hand, the convergence features are quite strong. It is not required that plant mapping or its mathematical model be convex. Note that if the model performance function $q(\cdot,\alpha)$ is convex for every $\alpha \in A$, then $\rho_1 \leq 0$ and the algorithm converges for every $\rho \geq 0$. The values of $\rho \geq \rho_1 > 0$ are necessary and sufficient to assure convergence in nonconvex cases. The feature of monotone decrease of the true plant performance function, see inequality (4.38), is extremely important from a practical point of view — it means that each step of the algorithm yields improvement, in spite uncertainty. Therefore, application of the algorithm may be justified even for very few steps, then interrupting the algorithm before the termination criterion is satisfied.

The ISOPE algorithm converges to points from the solution set Ω, see (4.30). From the construction of the algorithm, these points should satisfy at least necessary optimality conditions for the original optimizing control problem (OCP). The theoretical result which follows formally proves this statement.

Theorem 4.2 *Assume that assumptions of the Theorem 4.1 are satisfied. Then at each point $c \in \Omega$ the first-order necessary optimality conditions for the OCP are satisfied. If, additionally, the function q_* is convex on C, then each $c \in \Omega$ is a solution to the OCP.*

Proof. Since $c \in \mathcal{A}(c)$ then there is $\alpha \in \hat{\alpha}_S(c)$ and $c = \hat{v}(c, \alpha) \in \hat{v}_S(c, \alpha)$. Due to convexity of C the first order necessary optimality conditions for the solution of the MMOP can be written as

$$[q_c'(\hat{v}(c, \alpha), \alpha) - \lambda(c, \alpha)^T][v - \hat{v}(c, \alpha)] \geq 0 \quad \text{for all } v \in C. \tag{4.39}$$

We have from (4.10) and the equality $F(c, \alpha) = F_*(c)$

$$\begin{aligned} \lambda(c, \alpha)^T &= Q_y'(c, F(c, \alpha))F_c'(c, \alpha) - Q_y'(c, F(c, \alpha))F_*'(c) \\ &= Q_y'(c, F(c, \alpha))F_c'(c, \alpha) - Q_y'(c, F_*(c))F_*'(c) \\ &= q_c'(c, \alpha) - q_*'(c). \end{aligned} \tag{4.40}$$

Therefore, when $\hat{v}(c, \alpha) = c$ (4.39) can be written in the form

$$q_*'(c)(v - c) \geq 0 \quad \text{for all } v \in C \tag{4.41}$$

which is precisely the necessary optimality condition for the OCP. If the function q_* is convex on C then this condition becomes sufficient for optimality. This completes the proof.

4.3 On-line Estimation of Process Mapping Derivatives

At each iteration of the ISOPE algorithm the derivative $F_*^{'}(c^i)$ of the real process output mapping F_* must be obtained, in order to evaluate the multiplier (4.10). The way this derivative is evaluated is a key point of each version of the ISOPE algorithm. The reason for that is the time and cost for the derivative evaluation may be significantly higher than that of the whole remaining measurements and calculations during each iteration. It is precisely the case when the technique of the basic ISOPE algorithm, i.e., additional perturbations around each set-point c^i to get finite difference approximation of the derivative, is applied (see Section 4.1). That is why many attempts have been made to overcome this weak point, to find more effective realizations of the ISOPE algorithms. These attempts can be divided into three groups:

- Attempts to approximate the process output mapping derivative from steady-state output measurements, but significantly fewer than in the case of the standard finite difference approach of the basic ISOPE algorithm.
- Evaluation of the steady-state process output mapping derivative from a linear dynamic model, obtained using dynamic information and some identification technique.
- Formulation of the Stochastic Optimizing Control Problem using, consequently, only current dynamic measurements.

The first approach has led to the development of the dual realization of the ISOPE algorithm, which will be presented in the next section. The remaining approaches rely on dynamic information measured on-line from the controlled plant. The second one will be discussed further in this section. The third approach will not be presented in this book since the underlying concept is still at a stage of a dispute and initial trials, the reader is referred to the relevant literature (Lin *et al.*, 1989, 1990).

The idea of extracting a steady-state model from a dynamic one, obtained from an on-line identification procedure, was originally proposed by (Bamberger and Isermann, 1978). This model was then used for on-line optimization of the process operating point using a gradient algorithm. (Zhang and Roberts, 1990) tried to incorporate this idea into the basic ISOPE algorithm structure, for calculation of the plant output derivative $F_*^{'}(c^i)$. The procedure is as follows. At each iteration a *linear dynamic* model of the controlled plant (i.e., the plant together with its feedback controllers) is first identified. To explain the approach let us assume the following example structure of the discrete-time dynamic model of the controlled plant (Zhang and Roberts, 1990)

$$y[k] = A(q^{-1})y[k] + q^{-d}B(q^{-1})c[k] + d \tag{4.42}$$

where $A(q^{-1})$ and $B(q^{-1})$ are polynomial matrices in the backward shift operator q^{-1}, $y[k]$ and $c[k]$ are dynamic plant outputs and controller set-points at time instant k, respectively.
Having identified the model parameters at the current set-point $c = c^i$, the steady-state information can be then extracted by setting $q = 1$. For the dynamic model (4.42) its steady-state counterpart is of the following form

$$[I - A(1)]y = B(1)c + d. \tag{4.43}$$

Hence, we get the desired derivative

$$F_*{}'(c^i) \cong [I - A(1)]^{-1}B(1). \tag{4.44}$$

The presented idea seems very promising. But serious troubles can be connected with the identification process and accuracy of the resulting steady-state derivatives. First, an adequate structure of the linear dynamic model must be found — as an important and not obvious task of each identification procedure. Second, to get sufficiently accurate identification results the plant must be appropriately excited. It can hardly be assumed that passive identification experiment based on measurement records collected during the transient process between the last set-points, c^{i-1} and c^i, can always deliver sufficiently reach data records. Therefore, an active identification experiment performed around each actual set-point c^i should be planned. This would mean additional, dynamical changes of the set-points around c^i, for the identification purposes only. Moreover, the identification experiment is by no means easy. One of the main reasons is that identification experiments based on dynamic data records are generally aimed at getting linear models sufficiently accurately reflecting local *dynamics* of the plant, not *statics*. This difficulty has been discovered in the above mentioned early paper of (Bamberger and Isermann, 1978), and was there found to be particularly significant in the presence of noise. They suggested for noisy processes the identification based on a special correlation technique, without any parameter estimation.

Therefore, approximating the steady-state plant output derivative from a linear dynamical model, identified on-line at each iteration of the ISOPE algorithm, may lead to an efficient algorithm realization — but this is by no means an easy and well formalized approach.

4.4 Dual ISOPE Algorithm

Although a method of extracting the information about static plant characteristics from dynamic instantaneous measurements could be an alternative, as shown in the previous section, a reliable and technically sound steady-state optimizing control algorithm should rather rely on steady-state measurements. The reason is that only in this case the measurement noise can be well filtered. Hence the importance of deriving a technique for approximating the plant output derivative from steady-state measurements, but using significantly less set-point changes than in the case of the standard finite difference approach of the basic ISOPE algorithm (see Section 4.1). The situation would be ideal if additional steady-state measurements were not needed at all. However, from theoretical point of view it is, generally, not possible since it cannot be guaranteed that past measurements gathered during the run of the ISOPE algorithm in its basic structure contain sufficient information to approximate the derivatives. That is the reason why, for a long time, no progress has been made in this direction. The only work claiming an achievement in this area (Liu and Roberts, 1989) is rather a failure, at least because it attempts to approximate partial derivatives of a multivariable process input-output mapping $F_* : \mathbb{R}^n \mapsto \mathbb{R}^m$ from two last points only.

A new idea delivering a break-through was proposed by (Brdyś and Tatjewski, 1994). Realizing that relying on *passive* past output measurements only during the ISOPE run cannot guarantee the necessary information, an algorithm with *active* measurement gathering was proposed. That is, at each ISOPE iteration the next set-point c^{i+1} is generated in a way not only to perform the ISOPE task to optimize the performance index but also, simultaneously, to assure that c^{i+1} will be appropriately located for the purpose of future derivative approximation. This may lead to some current loss of optimality, but at the same time anticipates future measurement needs. This is, obviously, a dual structure in the sense of control duality defined by (Feldbaum, 1965). Therefore, the resulting algorithm will be a *dual optimizing control* algorithm, precisely the *Integrated System Optimization and Parameter Estimation Dual* (ISOPED) algorithm. It will now be presented.

Let us assume that there is a collection of n+1 points $c^i, c^{i-1}, \ldots, c^{i-n}$ such that all vectors

$$s^{ik} = c^{i-k} - c^i, \quad k = 1, ..., n \tag{4.45}$$

are linearly independent, and formulate a (nonsingular) matrix

$$S^i = [c^{i-1} - c^i \ c^{i-2} - c^i \ \cdots \ c^{i-n} - c^i]^T. \tag{4.46}$$

Directional derivative $DF_{*j}(c^i; s^{ik})$ of the j-th plant output F_{*j} at a point c^i and in a direction $s^{ik} = c^{i-k} - c^i$ can be defined as

$$DF_{*j}(c^i; s^{ik}) = \lim_{\beta \to 0} \frac{F_{*j}(c^i + \beta \frac{s^{ik}}{\|s^{ik}\|}) - F_{*j}(c^i)}{\beta}. \tag{4.47}$$

If F_{*j} has continuous partial derivatives in a neighborhood of c^i then the directional derivative can be computed as

$$DF_{*j}(c^i; s^{ik}) = \frac{1}{\|s^{ik}\|}(s^{ik})^T \nabla F_{*j}(c^i) \tag{4.48}$$

which can be written in the form

$$\|s^{ik}\| \, DF_{*j}(c^i; s^{ik}) = (s^{ik})^T \nabla F_{*j}(c^i). \tag{4.49}$$

Writing the above n equations, i.e. for n directions s^{ik}, $k = 1, \ldots, n$, in a matrix form we get

$$S^i \, \nabla F_{*j}(c^i) = \begin{bmatrix} \|s^{i1}\| \, DF_{*j}(c^i; s^{i1}) \\ \vdots \\ \|s^{in}\| \, DF_{*j}(c^i; s^{in}) \end{bmatrix}, \quad j = 1, \ldots, m. \tag{4.50}$$

Denote $\beta = \gamma \|s^{ik}\|$. Then the definition (4.47) can be written in the following form

$$DF_{*j}(c^i; s^{ik}) = \lim_{\gamma \|s^{ik}\| \to 0} \frac{F_{*j}(c^i + \gamma s^{ik}) - F_{*j}(c^i)}{\gamma \|s^{ik}\|}. \tag{4.51}$$

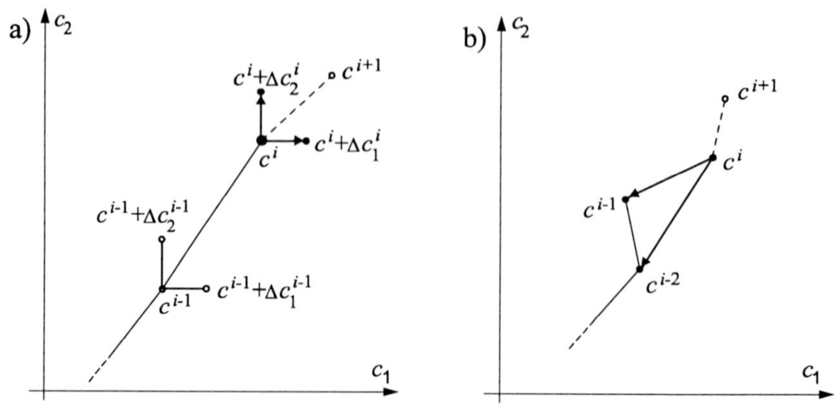

Fig. 4.1 Difference vectors used for estimation of the plant mapping derivatives: (a) based on n additional deviations at each main (basic) ISOPE algorithm point c^i, (b) based on main algorithm points only (dual algorithm).

If the points c^{i-k} are close enough to c^i, i.e., scalars $\|s^{ik}\|$ are small enough, then (assuming $\gamma = 1$) the following approximation is obtained

$$\|s^{ik}\| \, DF_{*j}(c^i; s^{ik}) \cong F_{*j}(c^i + s^{ik}) - F_{*j}(c^i). \tag{4.52}$$

The set of equations (4.50) can be now expressed in the form

$$S^i \, \nabla F_{*j}(c^i) \cong \begin{bmatrix} F_{*j}(c^{i-1}) - F_{*j}(c^i) \\ \vdots \\ F_{*j}(c^{i-n}) - F_{*j}(c^i) \end{bmatrix}, \quad j = 1, \ldots, m \tag{4.53}$$

In the ISOPED algorithm the consecutive points c^i are generated in such a way that the process mapping derivative estimation (4.53) can be employed at each iteration. The location of the subsequent points and vectors s^{i1}, s^{i2} in two-dimensional case ($n = 2$) is illustrated in Figure 4.1 (b). For comparison, in Fig. 4.1 (a) vectors of set-point deviations used in early ISOPE algorithms (as in the presented basic algorithm) are presented, i.e., vectors of differences generated using measurements at additional set-points creating, together with the main algorithm points c^i, linearly independent directions (of the Cartesian plane axes in the figure).

Calculation of the required derivative approximations from the sets of linear equations (4.53) will be practically useful only if the matrix S^i is not only non-singular, but also *sufficiently well conditioned.* The reason is that the right-hand side of these sets of equations is corrupted not only by the method error (approximation of the right-hand side of (4.50)) but also by plant outputs measurement errors. Good conditioning can be achieved only forcing appropriate location of the consecutive set-points in $\mathbb{R}^n$. It is achieved in the dual ISOPE algorithms by introducing a new inequality constraint to the modified model-based optimization problem MMOP, see (4.12), called *the conditioning constraint*, in the form

$$d(c^{i+1}(v), c^i, \ldots, c^{i-n+1}) = \frac{\sigma_{min}(S^{i+1}(v))}{\sigma_{max}(S^{i+1}(v))} \geq \delta \qquad (4.54)$$

where

$$c^{i+1}(v) = c^i + k_c(v - c^i) \qquad (4.55)$$

$$S^{i+1}(v) = [c^i - c^{i+1}(v) \quad \cdots \quad c^{i-n+1} - c^{i+1}(v)]^T \qquad (4.56)$$

and $\sigma_{min}(S^{i+1}(v))$, $\sigma_{max}(S^{i+1}(v))$ denote minimal and maximal singular values of $S^{i+1}(v)$. The value of δ, $1 > \delta > 0$, defines the required *conditioning* of the matrix $S^{i+1}(v)$. Notice that $\frac{\sigma_{min}(S^{i+1}(v))}{\sigma_{max}(S^{i+1}(v))}$ is the reciprocal of the standard condition number of the matrix $S^{i+1}(v)$.

The *ISOPED (ISOPE Dual) algorithm* is as follows:

Start. Given initial point c^0, parameter of the conditioning constraint $\delta > 0$, relaxation coefficient k_c, $0 < k_c \leq 1$, solution accuracy $\epsilon > 0$. Measure $y^0 = F_*(c^0)$.

Step 0 (initial phase). Set $c^{-n} := c^0$. Choose n points $c^{-n+1}, c^{-n+2}, \ldots, c^0$ such that the matrix S^0 is sufficiently well conditioned. Apply, consecutively, the points c^i to the process and measure $F_*(c^i)$, $i = -n+1, -n+2, \ldots, 0$. Set $i := 0$.

Step 1. Apply c^i to the process and measure $F_*(c^i)$, calculate $F_*^{'}(c^i)$ according to (4.53), i.e., solve m sets of n linear equations

$$S^i \cdot (F_{*j}^{'}(c^i))^T = \begin{bmatrix} F_{*j}(c^{i-1}) - F_{*j}(c^i) \\ \vdots \\ F_{*j}(c^{i-n}) - F_{*j}(c^i) \end{bmatrix}, \quad j = 1, \ldots, m. \qquad (4.57)$$

Step 2. *Parameter estimation problem* PEP: Using the obtained measurements update the parameters α of the model F under the restriction to match the process and model outputs at c^i. This yields $\alpha^i \in \hat{\alpha}(c^i)$ satisfying

$$y^i = F(c^i, \alpha^i) = F_*(c^i). \tag{4.58}$$

Step 3. Solve the *modified model-based optimization problem* MMOP:

$$\begin{aligned} &\text{minimize } \{q(v, \alpha^i) - \lambda(c^i, \alpha^i)^T v + \rho \|c^i - v\|^2\} \\ &\text{subject to}: \ v \in C = \{c \in \mathbb{R}^n : \ g(v) \leq 0\} \end{aligned} \tag{4.59}$$

denoting a solution by v^i. If

$$\|v^i - c^i\| \leq \epsilon \tag{4.60}$$

then terminate (solution found).

Step 4. If

$$d(c^{i+1}(v), c^i, \ldots, c^{i-n+1}) \geq \delta \tag{4.61}$$

then set $v_d^i := v^i$ and go to Step 5.
Else, solve the *conditioned modified model-based optimization problem* (CMMOP):

$$\begin{aligned} &\text{minimize } \{q(v, \alpha^i) - \lambda(c^i, \alpha^i)^T v + \rho \|c^i - v\|^2\} \\ &\text{subject to}: \ v \in C \cap D^i \end{aligned} \tag{4.62}$$

where

$$D^i = \{v \in \mathbb{R}^n : d(c^{i+1}(v), c^i, \ldots, c^{i-n+1}) \geq \delta\} \tag{4.63}$$

denoting a solution by v_d^i.

Step 5. Set

$$c^{i+1} := c^i + k_c(v_d^i - c^i) \tag{4.64}$$

set $i := i + 1$ and continue from Step 1.

The initial task in Step 4 is to check if v^i satisfies the conditioning constraint. If it does then $v_d^i := v^i$ and Step 5 can be immediately executed, without any change in the basic algorithm and, therefore, without any current loss of optimality. If it does not then v^i is suitably modified to satisfy the new constraint. This constraint reduces the feasible set of the modified model-based optimization problem. Hence, it might be that a current loss of optimality will be observed. However, the new generated set-point anticipates future needs of the next iteration and thus would enable to obtain better approximations of the derivatives, which in turn would lead to better optimality of the next generated set-point. This can be compared to what is known as active measurement gathering, or a *dual effect.*

Properties of the sets D^i, and thus of the whole CMMOP problem, are important from both theoretical and practical points of view. To present this properties let us denote

$$\begin{aligned} S^{i+1}(v) &= S^{i+1}[c^i - c^{i+1}(v) \quad \cdots \quad c^{i-n+1} - c^{i+1}(v)]^T && (4.65) \\ D^{i0} &= \{v \in \mathbb{R}^n : \det S^{i+1}(v) = 0\} && (4.66) \\ D^{i+} &= \{v \in \mathbb{R}^n : v \in D^i, \det S^{i+1}(v) > 0\} && (4.67) \\ D^{i-} &= \{v \in \mathbb{R}^n : v \in D^i, \det S^{i+1}(v) < 0\}. && (4.68) \end{aligned}$$

Then, taking into account the definition of D^i, it immediately follows that

$$\begin{aligned} D^i \cap D^{i0} = \emptyset && (4.69) \\ D^i = D^{i+} \cup D^{i-}. && (4.70) \end{aligned}$$

Moreover, D^{i+} and D^{i-} are symmetrical with respect to the hyperplane D^{i0}. Therefore, the feasible set $C \cap D^i$ of CMMOP consists of two separate sets $C \cap D^{i+}$ and $C \cap D^{i-}$. It turns out that the properties of the sets D^{i+} and D^{i-} are favorable from the CMMOP viewpoint. Due to the symmetry, only D^{i+} will further be considered.

It has been not managed, until now, to obtain a precise characterisation of the set D^{i+} defined by the function (5.55) in the general case. However, it has been shown (Brdyś and Tatjewski, 1994; Tadej and Tatjewski, 2001) for problem dimensionality $n = 2$ (for $n = 1$ the result is trivial) that D^{i+} is a closed, convex set. In particular, it is a disc with radius r^i, centered at a distance h^i from D^{i0} (i.e. the line defined by c^i and c^{i-1}), where

$$\begin{aligned} r^i &= \frac{1-\delta^2}{4k_c\delta} \left\| c^i - c^{i-1} \right\|_2 && (4.71) \\ h^i &= \frac{1+\delta^2}{4k_c\delta} \left\| c^i - c^{i-1} \right\|_2 && (4.72) \end{aligned}$$

as is depicted in Fig. 4.2. Therefore, if the original feasible set C is convex, then solving CMMOP can easily be performed by two optimizations on the (convex) sets $C \cap D^{i+}$ and $C \cap D^{i-}$ and selection of a better result (if

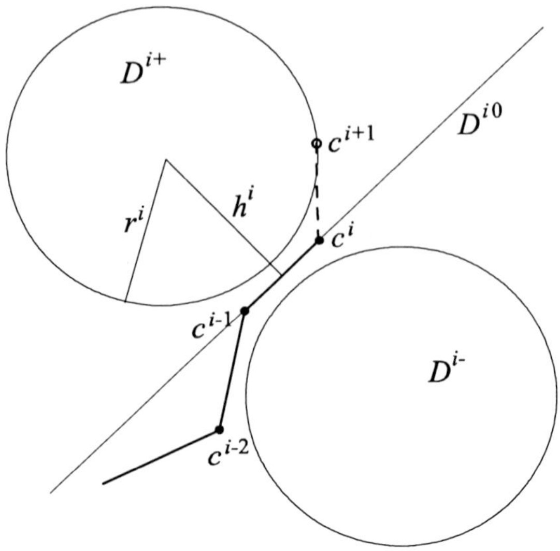

Fig. 4.2 Illustration of the conditioning set $D^i = D^{i+} \cup D^{i-}$ in the two-dimensional case.

they happen to be equal, the one closer to the point c^i should be chosen). Analytical expressions for D^{i+} in the case n=3 are not known until now. However, computer investigation shown that it is usually a convex, slightly deformed ball, for all possible locations of the points c^i, c^{i-1}, c^{i-2}. Moreover, the distance between D^{i+} and D^{i0}, and the volume of D^{i+} depend on δ and the points c^i, c^{i-1}, c^{i-2} similarly as in the previous case of n=2.

It has also been shown that if the feasible set $C \cap D^i$ of the problem CMMOP is not empty in the first iteration ($i = 0$) then it is not empty in the second and all subsequent iterations (Tadej and Tatjewski, 2001). An attempt to perform the theoretical convergence analysis is also reported in the cited paper. However, the results have been obtained for a slightly simplified version of the dual algorithm, namely with successive linearization of the original performance function at the beginning of each iteration (the ISOPEDL algorithm) and $k_c = 1$. Under reasonable assumptions, it has been shown that in the case $C = \mathbb{R}^n$ the gradient of the true plant performance function $Q(c^i, F_*(c^i))$ converges to zero for any $\delta \lesssim 0.284$. This result is important for practical applications because recommended (based on simulations mainly) values of δ are in the range $0.05 - 0.2$, with the larger values better for more erroneous measurements — hence are located in the guaranteed convergence region. Moreover, it has been shown on certain simulation examples that algorithms ISOPED and ISOPEDL react almost identically on changes in δ. Therefore, a hypothesis that the convergence limiting value of δ should be the same or very similar also for the ISOPED algorithm seems to be true. Unfortunately, the ISOPED (and ISOPEDL) algorithm has, due to the introduction of the conditioning constraint, a complicated nonlinear structure causing its not monotonic behaviour. Therefore, its theoretical analysis occurred to be extremely difficult even in the two-dimensional case (Tadej, 2001).

However, simulation results have shown excellent performance of the presented dual ISOPE algorithm.

4.5 Dual ISOPE Algorithm with Optimized Initial Phase

In practical applications any set-point change should be carefully planned at the optimization layer since it means a transient process in the plant and directly influences plant effectiveness. It is precisely the case of any ISOPED regular iteration (Steps 1 to 5), provided all data necessary to start these iterations have been gathered during the initial phase (Step 0). However, during the initial phase set-point changes are also necessary and should be planned as carefully as possible, with plant effectiveness taken into account. This is not the case in the formulation of the dual algorithm in the preceding section, a drawback important from a practical point of view. The aim of this section is to remove, or at least weaken this drawback.

Recall the goals of the initial phase:

- To gather points $c^{-n}, c^{-n+1}, \ldots, c^0$, spanning the space $\mathbb{R}^n$ and enabling then to create the matrix S^0 in the first regular iteration.
- To collect measurement information sufficient to evaluate the derivative $F_*'(c^0)$ of the real plant steady-state input-output mapping in the first regular iteration, i.e., to apply the points $c^{-n+1}, c^{-n+1}, \ldots, c^0$ to the controlled process and to measure corresponding steady-state output values.

Certainly, the two stated goals are closely interrelated and are performed simultaneously. The only way to achieve these goals is to make n set-point deviations, starting from the initial set-point. The easiest way is to make these deviations along axes of the Cartesian space and then estimate the derivative using finite differences. This is the technique taken from the basic ISOPE algorithm, where it was applied at the initial and every regular iteration point, making this algorithm hardly acceptable for practical applications.

The idea to make the initial phase of the dual ISOPE algorithm more efficient is to apply during this phase the iterative two-step (ITS) algorithm, see Section 2.3.1 in Chapter 2, suitably reformulated to suit our goals. Recall that ITS consists of iteratively repeated two steps:

(1) steady-state model update (adaptation) basing on the last output measurement added,
(2) optimization of the updated steady-state model, application of the resulted set-point to the plant.

It is known that this standard (i.e., not modified as in the ISOPE) two-step method converges generally (if at all) only to *suboptimal set-points*, the loss of optimality depending on the adequacy of the model, see Section 2.3.1 in Chapter 2. Nevertheless, it can lead to optimality improvement, especially in first steps. Therefore, it seemed to be a good idea to apply the general structure of this method during the initial phase of the ISOPED algorithm (Tatjewski, 1998). However, a reformulation leading to augmentation of the performance function and adding additional constraints is necessary to achieve the two stated original goals of the initial phase.

To formulate the new, optimized initial phase let us start with the definition of the matrix S^i for negative values of i, $-n < i \leq 0$:

$$S^i = [c^{i-1} - c^i \;\; c^{i-2} - c^i \;\; \cdots \;\; c^{-n} - c^i]^T \tag{4.73}$$

and the conditioning constraint function

$$d(v, c^i, \ldots, c^{-n+1}, c^{-n}) = \frac{\sigma_{min}(S^{i+1}(v))}{\sigma_{max}(S^{i+1}(v))} \tag{4.74}$$

where $\sigma_{min}(S^{i+1}(v))$ and $\sigma_{max}(S^{i+1}(v))$ denote minimal and maximal singular values of the matrix $S^{i+1}(v)$

$$S^{i+1}(v) = [c^i - c^{i+1}(v) \;\; c^{i-1} - c^{i+1}(v) \;\; \cdots \;\; c^{-n} - c^{i+1}(v)]^T$$

and where $c^{i+1}(v) = v$ if $k_c = 1$, as it will be assumed during the initial phase; compare (4.55) and (4.56). Notice that S^i is non-square for $i < 0$

and for $i = 0$ becomes precisely the square matrix S^0 needed to start the first regular iteration of the ISOPED algorithm in Step 1.

We are now in a position to formulate the *optimized initial phase of the ISOPED algorithm*:

Step 0 (Initial phase, optimized).

Start 0. Choose appropriately positive parameters ρ, γ, δ. Set $i := -n$, $c^{-n} := c^0$ — the initial point.

Step 0.1 Apply the set-point c^i to the controlled plant, measure the corresponding outputs after transient processes have died. Add the measurement to the data record and adapt the steady-state model (i.e., the parameters α).

Step 0.2 Solve the following *augmented model optimization problem* (AMOP)

$$\begin{aligned} &\text{minimize}_v\{q(v,\alpha) + \rho \left\|v - c^i\right\|^2\} \\ &\text{subject to}: \; g(v) \leq 0 \\ &\qquad\qquad\quad \left\|v - c^i\right\| \geq \gamma \end{aligned} \tag{4.75}$$

denoting the solution point by $c^{i+1} = \hat{v}(c^i)$. If

$$d(c^{i+1}, c^i, \ldots, c^{-n}) \geq \delta \tag{4.76}$$

then go to **Step 0.4**, else proceed.

Step 0.3 Solve the following *conditioned augmented model optimization problem* (CAMOP)

$$\begin{aligned} &\text{minimize}_v\{q(v,\alpha) + \rho \left\|v - c^i\right\|^2\} \\ &\text{subject to}: \; g(v) \leq 0 \\ &\qquad\qquad\quad \left\|v - c^i\right\| \geq \gamma \\ &\qquad\qquad\quad d(v, c^i, \ldots, c^{-n}) \geq \delta \end{aligned} \tag{4.77}$$

denoting the solution point by c^{i+1}.

Step 0.4 If $i \leq -1$ then set $i := i + 1$ and go to **Step 0.1**, else proceed with regular ISOPED iterations from **Step 1**.

It should be noticed that three elements have been added to the standard model optimization problem (i.e., minimize $q(v,\alpha)$ subject to $g(v) \leq 0$) to create the AMOP and CAMOP problems:

- First, the performance function has been augmented adding a penalty term as in the original ISOPED algorithm. The reasons are similar: regularization of the original optimization problem and prevention of too large deviations of a solution from the current point. The second reason may be important during the initial phase, where smaller deviations of the current set-point are recommended. Therefore, the value of ρ could be generally larger here than later during regular ISOPED iterations.
- Second, the constraint $\| v - c^i \| \geq \gamma$ has been added. Its role is to keep deviations of the set-point sufficiently large for the purpose of the later derivative estimation based on finite difference approximation using the matrix S^0. The minimal set-point deviations should be determined taking into account errors in the feedback (measurement) information and properties the of actual performance function. In particular, the discussed constraint is necessary if the algorithm starts from the model-optimal set-point, or a point close to it.
- Third, the conditioning constraint has been added to the AMOP in Step 0.3. The resulted CAMOP problem must be solved when the actual AMOP solution does not satisfy this constraint.

In conclusion, one can regard the presented optimized initial phase of the ISOPED algorithm as *standard initial phase improved by incorporation of the optimality factor*, expressed by augmented constrained model optimizations, into the process of generation of subsequent set-point deviations.

But, on the other hand, the presented optimized initial phase can be regarded as an ITS optimizing control algorithm augmented by introduction of the regularization term and the constraints to the model optimization problem, to include the goals of the ISOPED initial phase into the subsequent optimizations. Certainly, this augmentation slightly disturbs the optimization results. This should not be a real drawback, since the ITS algorithm leads generally only to suboptimal results, suboptimality depending on the accuracy of the model. Therefore, the optimality improvement during the initial phase can be awaited when the model is not too rough. Moreover, for relatively accurate models the whole ISOPED algorithm with the presented initial phase can be regarded as *an ITS approach modified (supplemented) by a second phase (regular ISOPED phase)* added to eliminate the suboptimality caused by model-reality differences.

Certainly, a choice of one from the two views presented above depends on the accuracy of the model. The latter point of view can be more suitable when the process model is sufficiently accurate to result in significant performance improvement under the standard two-step algorithm. On the other hand, the first view is more adequate when only rather rough process model is available — the optimized initial phase can then result only in achieving the ISOPED goals of the initial phase, even with no performance improvement at all. In such situations application of the standard initial phase still remains a reasonable alternative, in particular if directions of sufficiently significant change in the value of the true performance function are known and taken into account when choosing set-point deviations.

4.6 Comparative Simulation Study

In this section results of simulation studies of the ISOPE algorithms will be presented. First, simple examples, but appropriately chosen to show the features of the algorithms will be studied. Then, results of case-studies will be presented, in the next section.

Example 4.1

An example nonlinear plant described by the following (uncertain) input-output mapping is considered:

$$y = F_*(c_1, c_2) = 2c_1^{0.5} + c_2^{0.4} + 0.2c_1c_2. \tag{4.78}$$

The performance function (to be minimized) and the constraint set are

$$Q(c, y) = -y + (c_1 - 0.5)^2 + (c_2 - 0.5)^2 \tag{4.79}$$

$$C = \{c \epsilon \mathbb{R}^2 : \; 0 \le c_1 \le 1.5, \;\; 0 \le c_2 \le 1.5 \,\}. \tag{4.80}$$

The real optimal piont is $(\hat{c}_1, \hat{c}_2) = (1.067, 0.830)$, with the (real) optimal value of the performance function $Q(\hat{c}, F_*(\hat{c})) = -2.7408$. It should be noted that the optimal point is located in the interior of the constraint set, i.e., the inequality constraints are not active.

The following linear process model is assumed to be available

$$y = F(c_1, c_2, \alpha) = 0.6c_1 + 0.4c_2 + \alpha. \tag{4.81}$$

The process mapping (4.78) and its model (4.81) are visualized in Figs. 4.3 and 4.4, both seen from the directions of c_1 and c_2 axes, respectively.

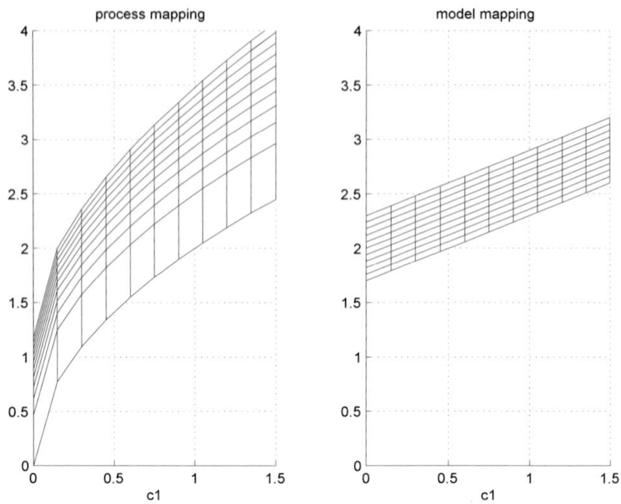

Fig. 4.3 Process and model mappings, projection on (y, c_1) plane (Ex. 4.1, 4.2, 4.3).

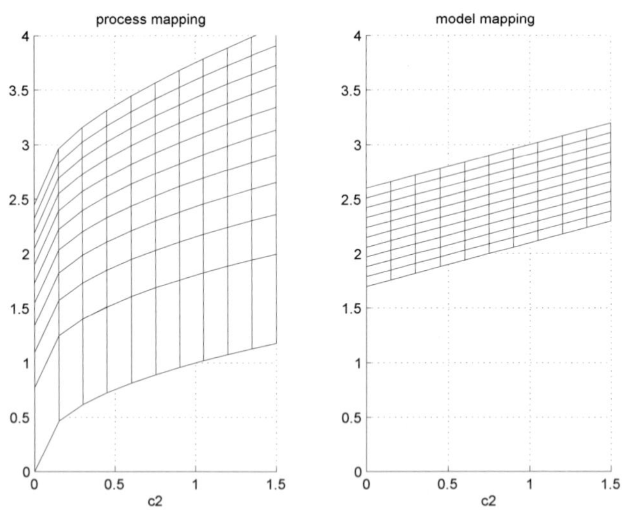

Fig. 4.4 Process and model mappings, projection on (y, c_2) plane (Ex. 4.1, 4.2, 4.3).

The modified model optimization problem (MMOP) takes for the presented example the form

$$\begin{aligned} &\min_{v}\{Q(v,F(v,\alpha))-\lambda(c^i)^T v+\rho\|c^i-v\|^2\}\\ &\text{subject to}:\ 0\le v_1\le 1.5\\ &\qquad\qquad\quad\ 0\le v_2\le 1.5 \end{aligned} \tag{4.82}$$

where $v=[v_1\ v_2]^T$,

$$Q(v,F(v,\alpha))=-0.6v_1-0.4v_2-\alpha+(v_1-0.5)^2+(v_2-0.5)^2 \tag{4.83}$$

$$\begin{aligned} \lambda(c^i)^T &= -\left([0.6\ \ 0.4]-[\frac{\partial F_*}{\partial c_1}(c^i)\ \ \frac{\partial F_*}{\partial c_2}(c^i)]\right)\\ &=[-0.6+\frac{\partial F_*}{\partial c_1}(c^i)\quad -0.4+\frac{\partial F_*}{\partial c_2}(c^i)] \end{aligned} \tag{4.84}$$

and the vector of partial derivatives $(F_*)'(c^i)=[\frac{\partial F_*}{\partial c_1}(c^i)\ \frac{\partial F_*}{\partial c_2}(c^i)]$ will be approximated during the run of the basic ISOPE algorithm using the finite difference method based od values of measured outputs $F_*(c)$.

In the ISOPED algorithm the CMMOP is solved instead of the MMOP, being its augmentation by adding the conditioning constraint

$$d(c^{i+1}(v),c^i,c^{i-1})=\frac{\sigma_{\min}(\mathbf{S}^{i+1}(v))}{\sigma_{\max}(\mathbf{S}^{i+1}(v))}\ge\delta \tag{4.85}$$

where $c^{i+1}(v)=c^i+k_c(v-c^i)$,

$$\mathbf{S}^{i+1}(v)=\begin{bmatrix} c_1^i-c_1^{i+1}(v) & c_1^{i-1}-c_1^{i+1}(v)\\ c_2^i-c_2^{i+1}(v) & c_2^{i-1}-c_2^{i+1}(v)\end{bmatrix}^T \tag{4.86}$$

and $\sigma_{\min}(\mathbf{S}^{i+1}(v))$ i $\sigma_{\max}(\mathbf{S}^{i+1}(v))$ denote eigenvalues of the matrix $\mathbf{S}^{i+1}(v)$. Elements of the derivative $(F_*)'(c^i)=[\frac{\partial F_*}{\partial c_1}(c^i)\ \frac{\partial F_*}{\partial c_2}(c^i)]$ will be in this case estimated from the solution of two sets of linear equations

$$\mathbf{S}^i\begin{bmatrix}\frac{\partial F_*}{\partial c_1}(c^i)\\ \frac{\partial F_*}{\partial c_2}(c^i)\end{bmatrix}=\begin{bmatrix}F_*(c^{i-1})-F_*(c^i)\\ F_*(c^{i-2})-F_*(c^i)\end{bmatrix} \tag{4.87}$$

where

$$\mathbf{S}^i=[\ c^{i-1}-c^i\quad c^{i-2}-c^i\ ]^T,\qquad i=1,2. \tag{4.88}$$

Three versions of the ISOPE algorithm were simulated for the above example problem:

ISOPEB - Basic ISOPE algorithm as presented in Section 4.1, i.e., with n ($n = 2$) additional set-points perturbations along Cartesian axes at every iteration, to generate finite difference approximations of the process mapping derivatives.

ISOPEDstd - Dual algorithm ISOPED as presented in Section 4.4, with standard simple initial phase consisting of $n = 2$ set-points perturbations along Cartesian axes at the initial point, to generate first matrix S^0 and first finite difference approximation of the process mapping derivatives.

ISOPEDopt - Dual algorithm ISOPED with optimized initial phase, as presented in Section 4.5.

Initial point $c^0 = (0.5\ 0.5)$ has been chosen and the following nominal values of the parameters of the considered algorithms:

- relaxation coefficient, see (4.21), $k_c = 0.5$,

- quadratic convexification (penalty) coefficient $\varrho = 0.1$,

- final accuracy (termination) parameter (i.e., norm of smallest acceptable change in the set-point c) $\varepsilon = 0.04$,

- set-point perturbations for finite difference approximations of the derivative for the ISOPEB algorithm and the initial phase of the ISOPEDstd algorithm, $\Delta c_j = \varepsilon$,

- right-hand side value of the conditioning constraint for both ISOPED algorithms, $\delta = 0.05$,

- right-hand side value of the additional constraint in the AMOP and CAMOP problems of the optimized initial phase of the ISOPEDopt algorithm, $\gamma = \varepsilon$,

- penalty coefficient in the AMOP and CAMOP problems of the optimized initial phase of the ISOPEDopt algorithm, $\varrho 1 = 4 * \varrho$.

Values of certain parameters were then appropriately changed, to investigate influence of their values on simulation results, i.e., on the algorithm behaviour. It should be noted that a relatively large, from a pure mathematical viewpoint, value of the termination accuracy parameter ε has been chosen. However, our aim is not to investigate mathematical optimization algorithms (where far smaller termination accuracies are usually applied), but *to simulate* operation of industrial process optimizing control algorithms. Termination of the algorithm iterations, i.e., termination of the set-point changes, is here defined by minimal reasonable change of the set-point values at given process operating conditions, determined by accuracy of measurements, actuators positioning and curvature of the plant performance function. Therefore, also the additional set-point perturbations used for estimation of process derivatives Δc_j cannot be less than the stated termination accuracy, as well as the minimal admissible change γ in set-points during the optimized initial phase ($\Delta c_j = \varepsilon$, $\gamma = \varepsilon$ was set).

All simulation runs have been visualized in the form of two curves:

- trajectories of the plant performance function values $Q(c^i, F_*(c^i))$ as function of all set-point changes, i.e., including also additional set-point changes applied for derivative estimation,
- trajectories of the set-points on the plane (c_1, c_2).

Trajectories of all three algorithms for nominal parameter values has been shown in Figs. 4.5 and 4.6. Looking at the set-point trajectories in Fig. 4.6 one can easily see two additional set-point changes proceeding every main iteration of the ISOPEB algorithm and a characteristic "zig-zag like" nature of the ISOPED trajectories resulting from activity of the conditioning constraint. As awaited, the most efficient is the ISOPEDopt algorithm due to the optimized initial phase.

Trajectories obtained after enlarging the conditioning constraint right-hand value δ from 0.05 to 0.2 are shown in Figs. 4.7 and 4.8. It results in stronger "zig-zag" behaviour of the dual algorithms, easily seen on the set-point plane. However, efficiency of the algorithms is still similar. □

Example 4.2

Example 4.1 but with the following tighter constraint set

$$C = \{c \in \mathbb{R}^2 : \ 0 \le c_1 \le 0.8, \ \ 0 \le c_2 \le 1.5 \,\} \tag{4.89}$$

will be now considered. Therefore, the optimal point lies on the feasible set boundary, the constraint $c_1 \le 0.8$ is active at the optimal point. The example tests behaviour of the algorithms in such situation.

Trajectories of all three algorithms, ISOPEB, ISOPEDstd and ISOPEDopt are shown in Figs. 4.9 and 4.10, for nominal values of the algorithm parameters as given in Example 4.1. Note the especially high efficiency of the ISOPEDopt algorithm in this example case, and the necessity to control

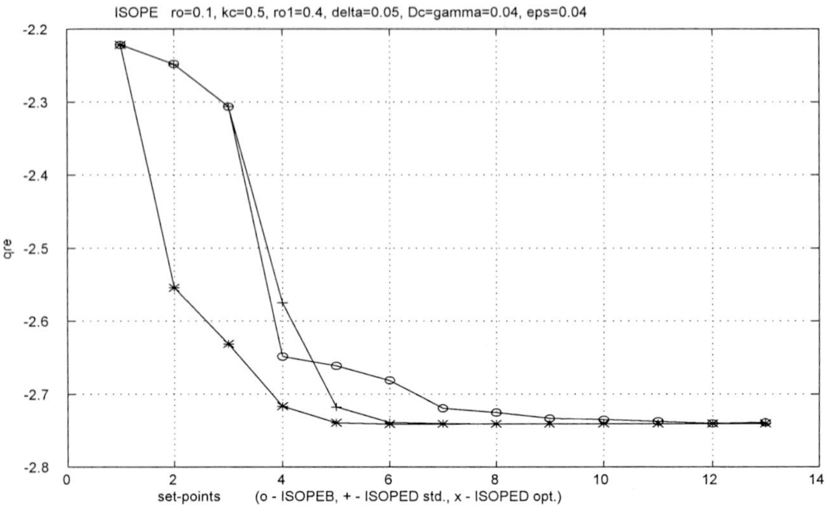

Fig. 4.5 Performance function trajectories for nominal parameter values (Ex. 4.1).

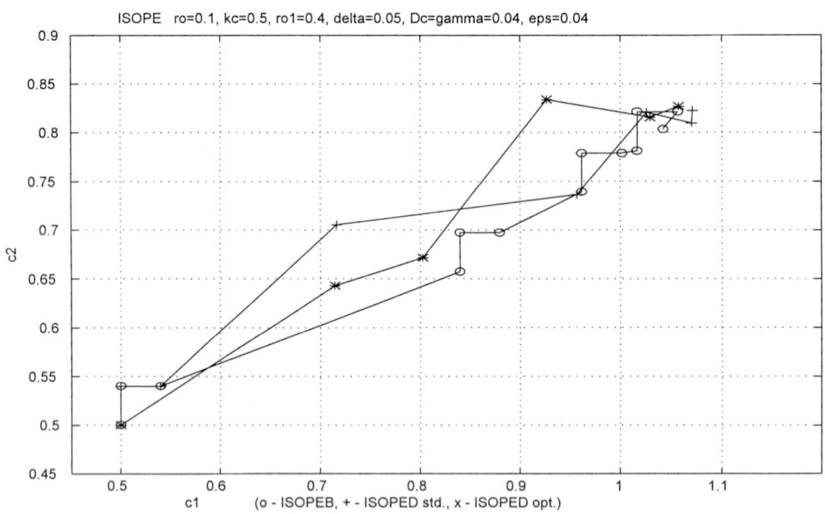

Fig. 4.6 Set-point trajectories for nominal parameter values (Ex. 4.1).

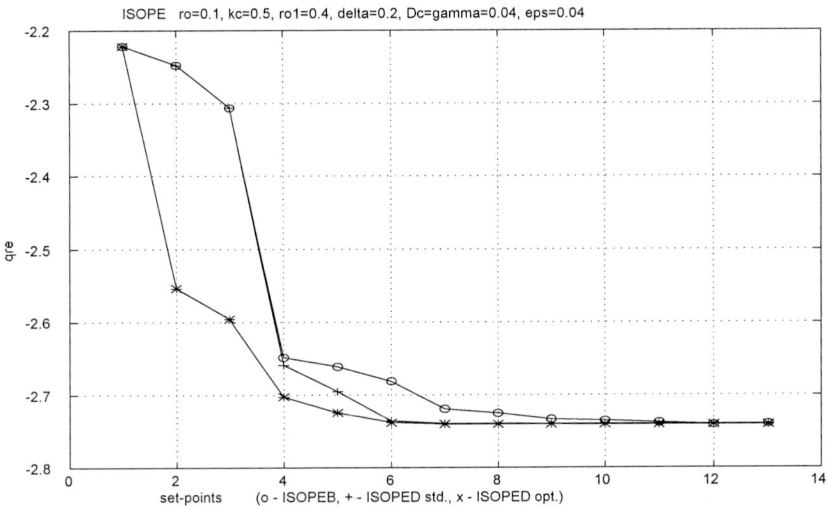

Fig. 4.7 Performance function trajectories with $\delta = 0.2$ (Ex. 4.1).

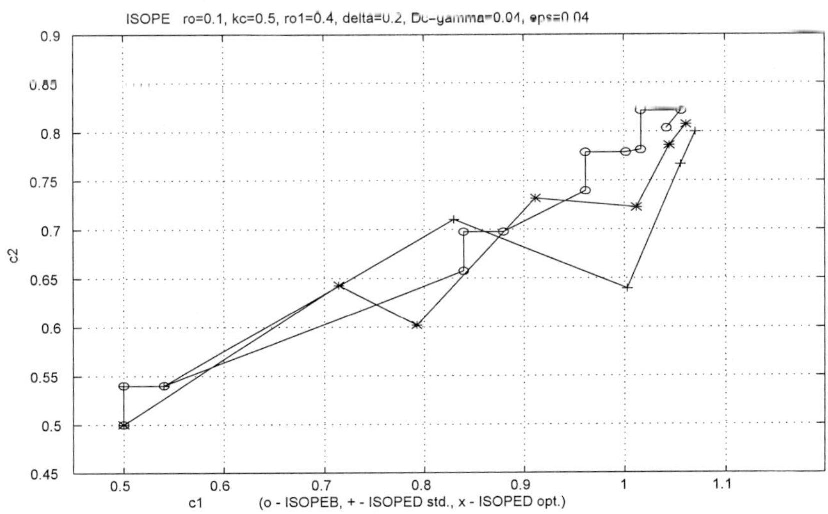

Fig. 4.8 Set-point trajectories with $\delta = 0.2$ (Ex. 4.1).

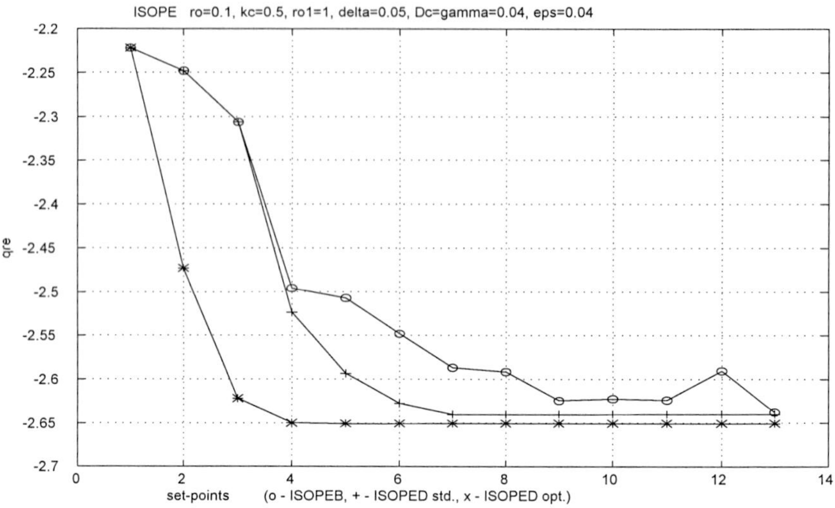

Fig. 4.9 Performance function trajectories (Example 4.2).

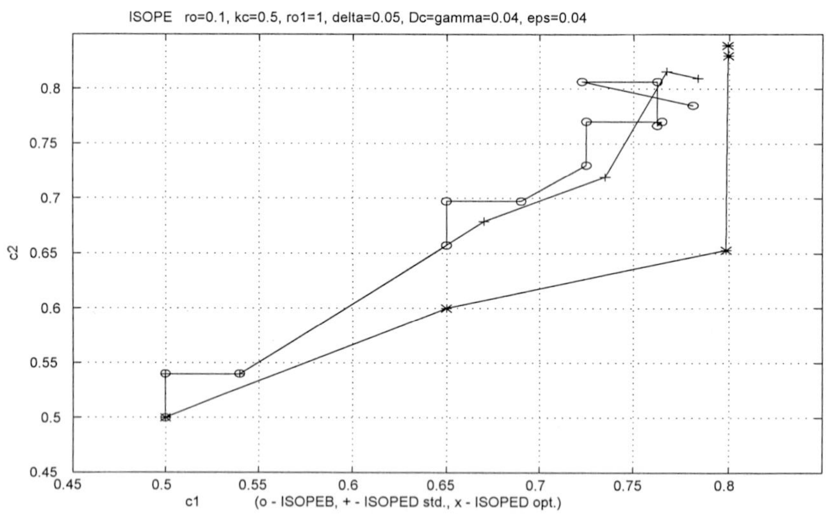

Fig. 4.10 Set-point trajectories (Example 4.2).

and appropriately change the sign of the additional set-point c_2 deviations in the vicinity of the constraint boundary in the ISOPEB algorithm. □

Example 4.3

It is still a modified Example 4.1, this time with measurement error added. That is, instead of the process output mapping (4.78) we have

$$y = F_*(c_1, c_2) = 2c_1^{0.5} + c_2^{0.4} + 0.2c_1c_2 + \nu \tag{4.90}$$

where ν is a pseudo-random signal uniformly distributed in $[-0.03\ +0.03]$. It corresponds to errors not exceeding 3% of the actual range of the output variable $y \cong 2 \div 3$ (and 1% of the whole range $y = 1 \div 3$). This can result in possible error in a difference of two output values equal up to $\Delta y_{err} = 0.06$, i.e. 6% (2%).

One should always remember the practical importance of an adequate choice of the minimal admissible change in the set-point values, directly influencing the choice of the parameters ε, Δc_j i γ, for every real case at hand with concrete output measurements accuracy. The reason is in all ISOPE algorithms estimation of the process mapping derivatives is based on differences of (measured) process output values. Therefore, differences between set-points should imply differences in resulting output values sufficiently larger than possible errors in differences of measured outputs. In the case of our example, the following approximate relation between set-points c and outputs y is valid for the values occurring during simulation runs:

$$\Delta y \cong (0.6 \div 1.2)\Delta c. \tag{4.91}$$

Therefore, taking, e.g., $\Delta c_j = 0.06$ results in

$$\Delta y \cong 0.036 \div 0.072 = (0.6 \div 1.2) \cdot \Delta y_{err} \tag{4.92}$$

and taking $\Delta c_j = 0.12$ results in

$$\Delta y \cong 0.072 \div 0.144 = (1.2 \div 2.4) \cdot \Delta y_{err} \tag{4.93}$$

where $\Delta y_{err} = 0.06$. These two cases correspond, therefore, to very large possible errors in output differences.

Due to the random nature of measurement errors the results for the same algorithm are different for different simulation runs. Therefore, representative exemplary simulation results for all three algorithms for the considered example with output errors are shown in Figs. 4.11 and 4.12, for nominal parameter values as stated in Example 4.1 except for $\varepsilon = \Delta c_j = \gamma = 0.12$.

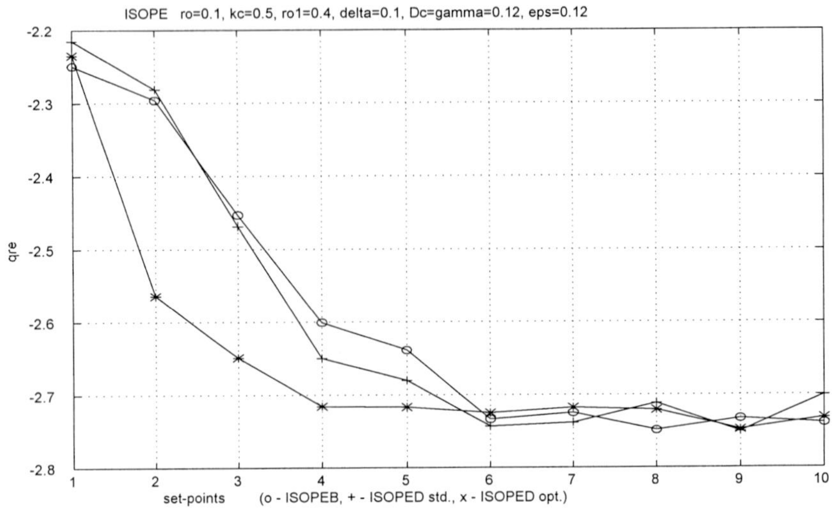

Fig. 4.11 Performance function trajectories (smaller error, Ex. 4.3).

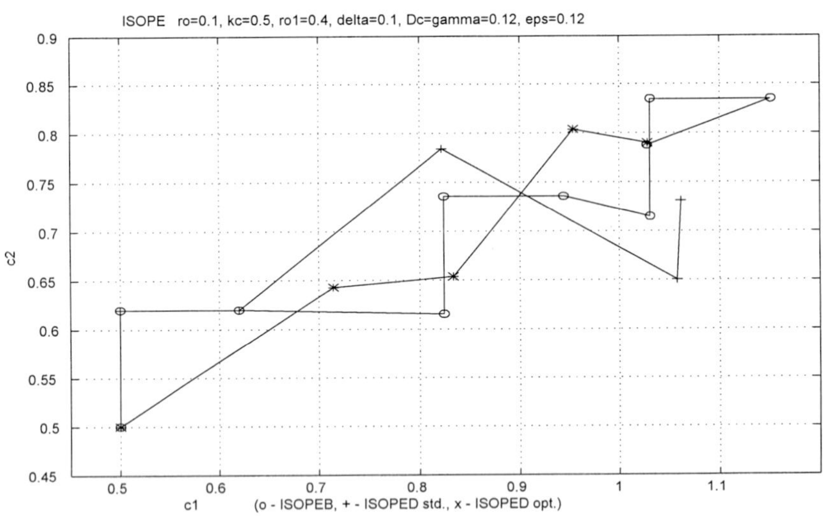

Fig. 4.12 Set-point trajectories (smaller error, Ex. 4.3).

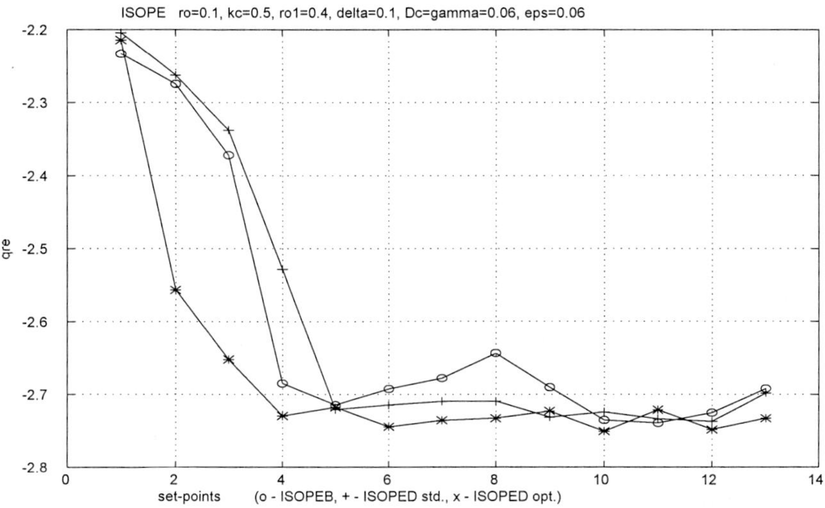

Fig. 4.13 Performance function trajectories (larger error, Ex. 4.3).

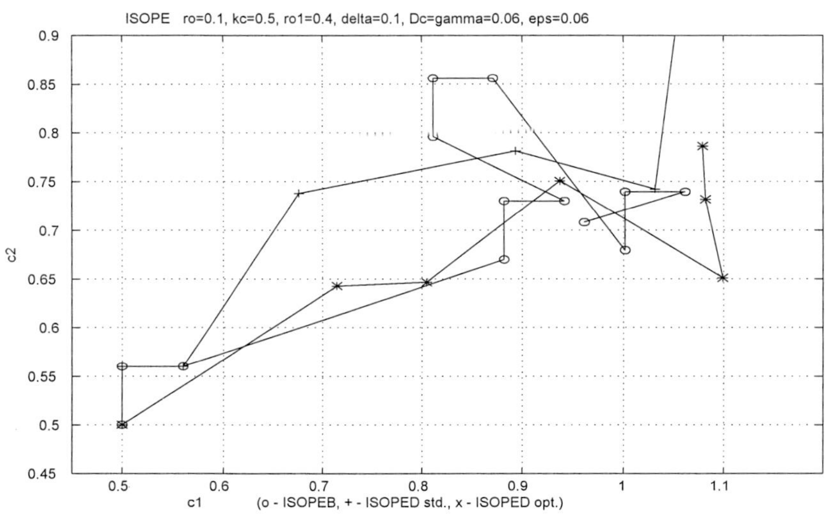

Fig. 4.14 Set-point trajectories (larger error, Ex. 4.3).

The right-hand side value of the conditioning constraint was also slightly enlarged to $\delta = 0.1$, to cope better with the errors. Further, simulation results for all three algorithms for $\varepsilon = \Delta c_j = \gamma = 0.06$ are shown in Figs. 4.13 and 4.14. In spite of large errors in the output information all algorithms behave unexpectedly well, quickly approaching a neighborhood of the optimal point, even in the worst case of the last simulations. Certainly, the trajectories are randomized when compared to the cases of previous examples without errors. In particular, when approaching a vicinity of the optimum and fulfilling the termination criterion, further variations of the performance function value are only due to random errors while measuring the process steady-state output value. In practical applications a human operator or an automatic supervisory algorithm should always stop the optimizing control algorithm after reaching the vicinity of the optimum, if the algorithm does not stop by itself. □

4.7 Case Study Examples

Example 4.4

An ethylene distillation column depicted in Fig. 4.15 will be considered, where ethylene is distilled from a mixture containing mainly ethane and ethylene (a case-study taken from (Tatjewski, 2002)).

The output vector $y = [y_1 \; y_2]^T$ represents values (in steady states) of the following variables:

y_1 – concentration of ethane in the product stream P, measured in *ppm* (*parts per million*),

y_2 – concentration of ethylene in the bottom product W, given as a molar fraction.

The vector of decision variables $c = [c_1 \; c_2]^T$ represents set-points for feedback controllers stabilizing:

c_1 – ratio of reflux flow R to product flow P (forced through the control of P with respect to measured value of R),

c_2 – ethylene concentration on the control shelf.

Models of the column which were considered during the optimizing control simulations had been taken from (Tadej, 2001), where on the basis of input-output data records two models have been developed: a "true plant" model

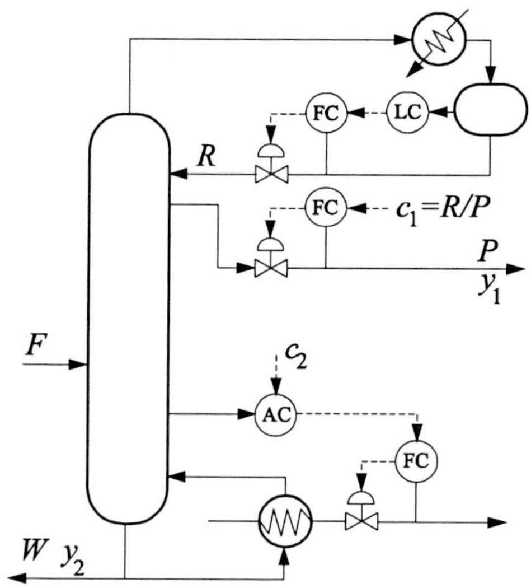

Fig. 4.15 Distillation column for Example 4.4 ($R-$ reflux, $P-$ product, $W-$ bottom product).

for simulation purposes and a simplified model for control purposes. The true plant model will serve as a true plant mapping during the optimizing control simulations, it is of the form

$$y_1 = F_{*1}(c_1, c_2) = \exp(-s_{11}(c_1 - \bar{c}_{11})) \cdot \exp(-s_{12}(c_2 - \bar{c}_{12})) \quad (4.94a)$$
$$y_2 = F_{*2}(c_1, c_2) = \exp(-s_{21}(c_1 - \bar{c}_{21})) \cdot \exp(-s_{22}(c_2 - \bar{c}_{22})) \quad (4.94b)$$

where

$$s_{11} = 12.7049, \quad \bar{c}_{11} = 4.6816$$
$$s_{12} = 0.2536, \quad \bar{c}_{12} = 0.3252$$
$$s_{21} = 0.3340, \quad \bar{c}_{21} = 2.5544$$
$$s_{22} = -5.3719, \quad \bar{c}_{22} = 1.1838.$$

The shape of the plant mappings is visualized in Fig. 4.16. The simplified model (for control purposes) is of the form

$$y_1 = F_1(c_1, c_2, a) = m_1(c_1 - \bar{c}_{11m})^4 \cdot (c_2 - \bar{c}_{12m})^4 + \alpha_{a1} \quad (4.95a)$$
$$y_2 = F_2(c_1, c_2, a) = m_{21}c_1 + m_{22}c_2 + \alpha_{a2} \quad (4.95b)$$

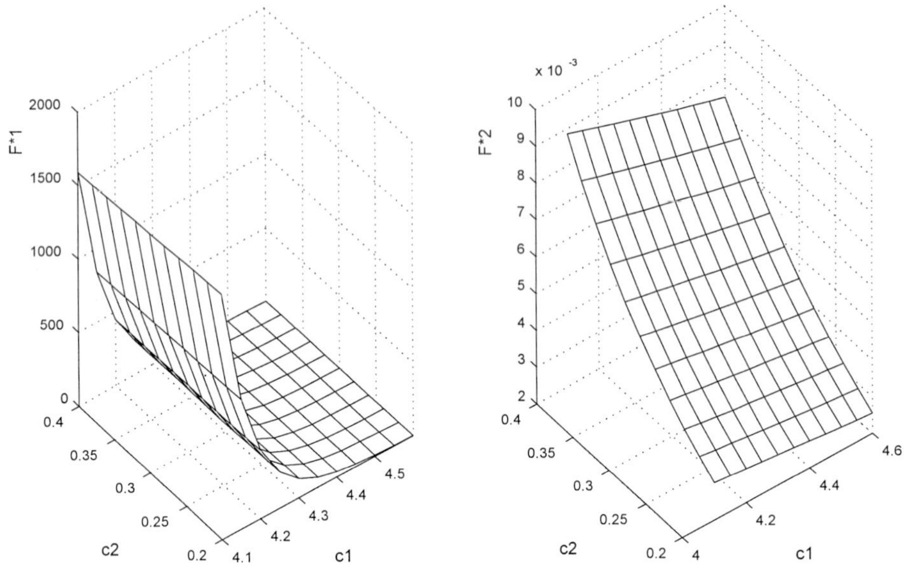

Fig. 4.16 Surfaces of the plant mappings $F_{*1}(c_1, c_2)$ and $F_{*2}(c_1, c_2)$.

where

$$m_1 = 999.9923, \quad \bar{c}_{11m} = 4.6261, \quad \bar{c}_{12m} = 2.6481, \quad \alpha_{a1} = 0$$
$$m_{21} = -0.0016, \quad m_{22} = 0.0254, \quad \alpha_{a2} = 0.0043$$

and $\alpha_a = [\alpha_{a1}, \alpha_{a2}]$ is the vector of additive model parameters. These parameters will be updated when performing parameter estimation (PEP) during runs of the ISOPE algorithms. It is assumed that it is enough to update other (non-additive) model parameters $\alpha_n = (m_1, \bar{c}_{11m}, \bar{c}_{12m}, m_{21}, m_{22})$ before initialization of the ISOPE algorithm and that they are kept constant during the single run of this algorithm. The shape of the model mappings is visualized in Fig. 4.17. The goal of the set-point optimization is to achieve desired steady-state values y_{r1} and y_{r2} of the plant outputs. Therefore, the performance function is formulated as

$$Q(c, y) = Q(y) = \mu_1(y_1 - y_{r1})^2 + \mu_2(y_2 - y_{r2})^2 \tag{4.96}$$

where $\mu_1 = 2 \cdot 10^{-5}$ and $\mu_2 = 10^6$ are scaling coefficients needed due to very different scales of the outputs y_1 (hundreds) and y_2 (in the range to

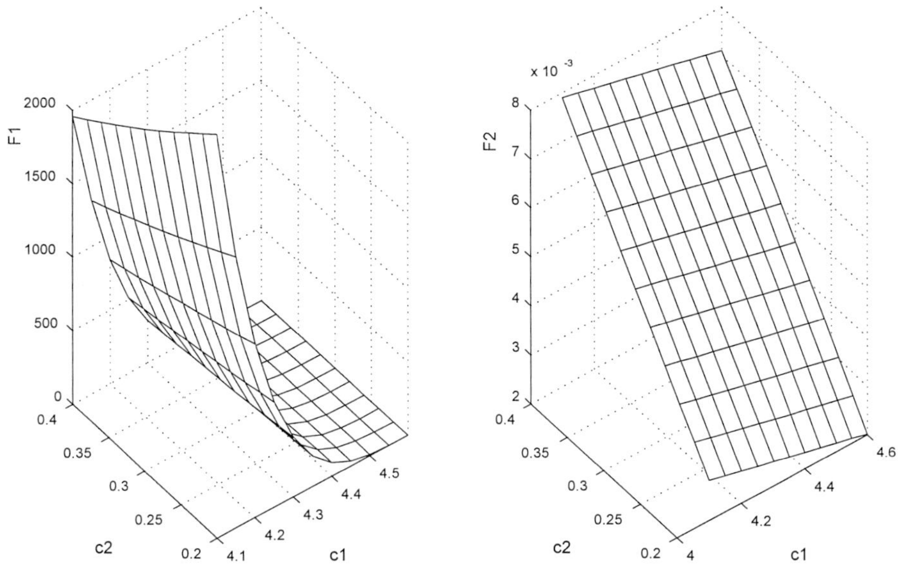

Fig. 4.17 Surfaces of the plant model mappings $F_1(c_1, c_2)$ and $F_2(c_1, c_2)$.

10×10^{-3}). Typical desired set-point values are $y_{r1} = 500$, $y_{r2} = 0.005$, and these values were assumed in the simulations which are to be presented. The set-points $c \in \mathbb{R}^2$, and therefore decision variables $v \in \mathbb{R}^2$ of the optimization problems, are constrained to the set

$$C = \{ v \in \mathbb{R}^2 : \ 4.1 \leq v_1 \leq 4.6, \quad 0.2 \leq c_2 \leq 0.4 \}. \tag{4.97}$$

The presented performance function (4.96) has an indirect economical sense. Namely, the more polluted the product the cheaper its production (less energy consumed for distillation). Therefore, it is optimal to run the distillation at the desired value y_{r1} (ethane concentration) located as close as possible to the largest admissible value, but at a distance of a safety zone of a width resulting from the uncertainty level in the control system. On the other hand, losses of ethylene should not be too high in the bottom product — therefore, a soft constraint on its concentration in this product expressed as a desired value y_{r2}.

First, the performance function (4.96) was optimized subject to the model constraints (4.95a), (4.95b) and inequality constraints yielding the model-optimal point $(\hat{c}_{m1},\ \hat{c}_{m2}) = (4.2685,\ 0.2964)$. The corresponding values of the plant mapping (4.94a), (4.94b) are $F_*(\hat{c}_{m1},\ \hat{c}_{m2}) = [191.6\ \ 0.0048]^T$ — and are obviously far from the desired values $y_{r1} = 500$, $y_{r2} = 0.005$, especially from the first one. Then, simulation of the optimizing control using the ISOPEDstd and ISOPEopt algorithms (see Example 4.1 for definitions) were performed. The conditioned modified model optimization problem (CMMOP) takes for the presented problem the form

$$\begin{aligned} &\min_v \{Q(F(v,\alpha^i)) - \lambda(c^i,\alpha^i)^T v + \rho\|c^i - v\|^2\} \\ &\text{subject to: } 4.1 \le v_1 \le 4.6 \\ &\qquad 0.2 \le v_2 \le 0.4 \\ &\qquad d(c^{i+1}(v), c^i, c^{i-1}) = \frac{\sigma_{\min}(\mathbf{S}^{i+1}(v))}{\sigma_{\max}(\mathbf{S}^{i+1}(v))} \ge \delta \end{aligned} \tag{4.98}$$

where $v = [v_1\ v_2]^T$,

$$c^{i+1}(v) = c^i + k_c(v - c^i) \tag{4.99}$$

and $\sigma_{\min}(\mathbf{S}^{i+1}(v))$ i $\sigma_{\max}(\mathbf{S}^{i+1}(v))$ denote singular values of the matrix $\mathbf{S}^{i+1}(v)$,

$$\mathbf{S}^{i+1}(v) = \begin{bmatrix} c_1^i - c_1^{i+1}(v) & c_1^{i-1} - c_1^{i+1}(v) \\ c_2^i - c_2^{i+1}(v) & c_2^{i-1} - c_2^{i+1}(v) \end{bmatrix}^T. \tag{4.100}$$

Further, in the formulation of the CMMOP we have :

$$Q(F(v,\alpha^i)) = \mu_1(F_1(v,\alpha^i) - y_{r1})^2 + \mu_2(F_2(v,\alpha^i) - y_{r2})^2 \tag{4.101}$$

$$\begin{aligned} \lambda(c^i,\alpha^i)^T &= [2\mu_1(F_1(c^i,\alpha^i) - y_{r1})\ \ 2\mu_2(F_2(c^i,\alpha^i) - y_{r2})] \times \\ &\times \left(\begin{bmatrix} \frac{\partial F_1}{\partial c_1}(c^i,\alpha^i) & \frac{\partial F_1}{\partial c_2}(c^i,\alpha^i) \\ \frac{\partial F_2}{\partial c_1}(c^i,\alpha^i) & \frac{\partial F_2}{\partial c_2}(c^i,\alpha^i) \end{bmatrix} - \begin{bmatrix} \frac{\partial F_{*1}}{\partial c_1}(c^i) & \frac{\partial F_{*1}}{\partial c_2}(c^i) \\ \frac{\partial F_{*2}}{\partial c_1}(c^i) & \frac{\partial F_{*2}}{\partial c_2}(c^i) \end{bmatrix} \right) \end{aligned} \tag{4.102}$$

where

$$\alpha_1^i = F_{*1}(c^i) - m_1(c_1^i - \bar{c}_{11m})^4 \cdot (c_2^i - \bar{c}_{12m})^4 \tag{4.103}$$

$$\alpha_2^i = F_{*2}(c^i) - m_{21}c_1^i + m_{22}c_2^i \tag{4.104}$$

because only the additive parameters were assumed to be updated during the runs of the ISOPED algorithms. Vectors of the model mapping derivatives $(F_j)'(c^i,\alpha^i) = [\frac{\partial F_j}{\partial c_1}(c^i,\alpha^i)\ \frac{\partial F_j}{\partial c_2}(c^i,\alpha^i)]$, $j = 1,2$ are calculated directly from the formulae (4.95a) i (4.95b), whereas vectors of the plant mapping derivatives $(F_{*j})'(c^i) = [\frac{\partial F_{*j}}{\partial c_1}(c^i)\ \frac{\partial F_{*j}}{\partial c_2}(c^i)]$, $j = 1, 2$ are calculated, during the runs of the algorithms, from solutions of linear equations

$$\mathbf{S}^i \begin{bmatrix} \frac{\partial F_{*j}}{\partial c_1}(c^i) \\ \frac{\partial F_{*j}}{\partial c_2}(c^i) \end{bmatrix} = \begin{bmatrix} F_{*j}(c^{i-1}) - F_{*j}(c^i) \\ F_{*j}(c^{i-2}) - F_{*j}(c^i) \end{bmatrix}, \quad j = 1, 2 \tag{4.105}$$

where

$$\mathbf{S}^i = \begin{bmatrix} c^{i-1} - c^i & c^{i-2} - c^i \end{bmatrix}^T. \tag{4.106}$$

Simulations of the optimizing control using the ISOPEDstd and ISOPEDopt algorithms were performed for the presented plant description and its model. The results shown in Figs. 4.18, 4.19 and 4.20 present trajectories generated by the algorithms starting from the set-point equal to the model-optimal one, $(\hat{c}_{m1}, \hat{c}_{m2}) = (4.2685,\ 0.2964)$, i.e., being the solution to the MOP problem. The following algorithm parameters were assumed: $\rho = 100$, $k_c = 1$, $\epsilon = 0.001$ and for the optimized initial phase of the ISOPEDopt algorithm: $\rho_1 = 200$ and $\gamma = 0.01$. A relatively small value of ϵ was assumed intentionally, to investigate convergence in a small neighborhood of the optimum (in practical optimizing control applications this value should be larger, adjusted to output measurement errors and differences). In Fig. 4.18 trajectories of the true performance function value $Q(c^i, F_*(c^i))$ are presented, whereas in Fig. 4.19 trajectories of the decision variables (set-points) plotted against the background of the performance function $Q(c, F_*(c))$ contour lines (dotted lines) and in Fig. 4.20 trajectories of the plant outputs.

It is remarkable that effectiveness of the first iterations was significantly improved when applying the algorithm with optimized initial phase (ISOPEDopt), in spite the fact that initial point was a model-optimal one. The reason is that the adaptation of the additive model parameters and relatively good modeling of the output trends by the simplified model mapping, in spite of their structural uncertainty.

Simulation runs were also performed from other, in particular more distant than the model-optimal one, initial points. Exemplary trajectories of the performance function, set-points and outputs are depicted in Figs. 4.21, 4.22 and 4.23.

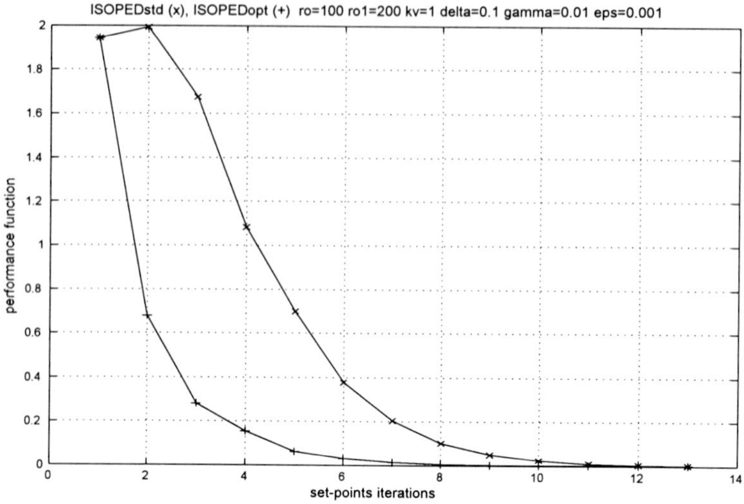

Fig. 4.18 Performance function trajectories for ISOPEDstd and ISOPEDopt algorithms, model-optimal initial point.

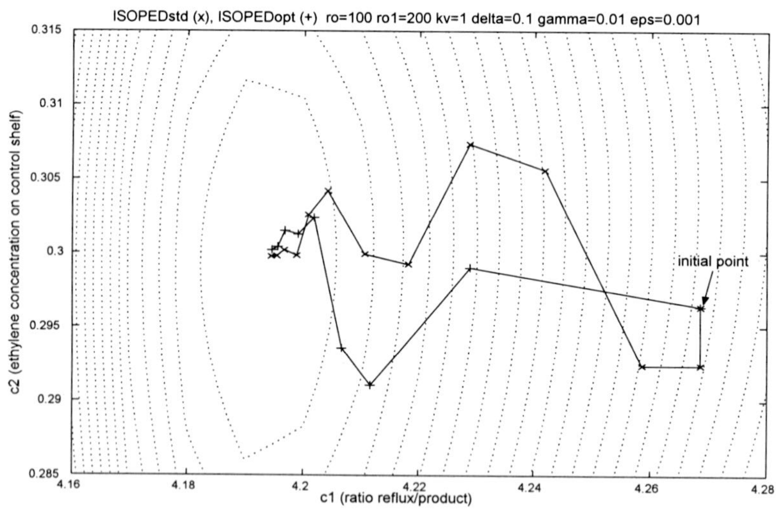

Fig. 4.19 Set-points trajectories for ISOPEDstd and ISOPEDopt algorithms, model-optimal initial point.

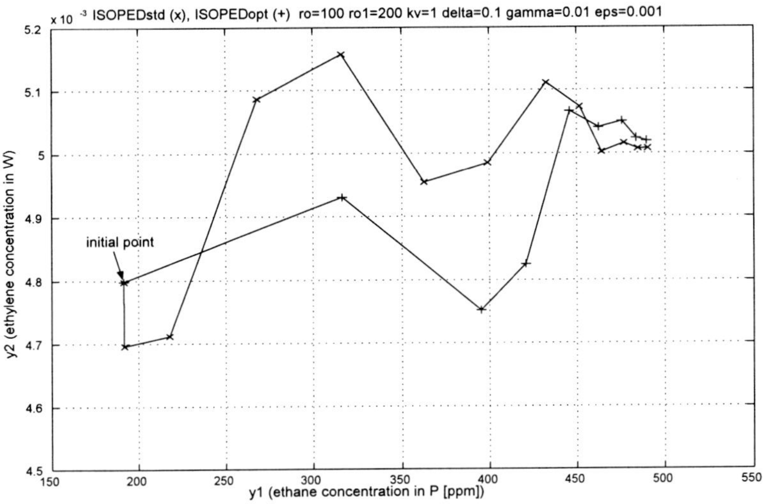

Fig. 4.20 Output trajectories for ISOPEDstd and ISOPEDopt algorithms, model-optimal initial point.

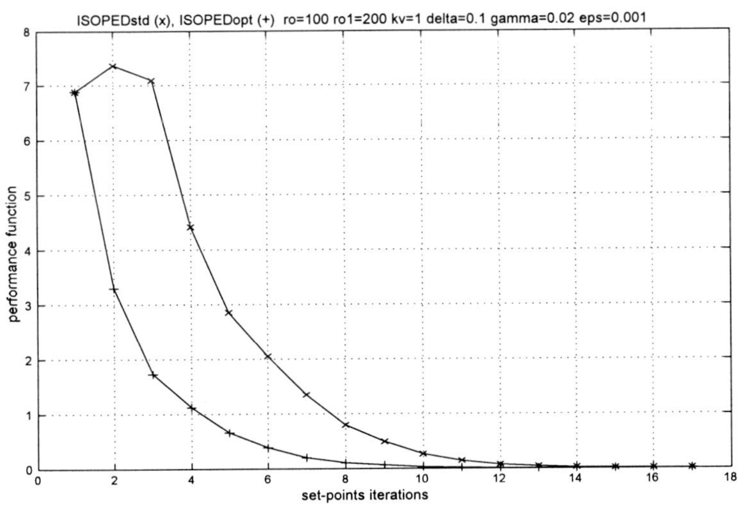

Fig. 4.21 Performance function trajectories for ISOPEDstd and ISOPEDopt algorithms, initial point not model-optimal.

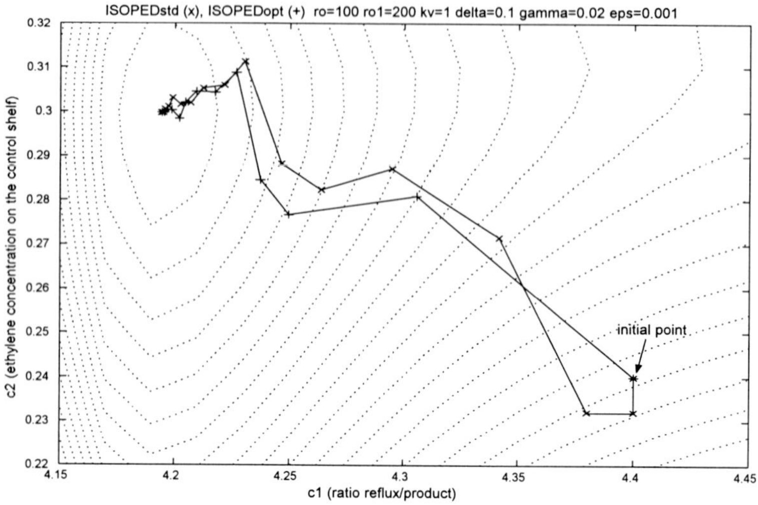

Fig. 4.22 Set-points trajectories for ISOPEDstd and ISOPEDopt algorithms, initial point not model-optimal.

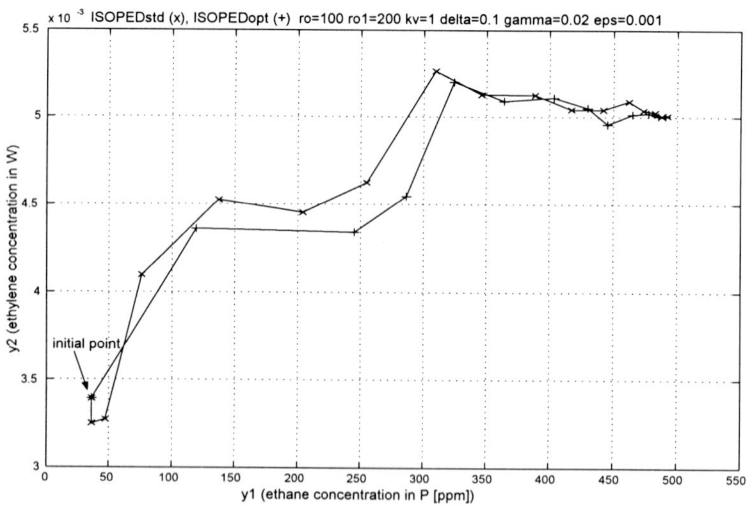

Fig. 4.23 Output trajectories for ISOPEDstd and ISOPEDopt algorithms, initial point not model-optimal.

Example 4.5

The ISOPE in the augmented version, see Section 4.1, is now applied to the pilot-scale chemical vaporiser process shown in Fig. 4.24, to test main features of the algorithm.

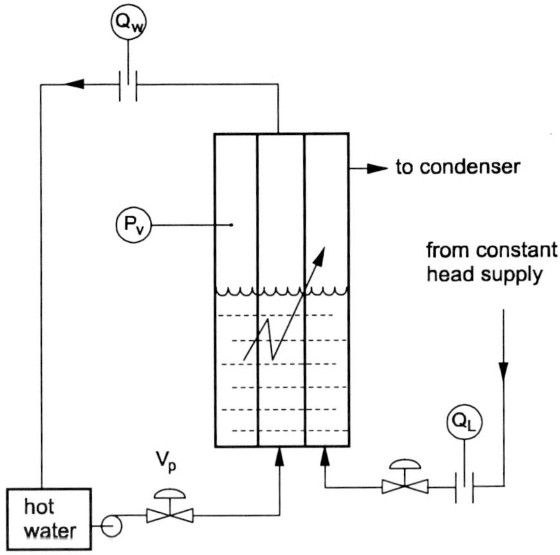

Fig. 4.24 Vaporiser pilot-scale system.

Freon liquid enters the base of the vaporiser tube and undergoes a phase change after being supplied with heat from a central heating tube. The resulting freon vapour is then led away. Controls are vapour pressure, via the freon feed valve, and the hot water valve position.

The vaporiser is controlled from within a distributed computer network (Stevenson *et al.*, 1984). A local microprocessor-based controller contains a two-term control algorithm for the vapour-pressure loop and also a digital valve positioner. The local controller also logs the process measurements and communicates with a supervisory host. The ISOPE algorithm is implemented at the host, which receives the process measurements from the local controller, to be used for updating the set-points which are then transmitted to the local controller to be applied to the process.

Process Objectives

The process performance is measured with the quadratic objective function

$$Q(c,y) = 2000(V_p - 0.25)^2 + 5(Q_w - 66)^2 + 40000(Q_L - 1.17)^2 \quad (4.107)$$

where V_p is the hot water valve position and Q_L and Q_w are respective flows of liquid and hot water. The choice of objective is justified on the grounds of possible costs incurred in departures from the desired conditions.

The objective is subject to the simple bounds:

$$5.0 \leq P_v \leq 30.0 \quad (4.108a)$$

$$0.0 \leq V_p \leq 2.5 \quad (4.108b)$$

where P_v is the vapour pressure (in WG (water gauge)) and V_p is measured in volts.

Process Model

The modeling of any process which involves heat transfer is inevitably problematic and this is particularly true when boiling, such as that which occurs in the vaporiser, is evident. However, from physical equilibrium and mass and heat transport relations, the best process description, in the sense of closeness of the model and process output responses, has been shown to be severely nonlinear (Ellis and Roberts, 1978). This nonlinear descriptions are still poor from a predictive viewpoint. Because of this, and because of a desire to make greater demands on the algorithm, a simple linear model is adopted which possesses approximate process gains:

$$Q_L = 0.0025 P_v + \alpha_1 \quad (4.109a)$$

$$Q_w = 3.0 V_p + \alpha_2. \quad (4.109b)$$

The model clearly has the required property of being point-parametric, if measurements of Q_L and Q_w are available.

The control is given as

$$c = (P_v, V_p) \quad (4.110)$$

and, because of the outputs appearing in the objective (4.107), it is required to have $y \in \mathbb{R}^2$ where

$$\begin{aligned} \alpha &= (\alpha_1, \alpha_2)^T && (4.111a)\\ y_1 &= Q_L = F_1(c, \alpha) = 0.025c_1 + \alpha_1 && (4.111b)\\ y_2 &= Q_w = F_2(c, \alpha) = 3.0c_2 + \alpha_2. && (4.111c) \end{aligned}$$

We shall verify that the assumptions of Theorem 4.1 are satisfied. Indeed, owing to inequalities (4.108a) and (4.108b), assumption (i) is satisfied. The equalities (4.109a) and (4.109b) yield that

$$\begin{aligned} \widehat{\alpha}_1(P_v, V_p) &= F_{*1}(c_1, c_2) = Q_{L*}(P_v, V_p) - 0.025P_v && (4.112)\\ \widehat{\alpha}_2(P_v, V_p) &= F_{*2}(c_1, c_2) = Q_{w*}(P_v, V_p) - 3.0V_p && (4.113) \end{aligned}$$

where $Q_{L*}(\cdot,\cdot)$ and $Q_{w*}(\cdot,\cdot)$ denote real process continuous input-output relationships, which are not known.

The above formulae imply that $\widehat{\alpha}(\cdot,\cdot)$ is a continuous function on C that is defined by inequalities (4.108a) and (4.108b), and thus assumption (vi) is fulfilled. Moreover, this continuity together with compactness of the set C yield compactness of the parameter value set A. Hence, the assumption (ii) is also satisfied. The smoothness assumptions (iii) and (iv) regarding the unknown mappings $F_*(\cdot)$ and $q_*(\cdot)$ have been verified experimentally by applying different values of controls P_v, V_p to the plant, measuring plant responses and performing suitable plots. Clearly, all the smoothness assumptions regarding only the plant mathematical model (see (4.111a), (4.111b) and (4.111c)) are met.

Algorithmic Performance

The optimized trajectory of the ISOPE method using the above presented objective, the above presented model and with convexifying penalty coefficient $\rho > 0$ are shown in Fig. 4.25, as a dashed line denoted by AISOPE (augmented ISOPE), along with contours of the process performance. It should be noted, however, that these contours are obtained using the nonlinear model (Ellis and Roberts, 1978) which had its parameters obtained corresponding to a defined optimum point $c^{opt} = (20.0, 1.0)^T$. As the nonlinear model still contains model-reality differences, the contours can only be regarded as approximate, but do serve as a useful guide. This is also before other process disturbances have been taken into consideration. These

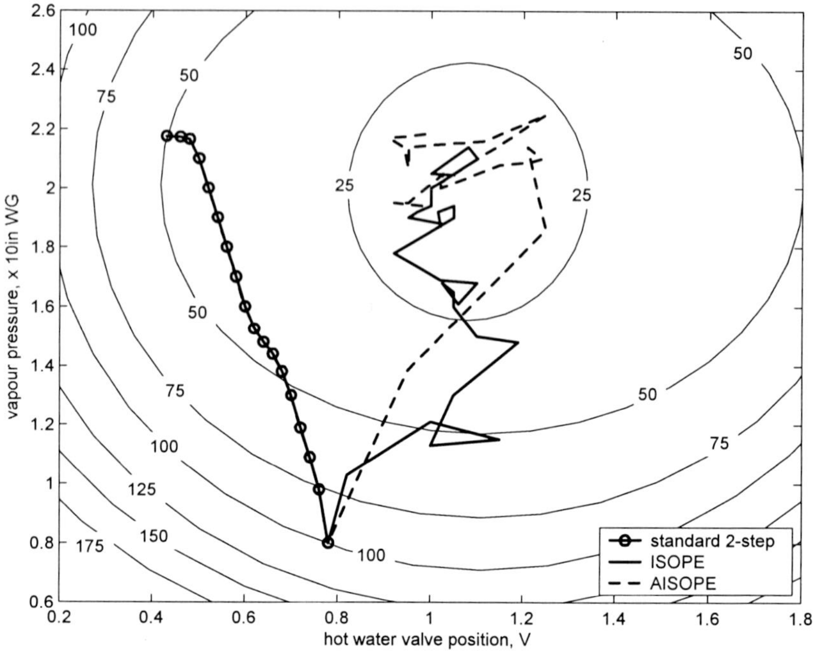

Fig. 4.25 Vaporiser optimizing control trajectories.

can be stochastic in nature, due to the boiling processes within the vaporiser, or more deterministic, due to, for example ambient condition changes. These factors could cause the optimum to depart from the c^{opt} point. From a practical viewpoint, the best that can be said is that the optimum lies in the region $\overline{B}(c^{opt}, \varepsilon)$, where ε is some suitable radius.

Also shown in Fig. 4.25, for comparison, are optimizing control trajectories for the ISOPE method with $\rho = 0$ (called ISOPE in the figure) and the standard two-step algorithm (the ITS method, see Subsection 2.3.1 in Chapter 2). Obviously, because the varporiser is a dynamic process, all methods require that a measurement time be taken into account. This is the time between making a set-point change and taking near steady-state measurements. It was set at 360 s. All methods use the gain coefficients: $k_{c_1} = 0.3$, $k_{c_2} = 0.2$. The standard two-step is seen to converge to a sub-optimal point, some distance from c^{opt}, while both ISOPE and AISOPE converge to within $\overline{B}(c^{opt}, \varepsilon)$. However, AISOPE can be seen to arrive to this region in fewer iterations.

Notice that both ISOPE and AISOPE methods require perturbation in controls, δc to obtain estimates of derivatives $F_*^{'}(c)$ in order that the modifier λ in (4.10) can be evaluated. Clearly, the estimation of these derivatives is fraught with difficulties when process noise is evident. The varporiser is certainly a process that possesses a high degree of process noise due to the boiling occurring, and this can result in severely incorrect estimates of the process output derivatives. This is the most likely cause for erratic behaviour of ISOPE when the optimizing control trajectory starts to wander in an incorrect direction. However, when at subsequent iterations more correct estimates of $F_*^{'}(c)$ are obtained, the algorithm recovers and the trajectory progresses towards $\overline{B}(c^{opt}, \varepsilon)$. The AISOPE algorithm does not suffer anywhere near as much from the erratic behaviour. The penalty term $\rho||c-\nu||^2$, because it is tending to convexify the optimization problem MMOP (see (4.12)), effectively exerts a smoothing influence on the algorithm probably not making it overly sensitive to errors in the estimates of $F_*^{'}(c)$.

One difficulty with AISOPE is that the penalty coefficient ρ must be determined appropriately. Too small a value will mean the performance of AISOPE will tend towards the performance of ISOPE, while too large a value will cause the penalty term to dominate the optimization problem (4.12), causing excessively slow movement of the algorithm iterations. □

[illegible] Systems [illegible]

[illegible] motivation.

[illegible]

[illegible] the variables [illegible]

[illegible] pressure noise is evident. [illegible] that exhibit a high degree of process [illegible] and this can result in severely incorrect [illegible]. The most likely cause for [illegible] control mode they [illegible] to [illegible]

Chapter 5

ISOPE for Problems with Output Constraints

The optimizing control problem with constraints on process outputs was preliminarily discussed in Chapter 2, see Section 2.3.2. It was shown there that constraints on process outputs are significantly more difficult to treat than the constraints on the set-points (optimizing controls) c. In practical cases when the process is under uncertainty, satisfying output constraints in the model does not imply satisfying these constraints by the real process outputs. As a result, an output constraint active in the model will generally be violated or inactive in the process, both situations are hardly acceptable for safety or optimality reasons. In particular, violation of the output constraint is usually too risky or unacceptable, leading either to dangerous operating conditions or an unacceptable decrease in product quality. Therefore, a practical solution to add *safety zones* to the constraints has been applied for years in practical applications to assure output constraint satisfaction. Usually, the larger the model uncertainty the bigger safety zone must be used. Obviously, appropriate values of the safety zones can be estimated only experimentally and, therefore, the zones are usually chosen in a rather conservative way.

Introduction of safety zones for active, critical constraints usually leads to decrease in process productivity and the loss may be significant. Therefore, there is a common interest in control algorithms that are capable to operate with smaller safety zones. The ISOPE iterative optimizing control algorithms, appropriately developed for the case with output constraints, are capable *to operate with significantly decreased safety zones,* thus *leading to increased process productivity.* These algorithms are the subject of this chapter. Two approaches will be developed:

- iterative algorithms for the *control structure with feedback controlled output constraints,* keeping the process output constraints satisfied at each set-point iteration with the accuracy of a feedback control error,
- iterative ISOPE algorithms *incorporating the output constraints into the ISOPE algorithm itself* in such a way that the constraints are satisfied at the end of iterations. However, during the iterations they may be violated and therefore more significant safety zones may be needed.

5.1 Feedback Controlled Output Constraints

5.1.1 *Process with constraint controllers*

Let us recall the multilayer control structure with feedback controlled output constraints discussed in Section 2.3.2, and repeated here for convenience as Fig. 5.1. The process is decomposed into two subprocesses: a fast actuating process and a slower optimized process representing the process description essential at the optimizing control layer, see Chapters 1 and 2. The actuating process together with the follow-up actuating controllers constitute a fast actuating system forcing the outputs y_m to follow the set-points c, regardless fast disturbances z (see Chapter 2). Unlike the basic multilayer structure of Fig. 2.1, the structure in Fig. 5.1 incorporates a dedicated upper-layer *constraint follow-up controller* (CFC) designed to keep the constraints on process outputs satisfied. The figure has been drawn for the output constraints in the most encountered simple form $y^d - y^d_r \leq 0$. Assuming only always active constraints are taken into account, the CFC provides for the equality $y^d - y^d_r = 0$, both in steady-states and during transient processes in the plant.

Introduction of the set-point controller CFC implies the situation where the output vector y is partitioned into two parts: the subvector of constrained outputs y^d and the subvector of free (unconstrained) outputs y^f, $y = (y^f, y^d)$. The optimizing control (set-point) vector is also appropriately partitioned, $c = (c^f, c^d)$, where $\dim c^d = \dim y^d$. The process mapping is obviously also partitioned in the way corresponding to the partitioning

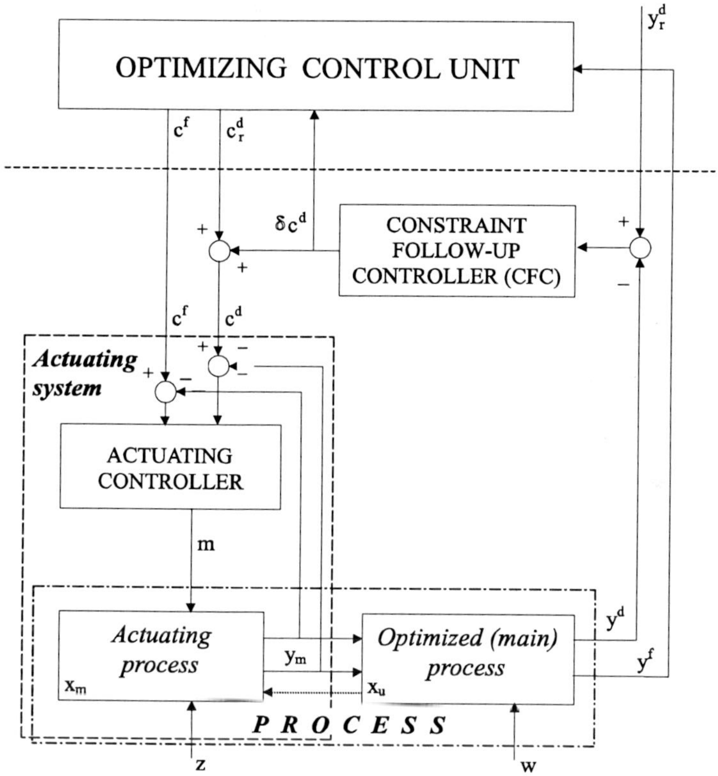

Fig. 5.1 Multilayer structure with feedback controlled output constraints.

of y, i.e., $F_* = (F_*^f, F_*)$,

$$y^f = F_*^f(c^f, c^d) \tag{5.1a}$$

$$y^d = F_*^d(c^f, c^d). \tag{5.1b}$$

The first mapping F_*^f describes the "free" outputs, not constrained. The second one corresponds to the "demanded" outputs which values are to be kept on the prescribed values y_r^d. Similarly as in Chapter 4, we shall write $F_*^f(c^f, c^d)$, $F_*^d(c^f, c^d)$ instead of $F_*^f(c^f, c^d, w)$, $F_*^d(c^f, c^d, w)$ *etc.*, to simplify the notation. This follows from the basic assumptions underlying the steady-state multilayer control, that changes in disturbance values affecting the main (optimized) process or changes of the process parameters itself are slow (or rare) when compared to the process dynamics. Therefore, the disturbance values and process parameters can be assumed constant (although uncertain, not known precisely) during a single application of the ISOPE iterative algorithm under given operating conditions. Taking it into account, symbols of slow disturbances w and their estimates w_e can be omitted in the process input-output mapping descriptions and models, as constant parameters (compare with Chapter 2, where this notation simplification is not used).

The primary *optimizing control problem* (OCP) can now be formulated in the following form OCP1:

$$\begin{aligned} &\text{minimize } Q(c^f, c^d, y^f, y_r^d) \\ &\text{subject to: } y^f = F_*^f(c^f, c^d) \\ &\qquad y^d = F_*^d(c^f, c^d), \quad y^d = y_r^d \\ &\qquad g(c^f) \leq 0. \end{aligned} \tag{5.2}$$

where general constraints on outputs have been simplified to the simpler form $y^d - y_r^d = 0$, as mentioned earlier, and activity of these constraints has been assumed, see Section 2.3.2 for the discussion on constraints activity. It should be noted that inequality constraints on free set-points c^f have only been assumed. In practical terms this means that the control system designer has managed to select such components c^d out of the vector c that do not enter their constraint boundaries during the process operation. Otherwise, due to the equality

$$c^d = c_r^d + \delta c^d \tag{5.3}$$

(see Fig. 5.1, where c_r^d is a constant value during every transient process) the CFC designer would have to consider constraints on δc^d, that is on this controller output, which would complicate the design significantly.

Applying a given constant value of c^f (and, simultaneously, c_r^d) to the system results in certain steady-state values of the CFC outputs δc^d, and thus in a steady-state value of c^d, after the transient processes in the whole dynamic part of the system have died. Therefore, due to the feedback action of the CFC, the set-points c^d are in fact dependent variables from the point of view of the steady-state optimizer. Clearly, they are dependent on c^f and this is emphasized by using the superscript "d". A functional relation between the steady-state values of c^d and c^f will be denoted by P_*,

$P_* : \mathbb{R}^{n_f} \times \mathbb{R}^{m_d} \mapsto \mathbb{R}^{n_d}$,

$$c^d = P_*(c^f, y_r^d). \tag{5.4}$$

The mapping P_* results, mathematically, from the steady-state relationship (5.1b). The vector of dependent set-points c^d should be chosen in such a way that P_* is a well-defined mapping (it is assumed throughout the paper).

Combining relations (5.1a) and (5.4) one can eliminate the dependence of the unconstrained outputs y^f on c^d, yielding the process mapping $\tilde{F}_*$,

$$y^f = \tilde{F}_*(c^f, y_r^d) = F_*^f(c^f, P_*(c^f, y_r^d)). \tag{5.5}$$

Using the introduced mappings the OCP1 problem can be transformed to the following form, which will be referred to as OCP2:

$$\begin{aligned} &\text{minimize } Q(c^f, c^d, y^f, y_r^d) \\ &\text{subject to: } \quad y^f = \tilde{F}_*(c^f, y_r^d) \\ &\qquad\qquad\qquad c^d = P_*(c^f, y_r^d) \\ &\qquad\qquad\qquad g(c^f) \leq 0. \end{aligned} \tag{5.6}$$

The optimizing control problem OCP1, and hence the equivalent problem OCP2 defined above, are formulations of the precise task of the steady-state optimizing control assuming an ideal operation of the CFC controller. This formulation is, however, not constructive because the true process mappings $\tilde{F}_*$ and P_* are unknown. However, it can be assumed that certain models $\tilde{F}$ and P of these mappings are available:

$$y^f = \tilde{F}(c^f, y_r^d, \alpha) \tag{5.7}$$

$$c^d = P(c^f, y_r^d, \beta) \tag{5.8}$$

where α and β are adjustable model parameters. Obviously, the models (5.7) and (5.8) can only be certain approximate descriptions of reality.

It may be difficult to model the rather complex mappings $\tilde{F}_*$ and P_* from the very beginning, due to the implicit relation between c^f and c^d involved. A reasonable alternative is to use first partitioned model F^f and F^d of the input-output process mappings F_*^f and F_*^d,

$$y^f = F^f(c^f, c^d, \alpha^f) \tag{5.9a}$$

$$y^d = F^d(c^f, c^d, \alpha^c) \tag{5.9b}$$

where α^f and α^c are parameters. A model P of P_* can then be obtained as a solution, with respect to c^d, of the implicit system of equations

$$y_r^d = F^d(c^f, c^d, \alpha^c). \tag{5.10}$$

Having obtained the model P we can get a model of the mapping $\tilde{F}_*$ by substituting P into the model (5.9a),

$$\tilde{F}(c^f, y_r^d, \alpha) = F^f(c^f, P(c^f, y_r^d, \beta), \alpha^f). \tag{5.11}$$

Having presented the optimizing control problem in the form OCP2 and defined models of the unknown process mappings we are in a position to derive the ISOPE technique. Having defined the function

$$q(c^f, \alpha, \beta, y_r^d) = Q(c^f, P(c^f, \beta), \tilde{F}(c^f, \alpha), y_r^d) \tag{5.12}$$

the problem OCP2 can be reformulated as follows

$$\begin{aligned} &\text{minimize}_{c^f,\alpha,\beta}\ q(c^f, \alpha, \beta, y_r^d) \\ &\text{subject to: } \tilde{F}(c^f, y_r^d, \alpha) = \tilde{F}_*(c^f, y_r^d) \\ &\qquad\qquad\quad P(c^f, y_r^d, \beta) = P_*(c^f, y_r^d) \\ &\qquad\qquad\quad g(c^f) \le 0. \end{aligned} \tag{5.13}$$

This formulation can be read as follows. An optimizing control algorithm should vary c^f over the feasible set described by the third constraint in (5.13) and take corresponding values of c^d and y^f from measurements. The model parameters are calculated so that the steady-state model responses are exactly the same as the real values for any chosen value of c^f, $P(c^f, y_r^d, \beta) = c^d$, $\tilde{F}(c^f, y_r^d, \alpha) = y^f$, as formulated by the first and second constraint in (5.13). Certainly, the models must be point-parametric (Brdyś, 1983). This means that for any real free output y^f and dependent set-point c^d corresponding to $c^f \in C^f$,

$$C^f = \{c^f \in \mathbb{R}^{n_f} :\ g(c^f) \le 0\} \tag{5.14}$$

parameter values must exist such that the model responses match exactly the real values (a much weaker requirement on the model than the standard one demanding that there are parameter values good for all possible input values). There is no problems with designing point-parametric models — a model augmented by an additive parameter in each output mapping can always do the job.

5.1.2 Basic ISOPE algorithm

Parameter estimation and model based optimization problems will now be integrated so that the real optimum can be achieved, similarly as it was presented in Chapter 4 for the case without output constraints. To this end, additional variables $v \in \mathbb{R}^{n_f}$ will be first introduced. Adding additionally a convexifying term, the problem OCP2 (5.13) can be reformulated to the following equivalent form

$$\begin{array}{ll} \text{minimize}_{c^f,v,\alpha,\beta}\{q(v,\alpha,\beta,y_r^d) + \rho\|c^f - v\|^2\} & \\ \text{subject to:} \quad \tilde{F}(c^f, y_r^d, \alpha) = \tilde{F}_*(c^f, y_r^d) & \\ \qquad\qquad\quad P(c^f, y_r^d, \beta) = P_*(c^f, y_r^d) & \\ \qquad\qquad\quad g(v) \le 0 & \\ \qquad\qquad\quad c^f = v. & \end{array} \tag{5.15}$$

Let us formulate the following Lagrange function

$$\begin{aligned} &L(v, c^f, \alpha, \beta, \lambda, \xi, \zeta, \mu) = \\ &= q(v,\alpha,\beta,y_r^d) + \rho\|c^f - v\|^2 + \lambda^T(c^f - v) + \xi^T(\tilde{F}(c^f, y_r^d, \alpha) + \\ &\quad - \tilde{F}_*(c^f, y_r^d)) + \zeta^T(P(c^f, y_r^d, \beta) - P_*(c^f, y_r^d)) + \mu^T g(v) \end{aligned} \tag{5.16}$$

then the corresponding necessary optimality conditions are of the form

$$q_v'(v,\alpha,\beta,y_r^d)^T - 2\rho(c^f - v) - \lambda + g'(v)^T\mu = 0 \tag{5.17a}$$

$$g(v) \le 0, \quad \mu \ge 0, \quad \mu^T g(v) = 0 \tag{5.17b}$$

$$\begin{aligned} -2\rho(v - c^f) + \lambda + [\tilde{F}_{c^f}'(c^f, y_r^d, \alpha) - \tilde{F}_*'(c^f, y_r^d)]^T\xi + \\ +[P_{c^f}'(c^f, y_r^d, \beta) - P_*'(c^f, y_r^d)]^T\zeta = 0 \end{aligned} \tag{5.17c}$$

$$q_\alpha'(v,\alpha,\beta,y_r^d)^T + \tilde{F}_\alpha'(c^f, y_r^d, \alpha)^T\xi = 0 \tag{5.17d}$$

$$q_\beta'(v,\alpha,\beta,y_r^d)^T + P_\beta'(c^f, y_r^d, \beta)^T\zeta = 0 \tag{5.17e}$$

$$\tilde{F}(c^f, y_r^d, \alpha) - \tilde{F}_*(c^f, y_r^d) = 0 \tag{5.17f}$$

$$P(c^f, y_r^d, \beta) - P_*(c^f, y_r^d) = 0 \tag{5.17g}$$

$$c^f - v = 0. \tag{5.17h}$$

Multipliers ξ and ζ can be explicitly derived from (5.17d) and (5.17e), assuming the matrices in square brackets are of full rank,

$$\xi = [\tilde{F}'_{\alpha}(c^f, y_r^d, \alpha)\tilde{F}'_{\alpha}(c^f, y_r^d, \alpha)^T]^{-1}\tilde{F}'_{\alpha}(c^f, y_r^d, \alpha)q'_{\alpha}(v, \alpha, \beta, y_r^d)^T \quad (5.18a)$$

$$\zeta = [P'_{\beta}(c^f, y_r^d, \beta)P'_{\beta}(c^f, y_r^d, \beta)^T]^{-1}P'_{\beta}(c^f, y_r^d, \beta)q'_{\beta}(v, \alpha, \beta, y_r^d)^T. \quad (5.18b)$$

Using the relations

$$q'_{\alpha}(c^f, \alpha, \beta, y_r^d) = Q'_{y^f}(c^f, P(c^f, y_r^d, \beta), \tilde{F}(c^f, y_r^d, \alpha), y_r^d)\tilde{F}'_{\alpha}(c^f, y_r^d, \alpha) \quad (5.19a)$$

$$q'_{\beta}(c^f, \alpha, \beta, y_r^d) = Q'_{c^d}(v, P(c^f, y_r^d, \beta), \tilde{F}(c^f, y_r^d, \alpha), y_r^d)P'_{\beta}(c^f, y_r^d, \beta) \quad (5.19b)$$

and the equality (5.17h) one arrives at simpler formulae for the multipliers, putting these into (5.17c) we get the formula for the multiplier λ:

$$\begin{aligned}\lambda(c^f, \alpha, \beta) = {} & [\tilde{F}'_{c^f}(c^f, y_r^d, \alpha) - \tilde{F}'_{*}(c^f, y_r^d)]^T \times \\ & \times Q'_{y^f}(c^f, P(c^f, y_r^d, \beta), \tilde{F}(c^f, y_r^d, \alpha), y_r^d)^T + \\ & + [P'_{c^f}(c^f, \beta, y_r^d) - P'_{*}(c^f, y_r^d)]^T \times \\ & \times Q'_{c^d}(c^f, P(c^f, y_r^d, \beta), \tilde{F}(c^f, y_r^d, \alpha), y_r^d)^T \end{aligned} \quad (5.20)$$

and the following reformulation of the necessary optimality conditions:

$$\begin{aligned} q'_v(v, \alpha, \beta, y_r^d)^T - 2\rho(c^f - v) - \lambda(c^f, \alpha, \beta) + g'(v)^T\mu &= 0 \quad (5.21a)\\ g(v) \le 0, \quad \mu \ge 0, \quad \mu^T g(v) &= 0 \quad (5.21b)\\ \tilde{F}(c^f, y_r^d, \alpha) - \tilde{F}_*(c^f, y_r^d) &= 0 \quad (5.21c)\\ P(c^f, y_r^d, \beta) - P_*(c^f, y_r^d) &= 0 \quad (5.21d)\\ c^f - v &= 0. \quad (5.21e)\end{aligned}$$

The ISOPE algorithm can be derived as an iterative numerical solver of the set (5.21a) – (5.21e). Let c^f be given.

A *parameter estimation problem (PEP)* activated at the point c^f should be formulated in a way that yields improved values of the model parameters $\alpha(c^f)$ and $\beta(c^f)$, under the condition to satisfy (5.21c) and (5.21d).

Putting $\alpha(c^f)$, $\beta(c^f)$ in place of α and β into (5.21a),(5.21b) converts them into necessary optimality conditions for the following *modified model-*

based optimization problem (MMOP):

$$\begin{aligned}&\text{minimize}_v\{q(v,\alpha(c^f),\beta(c^f),y_r^d)-\lambda(c^f,\alpha(c^f),\beta(c^f))^Tv+\rho\|c^f-v\|^2\}\\&\text{subject to: } g(v)\le 0. \qquad (5.22)\end{aligned}$$

Let $v^f = \hat{v}(c^f, \alpha(c^f), \beta(c^f))$ be a solution to the MMOP. If $v^f = c^f$ then (5.21e) is satisfied and the whole set of equations (5.21a) – (5.21e) is solved yielding optimal solution of the optimizing control problem, in the sense that its necessary optimality conditions are satisfied at this solution. If $v^f \neq c^f$, the value of c^f must be further adjusted according to a suitable iterative strategy. Therefore, any iterative algorithm finding a solution c^f of the equation

$$c^f = \hat{v}(c^f, \alpha(c^f), \beta(c^f)) \qquad (5.23)$$

can be viewed as an ISOPE *coordination strategy* (coordinating solutions of the parameter estimation problem (PEP) and solutions of the MMOP).

The following *relaxation iterative formula* will be considered as the way to adjust the set-points:

$$c^{f(i+1)} := c^{f(i)} + k_c(v^{f(i)} - c^{f(i)}) \qquad (5.24)$$

where k_c, $0 < k_c \le 1$ is a relaxation coefficient affecting the convergence rate.

Then one finally arrives at the following formulation:

The ISOPE basic algorithm for processes with feedback controlled output constraints:

Start. Given initial point $c^{f(0)}, c_r^{d(0)}$, solution accuracy $\epsilon > 0$. Set iteration counter $i := 0$.

Step 1. Apply $c^{f(i)}, c_r^{d(i)}$ to the controlled plant (see Fig. 5.1) and measure steady-state values of the outputs y^f

$$y^{f(i)} = \tilde{F}_*(c^{f(i)}, y_r^d). \qquad (5.25)$$

Read the CFC output $\delta c^{d(i)}$, where $\delta c^{d(i)}$ denotes steady-state value of the correction to $c_r^{d(i)}$ generated by the CFC that gives constraint tracking (see Fig. 5.1). Evaluate the set-points $c^{d(i)}$

$$c^{d(i)} := c_r^{d(i)} + \delta c^{d(i)} = P_*(c^{f(i)}, y_r^d). \qquad (5.26)$$

Evaluate derivatives $(\tilde{F}_*)'_{c^f}(c^{f(i)}, y_r^d)$ of the plant free output mapping and $(P_*)'_{c^f}(c^{f(i)}, y_r^d)$ of the CFC output mapping, applying a chosen technique. E.g., perform additional linearly independent perturbations around $c^{f(i)}$, measure corresponding values $y^{f(i)}$ and calculate values $c^{d(i)}$, then calculate finite difference approximation of the required derivatives.

Step 2. The *parameter estimation problem* (PEP):
Using the obtained new measurements $y^{f(i)}$ and $c^{d(i)}$ update parameters α and β of the models $\tilde{F}$ and P, under the restriction that model outputs match the actual controlled plant outputs. This yields $\alpha^{(i)} = \hat{\alpha}(c^{f(i)})$ satisfying

$$\tilde{F}(c^{f(i)}, y_r^d, \alpha^{(i)}) = y^{f(i)} \tag{5.27}$$

and $\beta^{(i)} = \hat{\beta}(c^{f(i)})$ satisfying

$$P(c^{f(i)}, y_r^d, \beta^{(i)}) = c^{d(i)} \tag{5.28}$$

compare with (5.21c), (5.21d).

Step 3. For $c^f = c^{f(i)}$, $\alpha = \alpha^{(i)}$, $\beta = \beta^{(i)}$ and, therefore, $\lambda(c^f, \alpha, \beta) = \lambda(c^{f(i)}, \alpha^{(i)}, \beta^{(i)})$ solve the *modified model-based optimization problem* (MMOP):

$$\begin{aligned} &\text{minimize}_v \{q(v, \alpha^{(i)}, \beta^{(i)}, y_r^d) - \lambda(c^{f(i)}, \alpha^{(i)}, \beta^{(i)})^T v + \\ &\qquad\qquad\qquad\qquad + \rho \|c^{f(i)} - v\|^2\} \\ &\text{subject to: } g(v) \le 0. \end{aligned} \tag{5.29}$$

Let $v^{(i)} = \hat{v}(c^{f(i)}, \alpha^{(i)}, \beta^{(i)})$ be a solution. If

$$\|v^{(i)} - c^{f(i)}\| \le \epsilon \tag{5.30}$$

then terminate (solution found). Else

Step 4. Set

$$c^{f(i+1)} \; := \; c^{f(i)} + k_c(v^{(i)} - c^{f(i)}) \tag{5.31}$$

$$c_r^{d(i+1)} \; := \; P(c^{f(i+1)}, y_r^d, \beta^{(i)}) \tag{5.32}$$

set $i := i + 1$ and continue from Step 1.

It should be realized that the number of the model parameters (α and β) is usually larger than the number of the model outputs. Therefore, there is more freedom in the parameter estimation problem than simply choosing the parameters to match current process outputs. This freedom can be utilized to satisfy certain model adaptation criteria.

The key element of the presented algorithm, important from both practical and theoretical points of view, is the *technique to evaluate appropriately estimates of the derivatives* $(\tilde{F}_*)'_{c^f}(c^{f(i)}, y_r^d)$ of the plant free output mapping and $(P_*)'_{c^f}(c^{f(i)}, y_r^d)$ of the CFC output mapping, in Step 1. The additional perturbation approach described in Step 1 above is not an efficient strategy, compare with Chapter 4. But, before passing to a more efficient dual algorithm, theoretical results concerning optimality and convergence of the presented basic ISOPE structure will be presented.

The theoretical analysis can be performed using the technique analogous to that applied in Chapter 4, see Section 4.2 therein. The possibility that the parameter estimation performed in Step 2 may lead to not unique results is taken into account, therefore the following sets need to be defined

$$\hat{\alpha}_S(c^f) = \{\alpha \in \mathbb{R}^{s_\alpha} : \ \tilde{F}(c^f, y_r^d, \alpha) = \tilde{F}_*(c^f, y_r^d) \ \} \tag{5.33}$$

$$\hat{\beta}_S(c^f) = \{\beta \in \mathbb{R}^{s_\beta} : \ P(c^f, y_r^d, \beta) = P_*(c^f, y_r^d) \ \} \tag{5.34}$$

$$A = \bigcup_{c^f \in C^f} \hat{\alpha}_S(c^f) \tag{5.35}$$

$$B = \bigcup_{c^f \in C^f} \hat{\beta}_S(c^f) \tag{5.36}$$

where $\hat{\alpha}_S(\cdot)$ and $\hat{\beta}_S(\cdot)$ are point-to-set mappings. Step 2 of the algorithm can now be described in the following way:

For a given $c^{f(i)}$ find $\alpha^{(i)}$ and $\beta^{(i)}$ such that
$$\alpha^{(i)} \in \hat{\alpha}_S(c^{f(i)}) \text{ and } \beta^{(i)} \in \hat{\beta}_S(c^{f(i)}).$$

The modified model based optimization performed in Step 3 may generally also be not unique, therefore the set of its solutions is defined as

$$\hat{v}_S(c^f, \alpha, \beta) = \operatorname{Argmin}_{v \in C^f} \{q(v, \alpha, \beta, y_r^d) - \lambda(v, \alpha, \beta)^T v + \rho \|c^f - v\|^2\}. \tag{5.37}$$

Step 3 of the algorithm can now be described as follows:

For given $c^{f(i)}$, $\alpha^{(i)}$ and $\beta^{(i)}$ find $v^{(i)}$ such that
$$v^{(i)} = \hat{v}(c^{f(i)}, \alpha^{(i)}, \beta^{(i)}) \in \hat{v}_S(c^{f(i)}, \alpha^{(i)}, \beta^{(i)}).$$

We are now in a position to define the algorithm solution set Ω,

$$\Omega = \{c^f \in \mathbb{R}^{n_f} : \exists\, \hat{\alpha} \in \hat{\alpha}_S(c^f), \hat{\beta} \in \hat{\beta}_S(c^f) \quad c^f \in \hat{v}_S(c^f, \hat{\alpha}, \hat{\beta}) \,\} \tag{5.38}$$

and definition of the true performance function will also be needed,

$$q_*(c^f, y_r^d) = Q(c^f, P_*(c^f), \tilde{F}_*(c^f), y_r^d). \tag{5.39}$$

The ISOPE algorithm should converge to points from the solution set Ω. Intuitively, following the construction of the algorithm, these points should satify at least necessary optimality conditions for the original optimizing control problem OCP. The optimality theorem which follows formally proves this statement.

Theorem 5.1 *Assume that, for a given constant value of* y_r^d :
(i) the set $C^f = \{c^f : g(c^f) \leq 0\}$ *is convex and compact,*
(ii) mappings $\tilde{F}_*(\cdot, y_r^d)$ *and* $P_*(\cdot, y_r^d)$ *are continuously differentiable on* C^f,
(iii) the function $Q(\cdot,\cdot,\cdot,y_r^d)$ *is continuously differentiable, the mappings* $\tilde{F}(\cdot, y_r^d, \cdot)$ *and* $P(\cdot, y_r^d, \cdot)$ *are continuous on* $C^f \times A$ *and* $C^f \times B$, *respectively, and differentiable with respect to* c^f *on* C^f *for any* $\alpha \in A$ *and* $\beta \in B$.
(iv) the model is point-parametric on C^f.
Then at each point $c^f \in \Omega$ *first-order necessary optimality conditions for the OCP are satisfied. If, additionally, the function* $q_*(\cdot, y_r^d)$ *is convex on* C^f, *then each* $c^f \in \Omega$ *is a solution to the OCP.*

Proof. Since $c^f \in \mathcal{A}(c^f)$ then there are $\alpha \in \hat{\alpha}_S(c^f)$, $\beta \in \hat{\beta}_S(c^f)$ and $c^f = \hat{v}(c^f, \alpha, \beta) \in \hat{v}_S(c^f, \alpha, \beta)$. Due to convexity of C^f first order necessary optimality conditions for the solution of MMOP can be written as

$$[q'_{c^f}(\hat{v}(c^f, \alpha, \beta), \alpha, \beta, y_r^d) - \lambda(c^f, \alpha, \beta)^T][v - \hat{v}(c^f, \alpha, \beta)] \geq 0 \quad \text{for all } v \in C^f. \tag{5.40}$$

We have from (5.20) :

$$\begin{aligned}
\lambda(c^f, \alpha, \beta)^T = \\
& Q'_{y^f}(c^f, P(c^f, y_r^d, \beta), \tilde{F}(c^f, y_r^d, \alpha), y_r^d)\tilde{F}'_{c^f}(c^f, y_r^d, \alpha) + \\
& + Q'_{c^d}(c^f, P(c^f, y_r^d, \beta), \tilde{F}(c^f, y_r^d, \alpha), y_r^d)P'_{c^f}(c^f, y_r^d, \beta) + \\
& - Q'_{y^f}(c^f, P(c^f, y_r^d, \beta), \tilde{F}(c^f, y_r^d, \alpha), y_r^d)(\tilde{F}_*)'_{c^f}(c^f, y_r^d) + \\
& - Q'_{c^d}(c^f, P(c^f, y_r^d, \beta), \tilde{F}(c^f, y_r^d, \alpha), y_r^d)(P_*)'_{c^f}(c^f, y_r^d).
\end{aligned} \tag{5.41}$$

Taking into account that $\tilde{F}(c^f, y_r^d, \alpha) = \tilde{F}_*(c^f, y_r^d)$, $P(c^f, y_r^d, \beta) = P_*(c^f, y_r^d)$ yields

$$\begin{aligned}
\lambda(c^f, \alpha, \beta)^T = \\
&= Q'_{y^f}(c^f, P(c^f, y_r^d, \beta), \tilde{F}(c^f, y_r^d, \alpha), y_r^d)\tilde{F}'_{c^f}(c^f, y_r^d, \alpha) + \\
&\qquad + Q'_{c^d}(c^f, P(c^f, y_r^d, \beta), \tilde{F}(c^f, y_r^d, \alpha), y_r^d)P'_{c^f}(c^f, y_r^d, \beta) + \\
&\quad - Q'_{y^f}(c^f, P_*(c^f, y_r^d), \tilde{F}_*(c^f, y_r^d), y_r^d)(\tilde{F}_*)'_{c^f}(c^f, y_r^d) + \\
&\qquad - Q'_{c^d}(c^f, P_*(c^f, y_r^d), \tilde{F}_*(c^f, y_r^d), y_r^d)(P_*)'_{c^f}(c^f, y_r^d) \\
&= q'_{c^f}(c^f, \alpha, \beta, y_r^d) - (q_*)'_{c^f}(c^f, y_r^d).
\end{aligned} \tag{5.42}$$

Therefore, if $c^f = \hat{v}(c^f, \alpha, \beta)$ then the inequality (5.40) can be written in the form

$$(q_*)'_{c^f}(c^f, y_r^d)(v - c^f) \geq 0 \quad \text{for all } v \in C^f \tag{5.43}$$

which is precisely the necessary optimality condition for the problem

$$\begin{aligned}
&\text{minimize}_{c^f}\, q_*(c^f, y_r^d) \\
&\text{subject to: } g(c^f) \leq 0
\end{aligned} \tag{5.44}$$

The problem formulation (5.44) is an equivalent formulation of the original optimizing control problem. It can be obtained from (5.6) after elimination of the variables y^f and c^d – by putting the equality constraints $y^f = \tilde{F}_*(c^f, y_r^d)$ and $c^d = P_*(c^f, y_r^d)$ into the performance function Q, see (5.39).
If the function $q_*(\cdot, y_r^d)$ is convex on C^f then the condition (5.43) becomes sufficient for optimality. That completes the proof.

Let us notice that the assumptions of the theorem are realistic. The feasible set C^f is bounded in all practical cases. Therefore, assumption (i) is satisfied if, additionally, the constraining functions $g_j(\cdot)$ are convex. The process input-output mappings F_*^f and F_*^d are continuously differentiable for the vast majority of continuous industrial processes. The mapping P_* is defined in an implicit way by the equation $y_r^d = F_*^d(c^f, c^d)$ (see 5.4). Hence, it is continuously differentiable based on the implicit function theorem. The mapping $\tilde{F}_*$ is then continuously differentiable as a composition of F_*^f and P_*. The mappings $\tilde{F}$ and P are model-based so that they can easily be constructed as continuous and differentiable. Hence, the smoothness assumptions (ii) and (iii) are not practically restricting.

Under certain additional assumptions convergence of the sequence of points $\{c^{f(i)}\}$ to the solution set can be assured, in an analogous way as it is given in Theorem 4.1 in Chapter 4 for the basic ISOPE algorithm. A related convergence theorem is given as follows.

Theorem 5.2 *Assume that the assumptions (i), (ii), (iii) of Theorem 1 are satisfied and, additionally:*
(v) the sets A *and* B *are compact,*
(vi) the function $q(\cdot,\alpha,\beta,y_r^d)$ *is twice (Frechet) differentiable with respect to* c^f *on* C^f*, for every* $(\alpha,\beta)\in A\times B$*, and* $q(\cdot,\cdot,\cdot,y_r^d), q'(\cdot,\cdot,\cdot,y_r^d), q''(\cdot,\cdot,\cdot,y_r^d)$ *are continuous on* $C^f\times A\times B$*,*
(vii) the point-to-set mappings $\hat{\alpha}_S(\cdot)$ *and* $\hat{\beta}_S(\cdot))$ *are open on* A *and* B*, respectively.*
Then, denoting

$$b(\alpha,\beta) = \min_{c^f\in C^f}\lambda_{min}(q''_{c^f c^f}(c^f,\alpha,\beta,y_r^d)) \tag{5.45}$$

where $\lambda_{min}(D)$ *denotes minimal eigenvalue of a symmetric matrix* D*, and*

$$\delta = \max_{c^f\in C^f}\|(q_*)'_{c^f}(c^f,y_r^d)\| \tag{5.46}$$

the following thesis holds:

There exist values ρ_1*,* τ*, function* $B(c^f)$ *and some scalar* $\epsilon>0$ *satisfying*

$$\rho_1 > -0.5\inf_{(\alpha,\beta)\in A\times B}b(\alpha,\beta) \tag{5.47}$$

$$0<\tau<\min\{1,\frac{2\inf_{(\alpha,\beta)\in A\times B}b(\alpha,\beta)+4\rho_1}{\delta+\epsilon}\} \tag{5.48}$$

$$B(c^f) = \min\{1,\frac{2\inf_{\alpha\in\hat{\alpha}(c^f),\beta\in\hat{\beta}(c^f)}b(\alpha,\beta)+4\rho_1}{\delta+\epsilon}\} \tag{5.49}$$

such that for every $\rho\ge\max\{0,\rho_1\}$*:*
(a) the algorithmic mapping is well defined on C^f*,*
(b) each point generated by the algorithm satisfies the plant constraints and for every $i=0,1,\ldots$

$$q_*(c^{f(i+1)},y_r^d) < q_*(c^{f(i)},y_r^d) \qquad if\ c^{f(i)}\notin\Omega, \tag{5.50}$$

(c) there is at least one cluster point of the sequence $\{c^{f(i)}\}$ *generated by the algorithm and each cluster point belongs to the solution set* Ω*.*

Proof is omitted, since it uses the reasoning analogous to that developed in (Brdyś *et al.*, 1987) and given in the Appendix A for the Theorem 4.1.

The discussion of the theorem assumptions would be also quite analogous to that given in Chapter 4 after Theorem 4.1, and is therefore not repeated here.

5.1.3 *Dual ISOPE algorithm*

A key element of any ISOPE algorithm is the technique for estimating the process mapping derivatives, as it was explained in Chapter 2 (see Section 2.3.1) and thoroughly discussed in Chapter 4 for the case without constraints on outputs, see Section 4.3. In the output constraint case considered in this chapter the required derivatives are: the derivative $(\tilde{F}_*)'_{c^f}(c^{f(i)}, y_r^d)$ of the process free output mapping $\tilde{F}_*$, and the derivative $(P_*)'_{c^f}(c^{f(i)}, y_r^d)$ of the biased (by c_r^d) CFC controller output mapping P_*. These derivatives, approximated in Step 2, are necessary for the evaluation of the modifier λ, see (5.20), which plays key role in creating the ability of the ISOPE algorithm to find the true process optimum.

As discussed in Section 4.3 of Chapter 4 the most practical is to use steady-state information for derivatives estimation, although an approach based on on-line dynamic information may also be possible. The concept of dual ISOPE algorithm, being a breakthrough in realization of the ISOPE method relying on steady-state information only, eliminated the serious drawback of additional set-point perturbations at each algorithm iteration as proposed in basic formulations. However, the algorithm presented up to now in this section must be appropriately redesigned to get its dual realization. The design is analogous to that of Section 4.4, but more derivatives must be now estimated.

Let us consider a collection of $n_f + 1$ points $c^{f(i)}, c^{f(i-1)}, \ldots, c^{f(i-n_f)}$ such that all vectors

$$s^{f(ik)} = c^{f(i-k)} - c^{f(i)}, \quad k = 1, ..., n_f \tag{5.51}$$

are linearly independent and formulate a (nonsingular) matrix

$$S^{(i)} = [s^{f(i1)} \; s^{f(i2)} \ldots \; s^{f(in_f)}]^T. \tag{5.52}$$

Then, reasoning analogously as in Section 4.4 of Chapter 4, one concludes that if the points $c^{f(i-k)}$ are close enough to $c^{f(i)}$, then for every $j =$

$1, \ldots, m_f$

$$S^{(i)} \nabla_{c^f} \tilde{F}_{*j}(c^{f(i)}, y_r^d) \cong \begin{bmatrix} \tilde{F}_{*j}(c^{f(i-1)}, y_r^d) - \tilde{F}_{*j}(c^{f(i)}, y_r^d) \\ \vdots \\ \tilde{F}_{*j}(c^{f(i-n_f)}, y_r^d) - \tilde{F}_{*j}(c^{f(i)}, y_r^d) \end{bmatrix} \tag{5.53}$$

where $\nabla_{c^f} \tilde{F}_{*j}(c^{f(i)}, y_r^d) = (\tilde{F}_{*j})'_{c^f}(c^{f(i)}, y_r^d)^T$ (formula (5.53) corresponds to (4.53) from Section 4.4). An analogous formula applies to $P_{*j}(c^{f(i)}, y_r^d)$, for every $j = 1, \ldots, n_d$

$$S^{(i)} \nabla_{c^f} P_{*j}(c^{f(i)}, y_r^d) \cong \begin{bmatrix} P_{*j}(c^{f(i-1)}, y_r^d) - P_{*j}(c^{f(i)}, y_r^d) \\ \vdots \\ P_{*j}(c^{f(i-n_d)}, y_r^d) - P_{*j}(c^{f(i)}, y_r^d) \end{bmatrix}. \tag{5.54}$$

To make an application of the evaluated derivative estimates possible and practical, the matrix $S^{(i)}$ must not only be nonsingular but, first of all, it should be *sufficiently well conditioned.* This can be fulfilled if the consecutive set-points $c^{f(i)}$ are appropriately located in their space. To this end, the following additional constraint called *conditioning constraint* is introduced (see Section 4.4)

$$d(c^{f(i+1)}(v), c^{f(i)}, \ldots, c^{f(i-n_f+1)}) = \frac{\sigma_{min}(S^{(i+1)}(v))}{\sigma_{max}(S^{(i+1)}(v))} \geq \delta \tag{5.55}$$

where $\sigma_{min}(S)$ and $\sigma_{max}(S)$ denote minimal and maximal singular values of a matrix S (thus (5.55) is the reciprocal of the condition number of the matrix $S^{(i+1)}(v)$, in 2-norm), $\delta > 0$ is a constant and

$$c^{f(i+1)}(v) = c^{f(i)} + k_c(v - c^{f(i)}) \tag{5.56}$$

(see Section 4.4 in Chapter 4 for detailed discussion of the features of the constraint (5.55) and the sets it defines).

The inequality (5.55) is introduced into the ISOPE algorithm as a new constraint to be satisfied by the set-points. This constraint reduces a feasible set of the modified model-based optimization problem. Hence, it might happen that a current loss on optimality will be observed. However, the new set-point generated will then enable us to obtain better approximations of the derivatives in the next iteration which in turn should lead to better optimality of the next set-point. This can be compared to what is known as a *dual effect*.

We are now in a position to present the formulation of the ISOPED (ISOPE Dual) algorithm for the considered case.

The ISOPED algorithm for processes with feedback controlled output constraints:

Start. Given initial point $c^{f(0)}, c_r^{d(0)}$, solution accuracy $\epsilon > 0$, conditioning constraint threshold $\delta > 0$. Measure free output $y^{f(0)} = \tilde{F}_*(c^{f(0)}, y_r^d)$ and evaluate set-point $c^{d(0)}$

$$c^{d(0)} := c_r^{d(0)} + \delta c^{d(0)} = P_*(c^{f(0)}, y_r^d) \tag{5.57}$$

where $\delta c^{d(0)}$ is the output of the CFC, in the steady-state (see Fig. 5.1).

Initial phase (Step 0). Set $c^{f(-n_f)} = c^{f(0)}$, $c^{d(-n_f)} = c^{d(0)}$. Choose n_f points $c^{f(-n_f+1)}, c^{f(-n_f+2)}, \dots, c^{f(0)}$ such that $S^{f(0)}$ is sufficiently well conditioned. Apply the points $(c^{f(i)}, c_r^{d(i)})$, where $c_r^{d(i)} = P(c^{f(i)}, y_r^d, \beta^{(i)})$, to the controlled plant (see Fig. 5.1), measure steady-state values of the free outputs $y^{f(i)}$

$$y^{f(i)} = \tilde{F}_*(c^{f(i)}, y_r^d) \tag{5.58}$$

and evaluate steady-state values of the set-points $c^{d(i)}$

$$c^{d(i)} := c_r^{d(i)} + \delta c^{d(i)} = P_*(c^{f(i)}, y_r^d) \tag{5.59}$$

$i = -n_f + 1, -n_f + 2, \dots, 0$. Set iteration counter $i := 0$.

Step 1. Apply $c^{f(i)}, c_r^{d(i)}$ to the controlled plant and measure steady-state values of the outputs $y^{f(i)} = \tilde{F}_*(c^{f(i)}, y_r^d)$. Read the value $\delta c^{d(i)}$ of the output of the CFC, in the steady-state and evaluate the set-points $c^{d(i)}$, according to (5.59). Apply the formulae (5.53) and (5.54) to evaluate the derivatives $(\tilde{F}_*)'_{c^f}(c^{f(i)}, y_r^d)$ of the plant free output mapping and $(P_*)'_{c^f}(c^{f(i)}, y_r^d)$ of the CFC output mapping.

Step 2. The *parameter estimation problem (PEP)*:
Using the obtained new measurements $y^{f(i)}$ and $c^{d(i)}$ update parameters α and β of the models $\tilde{F}$ and P, under the restriction that model outputs match the actual controlled plant outputs. This yields $\alpha^{(i)} = \hat{\alpha}(c^{f(i)})$ satisfying

$$y^{f(i)} = \tilde{F}(c^{f(i)}, y_r^d, \alpha^{(i)}) \tag{5.60}$$

and $\beta^{(i)} = \hat{\beta}(c^{f(i)})$ satisfying

$$P(c^{f(i)}, y_r^d, \beta^{(i)}) = c^{d(i)} \tag{5.61}$$

compare with eqs.(5.21c), (5.21d).

Step 3. For $c^f = c^{f(i)}$, $\alpha = \alpha^{(i)}$, $\beta = \beta^{(i)}$ and, therefore, $\lambda(c^f, \alpha, \beta) = \lambda(c^{f(i)}, \alpha^{(i)}, \beta^{(i)})$ solve the *modified model-based optimization problem* (MMOP):

$$\begin{aligned} &\text{minimize}_v \{q(v, \alpha^{(i)}, \beta^{(i)}, y_r^d) - \lambda(c^{f(i)}, \alpha^{(i)}, \beta^{(i)})^T v + \\ &\qquad\qquad\qquad\qquad + \rho \| c^{f(i)} - v \|^2 \} \\ &\text{subject to}: g(v) \leq 0 \end{aligned} \tag{5.62}$$

Let $v^{(i)} = \hat{v}(c^{f(i)}, \alpha^{(i)}, \beta^{(i)})$ be a solution. If

$$\| v^{(i)} - c^{f(i)} \| \leq \epsilon \tag{5.63}$$

then terminate (solution found).

Step 4. If

$$d(c^{f(i+1)}(v^{(i)}), c^{f(i)}, \ldots, c^{f(i-n_f+1)}) \geq \delta \tag{5.64}$$

then set $v_d^{(i)} := v^{(i)}$ and go to Step 5.

Else, solve the *conditioned modified model-based optimization problem* (CMMOP):

$$\begin{aligned} &\text{minimize}_v \{q(v, \alpha^{(i)}, \beta^{(i)}, y_r^d) - \lambda(c^{f(i)}, \alpha^{(i)}, \beta^{(i)})^T v + \\ &\qquad\qquad\qquad\qquad + \rho \| c^{f(i)} - v \|^2 \} \\ &\text{subject to}: \; g(v) \leq 0, \\ &\qquad\qquad d(c^{f(i+1)}(v), c^{f(i)}, \ldots, c^{f(i-n_f+1)}) \geq \delta \end{aligned} \tag{5.65}$$

denoting the solution point by $v_d^{(i)}$. If

$$\| v_d^{(i)} - c^{f(i)} \| \leq \epsilon \tag{5.66}$$

then terminate (solution found).

Step 5. Set

$$c^{f(i+1)} := c^{f(i)} + k_c(v_d^{(i)} - c^{f(i)}) \tag{5.67}$$

$$c_r^{d(i+1)} := P(c^{f(i+1)}, y_r^d, \beta^{(i)}) \tag{5.68}$$

set $i := i + 1$ and continue from Step 1.

The first task performed in Step 4 is to check if $v^{(i)}$ satisfies the conditioning constraint. If it does then $v_d^{(i)} := v^{(i)}$ is set and Step 5 can be immediately executed, without any change in the basic algorithm and, therefore, without any possible loss of optimality. If it does not then $v^{(i)}$ is suitably modified, by solving the CMMOP problem. This may lead to some current loss of optimality, but at the same time anticipates future needs of the next iteration (mechanism of active measurement gathering — dual structure).

In practical applications any set-point change should be carefully planned at the optimization layer since it means a transient process in the plant and directly influences plant effectiveness. It is precisely the case of any ISOPED regular iteration (Steps 1 to 5), provided all data necessary to start these iterations have been gathered during the initial phase (Step 0). However, during the initial phase set-point changes are also necessary and should be planned as carefully as possible, with plant effectiveness taken into account. This is possible by incorporating into the initial phase a suitably reformulated iterative two-step method leading to an *optimized initial phase*, the idea presented in Section 4.5 of Chapter 4. The idea can also be analogously applied in the case of feedback controlled output constraints. The reformulation of the optimized initial phase in this case is as follows.

First, we recall the definition of the matrix $S^{f(i)}$ for negative values of i, $-n_f < i \leq 0$:

$$S^{(i)} = [c^{f(i-1)} - c^{f(i)} \quad c^{f(i-2)} - c^{f(i)} \quad \cdots \quad c^{f(-n_f)} - c^{f(i)}]^T \tag{5.69}$$

and the conditioning constraint function d

$$d(c^{f(i)}, \ldots, c^{f(i-n_f+1)}, c^{f(-n_f)}) = \frac{\sigma_{min}(S^{(i)})}{\sigma_{max}(S^{(i)})} \tag{5.70}$$

where $\sigma_{min}(S)$ and $\sigma_{max}(S)$ denote minimal and maximal singular values of a (generally non-square) matrix S. Notice that $S^{(i)}$ is non-square for $i < 0$ and for $i = 0$ becomes precisely the square matrix $S^{(0)}$ (5.52) needed to start the first regular iteration of the ISOPED algorithm in Step 1.

The optimized initial phase of the ISOPED algorithm:

Step 0 (Initial phase, optimized):

Start 0. Choose appropriately positive parameters ρ, γ, δ. Set $(c^{f(-n_f)}, c_r^{d(-n_f)}) := (c^{f(0)}, c_r^{d(0)})$, set $i := -n_f$.

Step 0.1 Apply the points $(c^{f(i)}, c_r^{d(i)})$ to the controlled plant. Measure steady-state values of the outputs y^f, $y^{f(i)} = \tilde{F}_*(c^{f(i)}, y_r^d)$, and the value $\delta c^{d(i)}$ of the output of the CFC, in the steady-state. Evaluate $c^{d(i)} := c_r^{d(i)} + \delta c^{d(i)} = P_*(c^{f(i)}, y_r^d)$.
Add the measurement to the data record and *adapt the steady-state models* (i.e., the parameters α and β).

Step 0.2 Solve the following *augmented model optimization problem* (AMOP)

$$\begin{aligned} &\text{minimize}_{c^f} \{q(c^f, \alpha, \beta, y_r^d) + \rho \|c^f - c^{f(i)}\|^2\} \\ &\text{subject to:} \quad g(c^f) \le 0, \\ &\qquad\qquad\quad \|c^f - c^{f(i)}\| \ge \gamma \end{aligned} \tag{5.71}$$

denoting the solution point by $c^{f(i+1)}$. If

$$d(c^{f(i+1)}, c^{f(i)}, \ldots, c^{f(-n_f)}) \ge \delta \tag{5.72}$$

then go to **Step 0.4**, else proceed.

Step 0.3 Solve the following *conditioned augmented model optimization problem* (CAMOP)

$$\begin{aligned} &\text{minimize}_{c^f} \{q(c^f, \alpha, \beta, y_r^d) + \rho \|c^f - c^{f(i)}\|^2\} \\ &\text{subject to:} \quad g(c^f) \le 0, \\ &\qquad\qquad\quad \|c^f - c^{f(i)}\| \ge \gamma, \\ &\qquad\qquad\quad d(c^f, c^{f(i)}, \ldots, c^{f(-n_f)}) \ge \delta \end{aligned} \tag{5.73}$$

denoting the solution point by $c^{f(i+1)}$.

Step 0.4 If $i \le -1$ then set $i := i + 1$ and go to Step 0.1, else proceed from Step 1.

It should be noted that three elements have been added to the standard model optimization problem (i.e., minimize$_{c^f}$ $q(c^f, \alpha, \beta, y_r^d)$ subj. to $g(c^f) \le 0$) to create the AMOP and CAMOP problems:

- First, the performance function has been augmented adding a penalty term as in the original ISOPED algorithm. The reasons are similar: regularization of the original optimization problem and prevention of too large deviations of a solution from the current point.

- Second, the constraint $\|c^f - c^{f(i)}\| \geq \gamma$ has been added. Its role is to keep deviations of the set-point sufficient for the purpose of the later derivative estimation which is based on finite difference technique using the matrix $S^{(0)}$. The minimal set-point deviations should be determined taking into account actuating and measurement errors of actual application.
- Third, the conditioning constraint has been added to the AMOP in Step 0.3. The resulted CAMOP problem must be solved when the actual AMOP solution does not satisfy this constraint.

In conclusion let us remind the reader that, from practical point of view, the presented optimized initial phase of the ISOPED algorithm is a significant improvement of this phase by incorporation of the optimality factor into the process of generation of subsequent set-point deviations. Certainly, the quality of the subsequent set-points depends on the quality of the models, as no feedback from the process is used until the first regular ISOPED iteration starts. However, all available information seems to be exploited as much as possible. For more detailed discussion the reader is referred to Section 4.5 in Chapter 4, where the optimized initial phase was first introduced for the problem without feedback controlled output constraints.

5.1.4 *Simulation studies*

Example 5.1

A very simple optimizing control problem will be first considered, to illustrate the approach. The initial problem is three-dimensional, $c \in \mathbb{R}^3$, with one-dimensional plant output, $y \in \mathbb{R}$. The steady-state dependence of the plant output on the set-points c is given by the equation

$$y = F_*(c_1, c_2, c_3) = 2c_1^{0.5} + c_2^{0.4} + c_3 + 0.2c_1c_2 + c_1c_3 \tag{5.74}$$

the performance function (to minimize) is given by

$$Q(c, y) = Q(c) = c_1 + c_2 + 2c_3 + (c_1 - 0.5)^2 + 0.5(c_2 - 0.5)^2 \tag{5.75}$$

and the constraint sets are as follows:

$$C = \{c \in \mathbb{R}^3 : \ 0 \leq c_1 \leq 1.5, \ \ 0 \leq c_2 \leq 1.5, \ \ -1 \leq c_3 \leq 2\} \tag{5.76}$$

$$Y = \{y : \ y \geq 2\}. \tag{5.77}$$

The output constraint is in nominal conditions always active and is assumed to be vital. The optimal point for the true plant is $\hat{c}_* = [0.655\ \ 0.445\ \ -0.242]^T$, with corresponding value of the true performance function equal to $Q(\hat{c}_*, F_*(\hat{c}_*)) = 0.6418$.

Assuming linear structure of the model F of the mapping (5.74) identification of this model has been performed yielding

$$y = F(c_1, c_2, c_3, \alpha) = 2.14c_1 + 0.84c_2 + 1.75c_3 + 0.22 + \alpha \tag{5.78}$$

(the parameters were calculated for $\alpha = 0$ using the least squares method, based on simulations of measurements at points uniformly distributed in the considered range, with random measurement error of 2.5%).

The constraint follow-up controller (CFC) has been applied to force fulfillment of the output constraint. Because there are no another outputs in the considered case, i.e. there are no free outputs, then $y = y^d$, $y_r^d = 2$. The set-point c_3 has been chosen as the output of the CFC, i.e., $c^f = [c_1\ c_2]^T$, $c^d = c_3$. The implicit equation $y_r^d = F_*(c_1, c_2, c_3)$ with respect to c_3 is

$$2 = 2c_1^{0.5} + c_2^{0.4} + c_3 + 0.2c_1c_2 + c_1c_3. \tag{5.79}$$

Therefore, the explicit mapping $c^d = P_*(c^f, y_r^d)$ is in the form

$$c_3 = P_*(c_1, c_2, 2) = \frac{2 - (2c_1^{0.5} + c_2^{0.4} + 0.2c_1c_2)}{1 + c_1} \tag{5.80}$$

and is shown graphically in Fig. 5.2. This mapping will be used in simulations as a true mapping of the controlled plant.

On the other hand, solving with respect to c_3 the implicit equation $y_r^d = F(c_1, c_2, c_3, \alpha)$, i.e.,

$$2 = 2.14c_1 + 0.84c_2 + 1.75c_3 + 0.22 + \alpha \tag{5.81}$$

we get the explicit model mapping $c^d = P(c^d, y_r^d, \beta)$ (with $\beta = 0$)

$$c_3 = P(c_1, c_2, 2, \beta) = -1.22c_1 - 0.48c_2 + 1.02 + \beta. \tag{5.82}$$

The optimizing control simulation was performed applying the ISOPED algorithm. The conditioned modified model-based optimization (CMMOP)

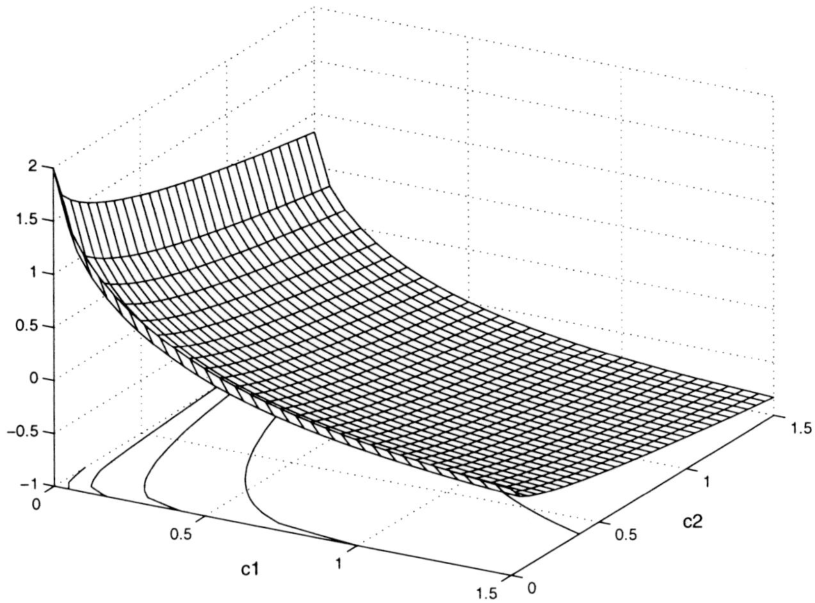

Fig. 5.2 Surface of the process mapping $P_*(c^f, y_r^d) = P_*(c_1, c_2, 2)$.

takes for our example problem the form

$$\begin{array}{ll} \text{minimize}_{c^f} \{Q(v, P(v, 2, \beta)) - \lambda(c^{f(i)})^T v + \rho \|c^{f(i)} - v\|^2\} & \\ \text{subject to: } 0 \leq v_1 \leq 1.5 & \\ \qquad\qquad\quad 0 \leq v_2 \leq 1.5 & \\ \qquad\qquad\quad d(c^{f(i+1)}(v), c^{f(i)}, c^{f(i-1)}) \geq \delta & \end{array} \tag{5.83}$$

where $v = [v_1 \; v_2]^T$,

$$\begin{aligned} Q(v, P(v, 2, \beta)) &= v_1 + v_2 + 2P(v_1, v_2, 2, \beta) + (v_1 - 0.5)^2 + 0.5(v_2 - 0.5)^2 \\ &= -1.44 v_1 + 0.04 v_2 + (v_1 - 0.5)^2 + 0.5(v_2 - 0.5)^2 + 2.04 + 2\beta \end{aligned} \tag{5.84}$$

$$\begin{aligned} \lambda(c^f)^T &= 2\left([-1.22 \; -0.48] - [\frac{\partial P_*}{\partial c_1}(c^f) \; \frac{\partial P_*}{\partial c_2}(c^f)]\right) \\ &= [-2.44 - 2\frac{\partial P_*}{\partial c_1}(c^f) \quad -0.96 - 2\frac{\partial P_*}{\partial c_2}(c^f)] \end{aligned} \tag{5.85}$$

$$d(c^{f(i+1)}(v), c^{f(i)}, c^{f(i-1)}) = \frac{\sigma_{\min}(\mathbf{S}^{f(i+1)}(v))}{\sigma_{\max}(\mathbf{S}^{f(i+1)}(v))} \geq \delta \tag{5.86}$$

$$c^{f(i+1)}(v) = c^{f(i)} + k_c(v - c^{f(i)}) \tag{5.87}$$

$$\mathbf{S}^{(i+1)}(v) = [\ c^{f(i)} - c^{f(i+1)}(v) \quad c^{f(i-1)} - c^{f(i+1)}(v)\ ]^T. \tag{5.88}$$

The derivative $(P_*)'(c^{f(i)}) = [\frac{\partial P_*}{\partial c_1}(c^{f(i)})\ \frac{\partial P_*}{\partial c_2}(c^{f(i)})]$ will be estimated during the ISOPED algorithm iterations using the finite difference approximation, as a solution to the set of linear equations

$$\mathbf{S}^{(i)} \begin{bmatrix} \frac{\partial P_*}{\partial c_1}(c^{f(i)}) \\ \frac{\partial P_*}{\partial c_2}(c^{f(i)}) \end{bmatrix} = \begin{bmatrix} P_*(c^{f(i-1)}) - P_*(c^{f(i)}) \\ P_*(c^{f(i-2)}) - P_*(c^{f(i)}) \end{bmatrix} \tag{5.89}$$

where

$$\mathbf{S}^{(i)} = [\ c^{f(i-1)} - c^{f(i)} \quad c^{f(i-2)} - c^{f(i)}\ ]^T. \tag{5.90}$$

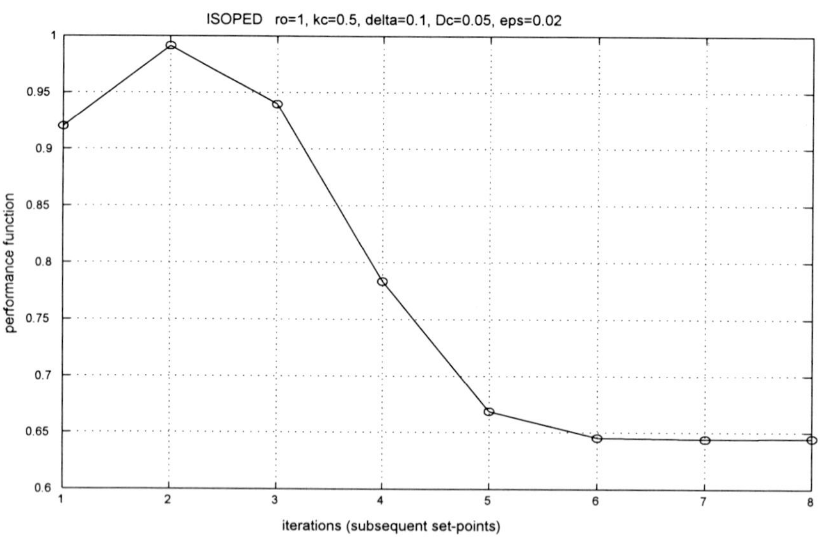

Fig. 5.3 ISOPED, performance function trajectory for example problem 5.1.

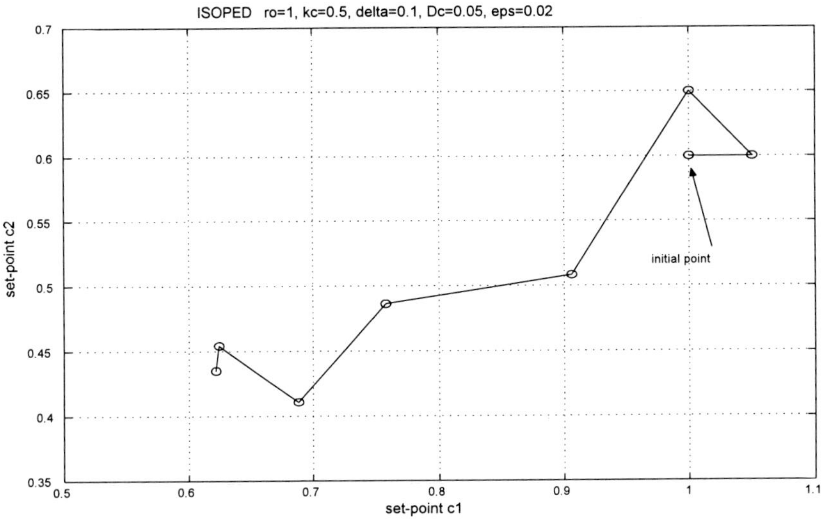

Fig. 5.4 ISOPED, set-point trajectory for example problem 5.1.

Results of an example simulation starting from the initial point $c^{f(0)} = [1.0 \;\; 0.6]^T$ are presented in Fig. 5.3 and Fig. 5.4. Trajectory of the performance function is shown in Fig. 5.3 whereas trajectory of the set points is shown in Fig. 5.4. The iterations were stopped by the stop test with $\epsilon = 0.02$. The first point ($i = 1$) represents the initial point, two subsequent points correspond to the standard (not optimized) initial phase (implemented as deviations from the initial point in directions of versors of the Cartesian space), next points present regular iterations of the ISOPED algorithm. □

Example 5.2

An application of the algorithms described in this section to a complex plant will be presented, see also (Tatjewski *et al.*, 2001). The plant is a production line consisting of three distillation columns connected in series, see Fig. 5.5.* It corresponds to a real industrial distillation unit aimed at separation of a mixture of five components: benzene, toluene, ethylobenzene, styrene and heavy polymers. The mixture is produced in a chemical

*Model of the three distillation column plant is due to J. T. Duda, see (Byrski and Duda, 1990; Tatjewski and Duda, 1997).

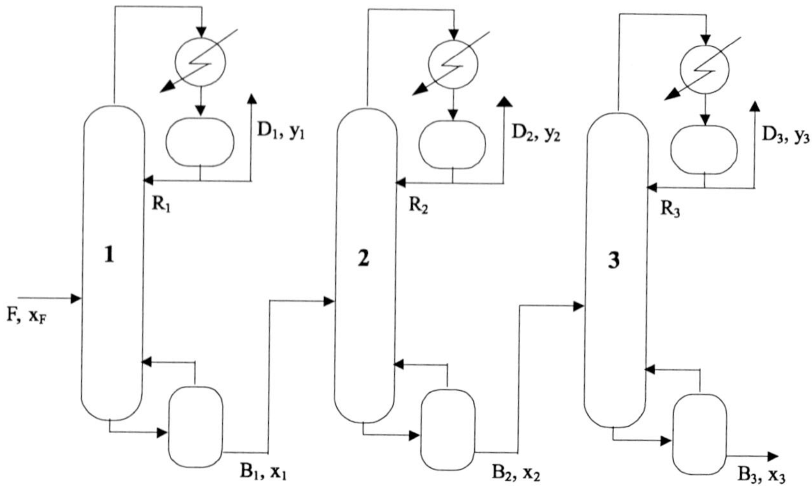

Fig. 5.5 Structure of the case study plant.

reactor converting ethylobenzene into styrene with benzene, toluene and heavy polymers as by-products. The goal of the distillation is to extract pure styrene being a main market product. The first column is fed with the crude mixture drawn from a buffer tank. The distillate from this column (benzene-toluene) is a lower price product and is not required to be of high purity. Therefore, no composition constraints have to be imposed on it. The second column, fed with the bottom product of the first one, separates ethylobenzene from styrene. Its top product (mainly ethylobenzene) is returned to the reactor and should not contain too much benzene being a harmful component. The third column improves the purity of the end-product by removing the heavy polymers drawn as the bottom (waste) product. High purity of the distillate (styrene) is demanded.

The production efficiency can be expressed quantitatively by the following objective function

$$\begin{aligned} Q &= c_{D1}D_1 + c_{D2}D_2 + c_{D3}D_3 + c_{B3}B_3 + \\ &\quad -c_F F - c_E(R_1 + R_2 + R_3 + D_1 + D_2 + D_3) \quad (5.91) \end{aligned}$$

where D_1, D_2, D_3, B_3, F, R_1, R_2, R_3 denote streams shown in Fig. 5.5, whereas c_{D1}, c_{D2}, c_{D3}, c_{B3}, c_F are prices of appropriated substances (c_{B3} being negative) and c_E is the unit evaporation and condensation cost. The quality demands, as discussed before, are expressed in terms of the constraints

$$y_{22} \leq y_{22max} \qquad (5.92a)$$
$$y_{33} \leq y_{33max} \qquad (5.92b)$$

where y_{22} is the concentration of benzene in the second column distillate and y_{33} concentration of the ethylobenzene in the end-product. The production goals can be affected mainly by proper control of the first two columns, especially the second one. The operation area of the third column is very restricted and hence it is led in practice in a prespecified way with its streams being kept at a proper ratio to the second column bottom product. This control structure was preserved when applying the optimizing control algorithms and during the simulations. Therefore, the first two columns were carefully examined in order to implement a computer control system. With this aim in mind, their steady-state nonlinear mathematical models were thoroughly elaborated and identified, see (Byrski and Duda, 1990). The models are based on a key components approach. The second, third and forth feed components are light key components in each of the consecutive columns respectively. Typical values for the stream flows and their relative prices are given in Table 5.1 and compositions in Table 5.2.

Col.	Feed		Reflux		Bottom prod.		Distillate	
no i	F_i	c_F/c_E	R_i	c_{Ri}/c_E	B_i	c_{Di}/c_E	D_i	c_{Di}/c_E
1	14.8	264	7.4	1	14.13		0.67	193
2	14.13		65	1	6.09		8.05	260
3	6.19		5.57	1	.3175	-2.9	5.77	310

Table 5.1 Typical values of flows (initial point) and their relative prices.

The description of each column is transformed into the pseudo-binary form by using mass-balance equations. For the first column, where the light key component is the second one, we have as follows

$$D_1 = F - B_1 \qquad (5.93a)$$
$$y_{11} = \frac{F \cdot x_{F1}}{D_1} = \frac{x_{F1}}{1 - b_1} \qquad (5.93b)$$
$$y_{14} = y_{15} = 0 \qquad (5.93c)$$

Stream	Symbol	benz. j=1	toluene j=2	ethylob. j=3	styrene j=4	heavy pol. j=5
F_1	x_{Fj}	0.0300	0.0400*	0.5100	0.4000	0.0200
D_1	y_{1j}	0.6663	0.2846*	0.0491	0.0000	0.0000
B_1=F_2	x_{1j}=x_{F_2j}	0.0000	0.0285	0.5317*	0.4189	0.0209
D_2	$y_{2,j}$	0.0000	0.0500	0.9328*	0.0172	0.0000
B_2=F_3	x_{2j}=x_{F_3j}	0.0000	0.0000	0.0014	0.9499*	0.0487
D_2	y_{3j}	0.0000	0.0000	0.0015	0.9985*	0.0000
B_3	x_{3j}	0.0000	0.0000	0.0000	0.0678	0.9322
* light key component, Controlled constraints: y_{22max}=0.05, y_{33max}=0.0015						

Table 5.2 Typical composition of streams (initial point for simulation).

$$x_{11} = 0 \tag{5.94a}$$

$$x_{14} = \frac{F \cdot x_{F4}}{B_1} = \frac{x_{F4}}{b_1} \tag{5.94b}$$

$$x_{15} = \frac{F \cdot x_{F5}}{B_1} = \frac{x_{F5}}{b_1} \tag{5.94c}$$

where y_{1j}, x_{1j} denote the j-th component fractions in the column top (distillate) and bottom streams, respectively, b_1 denotes the column relative bottom stream (i.e. B_1/F) and x_{Fj} the j-th component fraction in the feed stream. Pseudo-binary streams, denoted by the asterisk superscript, are defined as:

$$F^* = F \cdot (x_{F2} + x_{F3}) \tag{5.95a}$$

$$D_1^* = D_1 \cdot (y_{12} + y_{13}) = D_1 \cdot (1 - y_{11}) \tag{5.95b}$$

$$R_1^* = R_1 \cdot (y_{12} + y_{13}) = R_1 \cdot (1 - y_{11}) \tag{5.95c}$$

$$B_1^* = B_1 \cdot (x_{12} + x_{13}) = B_1 \cdot (1 - x_{14} - x_{15}) \tag{5.95d}$$

$$b_1^* = B_1^*/F^* \tag{5.95e}$$

$$r_1^* = R_1^*/F^*. \tag{5.95f}$$

The description of the second column has analogous form, but with B_1 taken as the feed flow F, and the components (3rd, 4th) as key components in (5.95a) – (5.95d), respectively. In particular, for the compositions of interest, see (5.92a) and (5.92b),

$$y_{22} = \frac{B_1 x_{12}}{D_2} = \frac{b_1 x_{12}}{b_1 - b_2} = \frac{x_{12}}{1 - b_2/b_1} \tag{5.96a}$$

$$y_{33} = \frac{B_2 x_{23}}{D_3} = \frac{b_2 x_{23}}{b_2 - b_3} = \frac{x_{23}}{1 - b_3/b_2}. \tag{5.96b}$$

As it can easily be seen, these relationships introduce additional interactions between the columns. The selection of the key components corresponding to technological goals is adequate provided that the following relations hold:

$$x_{F4} + x_{F5} < b_1 < 1 - x_{F1} = (x_{F2} + x_{F3}) + x_{F4} + x_{F5} \tag{5.97a}$$

$$x_{15} < b_2/b_1 < 1 - x_{11} - x_{12} = x_{13} + (x_{14} + x_{15}) \tag{5.97b}$$

$$0 < b_3/b_2 < x_{24} + x_{25}. \tag{5.97c}$$

The general model for binary distillation was found to be of the following form (Byrski and Duda, 1990)

$$x^*(r, b, x_F^*) = x_F^* - \frac{1 - x_{min}^*}{1 + \exp(S)} x_F^* \tag{5.98}$$

$$x_{min}^* = \begin{cases} 0 & , \quad A \leq 1 \\ \frac{A-1}{A} & , \quad A > 1 \end{cases} \tag{5.99}$$

$$A = \frac{b^*}{1 - b^*} \frac{x_F^*}{1 - x_F^*} \tag{5.100}$$

where x^* is the fraction of the light (key) component in the pseudo-binary bottom stream, x_{min}^* its physical minimum, and S denotes a separation function. The function S was found for each column by regression techniques based on measurement data. Its form is as follows

$$S_1 = -2.1004 r_1^* - 11.2073 (r_1^*)^2 \tag{5.101a}$$

$$S_2 = -1.4577 (r_2^*)^2 + 1.2246 A_2^2 (r_2^*)^2. \tag{5.101b}$$

The above strongly nonlinear and closely related to reality model of the plant has been used during simulations as the *true process representation.*

Taking into account mass balances:

$$D_1 = F - B_1 \tag{5.102a}$$

$$D_2 = B_1 - B_2 \tag{5.102b}$$

$$D_3 = B_2 - B_3 \tag{5.102c}$$

and applying prices relative to c_E the original performance function (5.91) can be converted to the following form

$$J = c_{Fe} + c_{b1} \cdot b_1 + c_{b2} \cdot b_2 + c_{b3} \cdot b_3 - (r_1 + r_2 + r_3), \tag{5.103}$$

with relative bottom product and reflux streams, b_1, b_2, b_3 and r_1, r_2, r_3 respectively, where

$$\begin{aligned} c_{Fe} &= (c_{D1} - c_F - c_E)/c_E & (5.104a)\\ c_{b1} &= (c_{D2} - c_{D1})/c_E & (5.104b)\\ c_{b2} &= (c_{D3} - c_{D2})/c_E & (5.104c)\\ c_{b3} &= (c_{B3} - c_{D3} + c_E)/c_E. & (5.104d) \end{aligned}$$

Let us now consider the optimizing control algorithms. The set-points for the relative reflux and bottom product streams of two first columns (r_1, r_2, b_1, b_2) will be the algorithm control variables, i.e., $c = [r_1, r_2, b_1, b_2]$. In order to meet the quality demands (5.92a) and (5.92b) a constraint follow-up controller CFC (see Fig. 5.1) has been applied in the control structure. It affects the bottom product flows B_1, B_2. So, in terms of the notation used in the paper we have

$$\begin{aligned} c^f &= [r_1, r_2], & c^d &= [b_1, b_2]\\ y^d &= [y_{22}, y_{33}], & y_r^d &= [y_{22max}, y_{33max}] \end{aligned} \tag{5.105}$$

the input-output mappings $y^d(c)$ being strongly nonlinear. There are no free outputs (5.1a) in the process, hence $y = y^d$. Therefore, a model P of the multivariable mapping P_*, $c^d = P_*(c^f, y_r^d)$, is only necessary.

The ISOPE algorithms have been investigated using the following second order polynomial model P :

$$\begin{aligned} b_1(r_1, r_2) &= \alpha_1 + 0.027r_1 - 0.0112r_2 - 0.0171(r_1)^2 + \\ &\qquad +0.0011(r_2)^2 & (5.106a)\\ b_2(r_1, r_2) &= \alpha_2 + 0.2456r_2 - 0.0244(r_2)^2 & (5.106b) \end{aligned}$$

with only shift parameters α_1 and α_2 updated during iterations, and with initial values: $\alpha_{10} = 0.9736$, $\alpha_{20} = -0.1911$.

The surfaces $b_1(r_1, r_2)$ and $b_2(r_1, r_2)$ of the controlled plant (i.e., the mapping P_*) are shown in Fig. 5.6, while those of the model P in Fig. 5.7.

The model P has been obtained from a set of input-output data, by a least-squares technique, in such a way as to approximate the surfaces

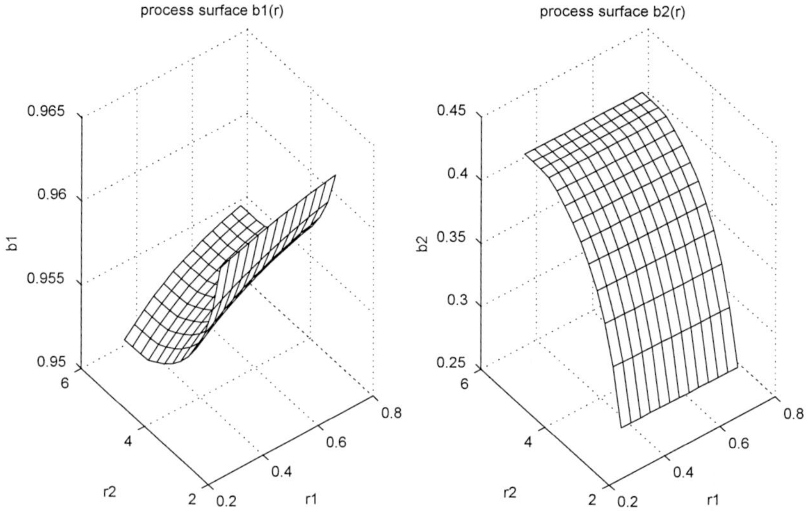

Fig. 5.6 Surfaces $b_1(r)$, $b_2(r)$ of the process mapping P_*.

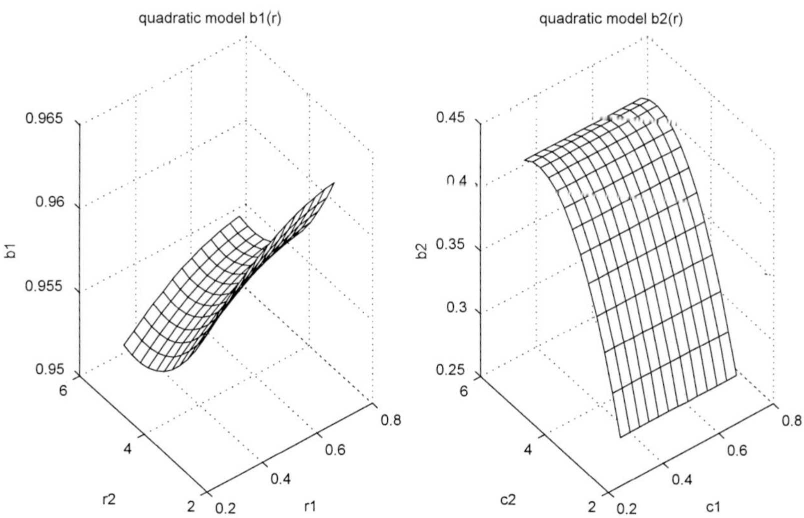

Fig. 5.7 Surfaces $b_1(r)$, $b_2(r)$ of the model mapping P.

P_* over the whole area admissible for (r_1, r_2). Looking at the figures, the surfaces P seem to be quite good models of P^* with similar shapes of nonlinearity, but with certain differences encountered. These differences are of important influence: the process and model optimal points differ significantly. Optimizing (off-line) the set-points with the original nonlinear model P_* gives the result $J_* = 0.5474$, while with the quadratic model P yields only $J = 0.4899$.

The following algorithms were tested in the presented case-study simulation:

- The ISOPEB algorithm, i.e., the basic ISOPE implementation as presented in Section 5.1.2, with derivative estimation at each iteration based on finite difference approximation resulting from additional set-point deviations along predefined directions (axes of the Cartesian space),
- The dual control algorithm ISOPED with standard initial phase, as presented in Section 5.1.3, where the first derivative is constructed as in the ISOPEB case based on additional set-point deviations along predefined directions;
- The dual control algorithm ISOPEDO with optimized initial phase, as presented in Section 5.1.3;
- The ITS algorithm of the standard iterative two-step method, for comparison with the ISOPE algorithms.

The goal of the simulation research was to investigate the effectiveness and practical applicability of the considered ISOPE algorithms. Apart from an obvious comparison of ISOPE and ITS approaches, comparing the ISOPE technique implementations was of primary interest, in particular:

(1) Investigation of the impact of duality on the effectiveness, i.e., comparison of the ISOPEB and ISOPED algorithms;
(2) Investigation of the impact of the optimized initial phase on the effectiveness of the dual control algorithm ISOPED, i.e., comparison of the IDSOPED and ISOPEDO algorithms;
(3) Checking the robustness of the ISOPE algorithms to errors corrupting the steady state feedback information from the controlled plant to the set-point optimizing algorithms, and sensitivity to user-adjusted algorithm parameters.

k	J	r_1	r_2	r_step	$d.act.$	Δb_1	Δb_2
0	0.4438	0.5000	4.3919	0.1081	0	-	-
1	0.4804	0.3963	4.3612	0.0384	1	0.001174	0.000596
2	0.4916	0.3876	4.3238	0.0790	1	0.000072	0.000918
3	0.5065	0.3263	4.2740	0.0856	0	0.000871	0.001160
4	0.5246	0.3263	4.1884	0.0382	1	0.000135	0.002329
5	0.5294	0.3079	4.1550	0.0653	0	0.000258	0.000909
6	0.5386	0.3335	4.0949	0.0425	1	0.000532	0.001847
7	0.5429	0.3305	4.0526	0.0384	0	0.000027	0.001306
8	0.5440	0.3534	4.0218	0.0319	0	0.000411	0.001042
9	0.5462	0.3364	3.9948	0.0184	1	0.000210	0.000844
10	0.5468	0.3349	3.9764	0.0187	0	0.000011	0.000612
11	0.5471	0.3150	3.9698	0.0134	0	0.000308	0.000172
12	0.5472	0.3134	3.9611	-	0	0.000001	0.000290

Table 5.3 Data of a single run of the ISOPEDO algorithm.

Several runs of all algorithms were simulated, first without errors and then with different levels of the errors. The iterations were stopped when the calculated set-point change was less than predefined tolerance value set to 0.01 ($\epsilon = 0.01$). A typical example taken from a collection of simulation runs of all the algorithms is depicted in 5.8 and 5.9, where trajectories of the performance function J (5.103) and the set-point $c^f = (r_1, r_2)$, respectively, are shown. Additionally, numerical data of the run for the ISOPEDO algorithm are presented in Table 5.3, where values of the performance function J, decision variables r_1,r_2 (i.e., c^f) and their change (r_step), activity of the dual constraint and, in the last two columns, changes in values of the dependent variables b_1 and b_2 (i.e, c^d) are shown.

As expected, all the ISOPE algorithms (ISOPEB, ISOPED and ISOPEDO) converged to the real optimal value $J_* = 0.5474$, while the ITS algorithm converged only to a suboptimal set-point resulting in much worse value of the performance function (see 5.8; all runs starting from the same initial point with performance function value $J_{st} = 0.4438$). As expected, the ISOPEB algorithm was less efficient than the dual algorithms, needing much more set-point changes due to two additional set-point deviations at each iteration. These can be easily seen in 5.9, as well as a characteristic "zig-zag" nature of the set-point trajectories of ISOPED and

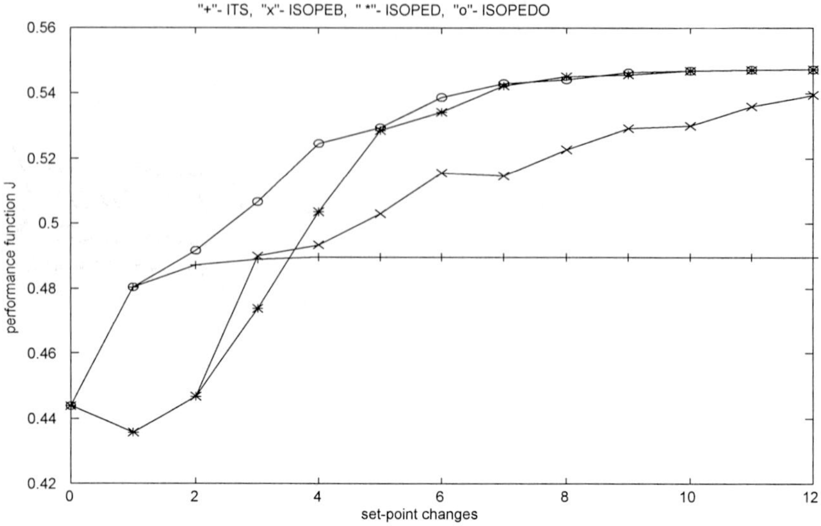

Fig. 5.8 Performance function trajectories, simulations without errors in the feedback information.

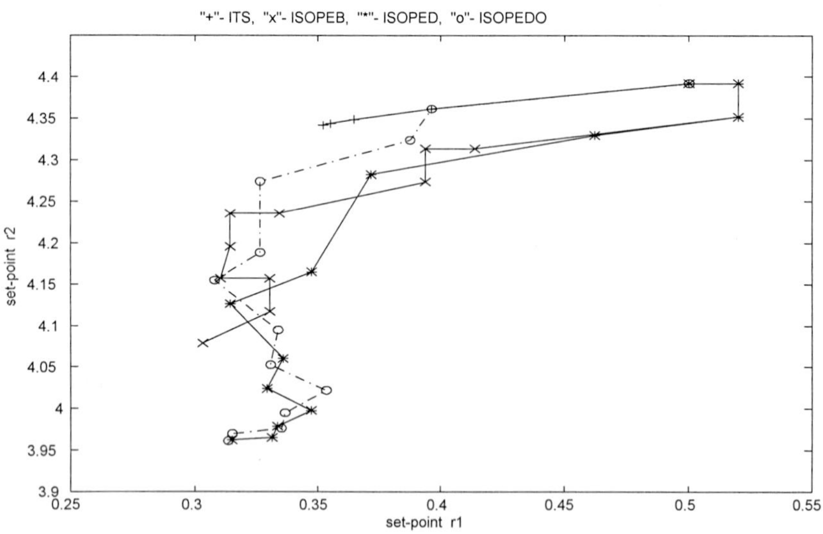

Fig. 5.9 Set point trajectories, simulations without errors in the feedback information.

ISOPEDO algorithms, forced by the dual constraint activity.

Further, introduction of the optimized initial phase significantly improved the initial phase (i.e., two first iterations counted $k = 1, 2$) of the ISOPED algorithm, leading to much better values of the performance function during first iterations. In spite of modeling errors the most developed ISOPEDO algorithm improves the performance function from the very beginning after each set-point change, rapidly arriving at the region of the true process optimum.

The discussed simulation runs were made for the following values of the parameters: $\rho = 0.5$, $k_c = 0.7$, $\delta = 0.1$ and $\gamma = 0.02$ (see Section 5.1.3). The algorithms converged for wide ranges of these parameter values, the values of $\rho \geq 0.2$ and $k_c = 0.3 - 0.9$ were tried, while $\delta = 0.1$ is a generally recommended value (Tatjewski and Duda, 1997) and was not adjusted. A very few trial runs were enough to adjust the parameters from their initial values $\rho = 1$, $k_c = 0.5$ (also resulting in a satisfactory convergence). It should be noted that the value of ρ influences not only the convexification of the original problem but also the convergence rate: the larger ρ the closer the solution point of the modified model optimization problems MMOP or MMOPD to the value of the current set-point c^f, due to the obvious influence of the quadratic penalty term $\rho\|c^f - v\|^2$. Therefore, to optimize the convergence rate, values of both ρ and k_c can be adjusted.

Before passing to simulation results with *errors in the feedback information* from the controlled plant to the considered optimizing control algorithms, let us discuss sources of these errors. Recall that the feedback information consists of steady-state values of the constraint follow-up controller (CFC) outputs, see Fig. 5.1. Therefore, reasons for the errors could be an imperfect action of the CFC and errors in process output measurements. But, the CFC is a computer controller acting precisely according to its algorithm. Next, it is an upper-layer controller (set-point controller) acting with much longer sampling interval than direct controllers. This interval can be in a range of minutes, typical for industrial controllers. Therefore, if the process output measurements are more frequent, the high frequency noise can be well filtered on the input to the CFC. Nevertheless, lower frequency noise (or other errors) can remain. But, optimizing control algorithms are interested in steady-state outputs of the CFC only, therefore filtering CFC outputs should further improve the feedback information. In conclusion, appropriate on-line processing of the measured process outputs and controller outputs can significantly reduce possible errors in the feedback information to the algorithms. Although, as discussed, relatively small values of errors in this information are to be expected, robustness properties of the algorithms should be investigated.

The errors generated by a random number generator (the one from MATLAB) were added to the feedback information, i.e., to the steady-state values of the outputs of the CFC, at each algorithm iteration. The controller acts on the flows b_1 and b_2, i.e., $c^d = b = (b_1, b_2)$, $c^d = c_r^d + \delta c^d$, δc^d being the controller output, comp. Fig. 5.1 and the algorithm structure description in Section 5.1.2. The erroneous signals of the following form were used

$$b_{1n} = b_1 + b_{1err} * 2 * (\text{rand} - 0.5) \tag{5.107}$$

$$b_{2n} = b_2 + b_{2err} * 2 * (\text{rand} - 0.5) \tag{5.108}$$

i.e., the values b_{1n} and b_{2n} were used in the ISOPE algorithms instead of b_1 and b_2 (uniformly distributed random numbers from [0, 1] were used). Two cases with errors were considered:

a) smaller error amplitude: $b_{1err} = 0.000025$, $b_{2err} = 0.0001$,
b) larger error amplitude: $b_{1err} = 0.00005$, $b_{2err} = 0.0002$.

To judge the amplitudes of the error signal let us have a look at Table 5.3, where changes of b_1 and b_2 during a typical simulation run are given (recall that relative values of flows are used, which results in relatively small numbers). It can be easily seen that even in case (a) the errors are significant, in most iterations of about $5 \div 15\%$ of the changes in b_i — and the derivative estimations are evaluated from these changes.

Simulation runs for case (a) are shown in Fig. 5.10, in terms of the performance function value behaviour. In spite the errors the algorithms are still working well. Simulations for case (b) are shown in Figs. 5.11 and 5.12, for different error signal realizations. Even for that large error level the algorithms are useful during first iterations, where the algorithm steps (set-point changes) are larger and most of the steady-state improvement is obtained. In further iterations, where the steps become smaller the errors are too large and the operator or supervisory algorithm should stop the action of the algorithm in a real application. In conclusion, the obtained results show *good robustness of the ISOPE algorithms against errors in the feedback information* from the controlled process. Certainly, the reported

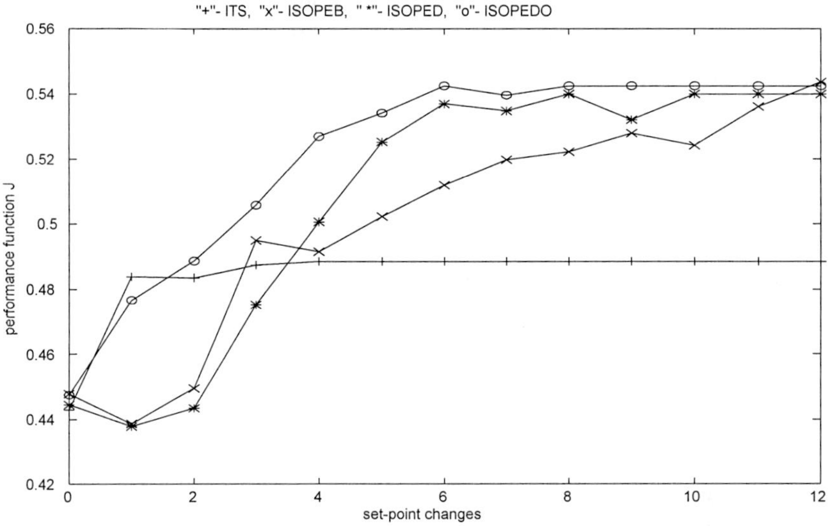

Fig. 5.10 Performance function trajectories, simulations with errors in the feedback information.

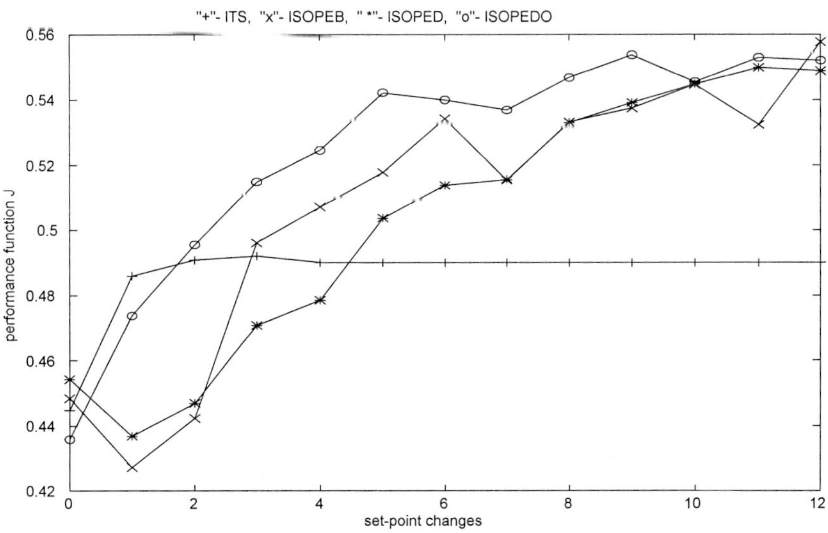

Fig. 5.11 Performance function trajectories, simulations with larger errors.

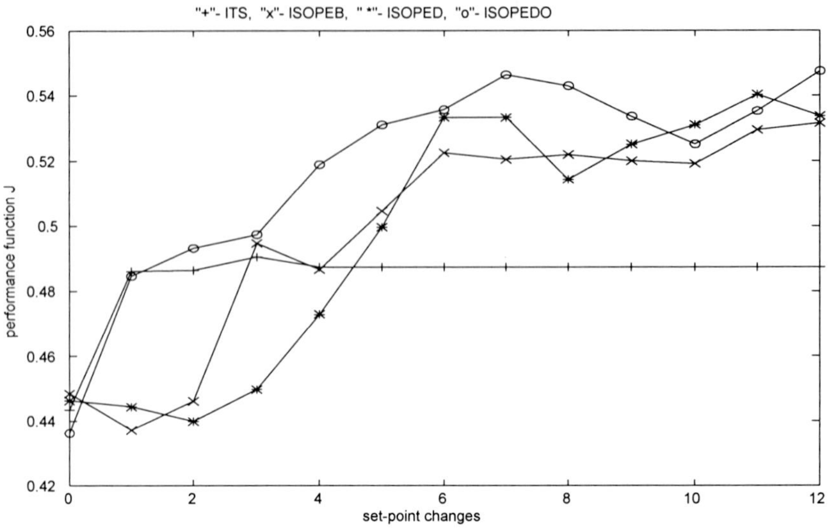

Fig. 5.12 Performance function trajectories, simulations with larger errors (other runs).

results have been obtained for the considered case study example, and cannot be generalized to plants of another nature. More extensive investigation is needed here. We should only add that similar results have been obtained during the tests on several simple (rather academic) examples.

Another important feature of the ISOPE algorithms was presented in (Tatjewski and Duda, 1997), where simulation results of the ISOPED algorithm for different process models are reported, for the same example problem. The conclusion is that the algorithm is robust also to model inaccuracies, in terms both of the model structure and parameter values. The result seems to be of primary practical value, since precise models are often not available or difficult to obtain in process industries. The result is not for nothing — it is due to the fact that the true plant mapping derivatives have to be estimated at each step and used to modify the models in the modified model-based optimization problems, giving good direction of improvement. In fact, the ISOPE technique can be seen as a kind of a sophisticated technique of an extremal search on the process, with extensive use of the process model to get rapid improvement. However, quality of the steady-state model may be important especially for the initial phase of the ISOPEDO algorithm, i.e., for the good improvement from the very first change of the set-point (in (Tatjewski and Duda, 1997) the initial phase was not optimized, ISOPED and not ISOPEDO algorithm was used). □

5.2 Algorithmic Implementation of Output Constraints

5.2.1 *Problem formulation and analysis*

Let us assume that the original steady-state optimizing control problem OCP is of the following form

$$\begin{aligned} &\text{minimize } Q(c,y) \\ &\text{subject to: } y = F_*(c) \\ &\qquad\qquad\quad g(c) \leq 0 \\ &\qquad\qquad\quad \psi(c,y) \leq 0 \end{aligned} \tag{5.109}$$

where $\psi : \mathbb{R}^n \times \mathbb{R}^m \mapsto \mathbb{R}^{r_y}$. The formulation (5.109) is a slight generalization of the initial optimizing control problem with output constraints OCP (2.18) introduced in Chapter 2: the function ψ defining constraints on outputs depends now additionally also on the set-points c. We introduced this generalization because it does not, in fact, influence further considerations in this section.

Comparing (5.109) with the basic formulation (4.1) used to derive the basic ISOPE algorithm in Section 4.1 one can see that they differ in the presence of additional inequality constraints $\psi(c,y) \leq 0$. Let us define

$$h(c,\alpha) = \psi(c, F(c,\alpha)). \tag{5.110}$$

Introducing now additional variables $v \in \mathbb{R}^n$ and proceeding analogously as with the OCP problem in Section 4.1 the problem (5.109) can be converted into the following equivalent form (compare with (4.4))

$$\begin{aligned} &\text{minimize}_{c,v,\alpha}\{q(v,\alpha) + \rho\|c - v\|^2\} \\ &\text{subject to: } F(c,\alpha) = F_*(c) \\ &\qquad\qquad\quad g(v) \leq 0 \\ &\qquad\qquad\quad h(v,\alpha) \leq 0 \\ &\qquad\qquad\quad c = v \end{aligned} \tag{5.111}$$

where

$$q(v,\alpha) = Q(c, F(c,\alpha)). \tag{5.112}$$

Let us write now the Lagrange function corresponding to (5.111)

$$L(c, v, \alpha, \lambda, \xi, \mu) = q(v, \alpha) + \rho\|c - v\|^2 + \\ +\lambda^T(c - v) + \xi^T(F(c, \alpha) - F_*(c)) + \mu^T g(v) + \eta^T h(v, \alpha). \quad (5.113)$$

Proceeding now as in Section 4.1 and taking, additionally, into account the relation

$$h'_\alpha(c, \alpha) = \psi'_y(c, F(c, \alpha))F'_\alpha(c, \alpha) \quad (5.114)$$

we can write necessary optimality conditions for the problem (5.111) in the following form (compare with (4.11a) - (4.11d))

$$\begin{aligned} q'_v(v, \alpha)^T - 2\rho(c - v) - \lambda(c, \alpha, \mu) + g'(v)^T\mu + \\ +h'_v(v, \alpha)^T\eta &= 0 \quad (5.115a) \\ g(v) \leq 0, \quad \mu \geq 0, \quad \mu^T g(v) &= 0 \quad (5.115b) \\ h(v, \alpha) \leq 0, \quad \eta \geq 0, \quad \eta^T h(v, \alpha) &= 0 \quad (5.115c) \\ F(c, \alpha) - F_*(c) &= 0 \quad (5.115d) \\ c - v &= 0 \quad (5.115e) \end{aligned}$$

where

$$\lambda(c, \alpha, \eta) = [F'_c(c, \alpha) - F'_*(c)]^T[Q'_y(c, F(c, \alpha))^T + \psi'_y(c, F(c, \alpha))^T\eta]. \quad (5.116)$$

The resulting set of equations (5.115a) - (5.116) is analogous to the set (4.10) – (4.11d) obtained in Section 4.1 for the basic formulation of the OCP problem with constraints on c only, except for obvious differences resulting from the presence of the constraint $h(c, \alpha)$. In particular, λ is now also dependent on the multiplier η corresponding to the inequality constraints depending on outputs. This causes a major change in the ISOPE algorithm — not only the set-points c but also *the multipliers η must now be treated as global (coordination) variables* and adjusted from iteration to iteration.

5.2.2 *Algorithm 1*

Assuming simple relaxation strategy analogous to (4.21) for set-point and multiplier adjustment, (Brdyś *et al.*, 1986) proposed the following structure of the ISOPE algorithm, denoted as ISOPEY1 in what follows:

ISOPE algorithm with algorithmic implementation of output constraints (ISOPEY1):

Start. Given initial set-point c^0 and multiplier η^0, coefficients k_c and k_η, $0 < k_c \leq 1,\ 0 < k_\eta$, solution accuracy $\epsilon > 0$. Set $i := 0$.

Step 1. Apply c^i to the controlled plant and measure $y^i = F_*(c^i)$. Apply a chosen technique for evaluating the plant output mapping derivative $F_*'(c^i)$.

Step 2. The *parameter estimation problem* PEP: using the obtained new measurements update parameters α under the restriction that the model outputs match the actual controlled plant outputs at c^i. This yields $\alpha^i = \hat{\alpha}(c^i)$ satisfying

$$y^i = F(c^i, \alpha^i) \tag{5.117}$$

Step 3. For $c = c^i$, $\eta = \eta^i$, $\alpha = \alpha^i$ and, therefore, $\lambda(c, \alpha, \eta) = \lambda(c^i, \alpha^i, \eta^i)$ solve the following *modified model-based optimization problem* (MMOPY1):

$$\begin{aligned} &\text{minimize}_v \{q(v, \alpha^i) - \lambda(c^i, \alpha^i, \eta^i)^T v + \rho \|c^i - v\|^2\} \\ &\text{subject to}: \ g(v) \leq 0 \\ &\qquad\qquad\quad\ h(v, \alpha) \leq 0 \end{aligned} \tag{5.118}$$

Let $v^i = \hat{v}(c^i, \alpha^i, \eta^i)$ be the solution, with the corresponding multipliers $\hat{\mu}^i = \hat{\mu}(c^i, \alpha^i, \eta^i)$, $\hat{\eta}^i = \hat{\eta}(c^i, \alpha^i, \eta^i)$.

Step 4. If

$$\begin{aligned} \|v^i - c^i\| &\leq \epsilon \\ \text{and } \|\hat{\eta}^i - \eta^i\| &\leq \epsilon \end{aligned} \tag{5.119}$$

then terminate (solution found).

Step 5. Adjust set-points and multipliers

$$c^{i+1} \ : \ = c^i + k_c(\hat{v}^i - c^i) \tag{5.120}$$

$$\eta_j^{i+1} \ : \ = \max\{0,\ \eta_j^i + k_\eta(\hat{\eta}_j^i - \eta_j^i)\},\ j = 1, \ldots, r_y \tag{5.121}$$

set $i := i + 1$ and continue from Step 1.

Notice that each element η_j^i of the multiplier vector η^i is adjusted independently in (5.121) in order to assure that $\eta \geq 0$.

The presented algorithm is the simplest and most natural generalization of the basic ISOPE algorithm, presented in Section 4.1, to the case with output dependent inequality constraints. It can be still regarded as an algorithm of fix-point type, with set-points v and multipliers η adjusted in order to fulfill the equations

$$\hat{v}(c, \hat{\alpha}(c), \eta) = c \tag{5.122a}$$

$$\hat{\eta}(c, \hat{\alpha}(c), \eta) = \eta \tag{5.122b}$$

compare with (4.22).

5.2.3 *Algorithm 2*

(Lin *et al.*, 1988) proposed a formulation of the ISOPE algorithm with constraints on the plant outputs treated also algorithmically, but in a different way – using a penalty function approach to these constraints. The structure of that algorithm is the same, only the modified model-based optimization problem and the strategy for iterating multipliers are different. This different formulation turned out to be better tractable theoretically, the authors managed to derive a rigorous convergence proof using the methodology analogous to that presented in Appendix A in the proof of Theorem 4.1 from Chapter 4. The algorithm uses the process-based form of the output constraint defined as

$$h_*(c) = \psi(c, F_*(c)) \tag{5.123}$$

ISOPE algorithm with algorithmic implementation of output constraints (ISOPEY2):

Start. Given initial set-point c^0 and multiplier η^0, coefficients k_c and k_η, $0 < k_c \leq 1$, $0 < k_\eta$, solution accuracy $\epsilon > 0$. Set $i := 0$.

Step 1. Apply c^i to the controlled plant and measure $y^i = F_*(c^i)$. Apply a chosen technique for evaluating the plant output mapping derivative $F_*'(c^i)$.

Step 2. The *parameter estimation problem* PEP: using the obtained new measurements update parameters α under the restriction that the model outputs match the actual controlled plant outputs at c^i. This yields $\alpha^i = \hat{\alpha}(c^i)$ satisfying

$$y^i = F(c^i, \alpha^i). \tag{5.124}$$

Step 3. For $c = c^i$, $\eta = \eta^i$, $\alpha = \alpha^i$ and, therefore, $\lambda(c, \alpha, \eta) = \lambda(c^i, \alpha^i, \eta^i)$ solve the following *modified model-based optimization* *problem* (MMOPY2):

$$\text{minimize}_v\{q(v,\alpha^i) - \lambda(c^i,\alpha^i,\eta^i)^T v + \rho\|c^i - v\|^2 + +[h_c'(c^i,\alpha^i)^T\eta^i]^T v\}$$

$$\text{subject to}: g(v) \le 0 \tag{5.125}$$

Let $v^i = \hat{v}(c^i,\alpha^i,\eta^i)$ be the solution, with the corresponding multipliers $\hat{\mu}^i = \hat{\mu}(c^i,\alpha^i,\eta^i)$, $\hat{\eta}^i = \hat{\eta}(c^i,\alpha^i,\eta^i)$.

Step 4. If

$$\begin{aligned} \|v^i - c^i\| &\le \epsilon \\ \text{and } \|\hat{\eta}^i - \eta^i\| &\le \epsilon \end{aligned} \tag{5.126}$$

then terminate (solution found).

Step 5. Adjust set-points and multipliers

$$c^{i+1} \; : = c^i + k_c(\hat{v}^i - c^i) \tag{5.127}$$

$$\eta^{i+1} \; : = \eta^i + k_\eta^i P\{h_*(c^i) + h_*'(c^i)(v^i - c^i)\} \tag{5.128}$$

where P is a projection operator defined by (5.129), set $i := i+1$ and continue from Step 1.

The projection operator P applied in Step 5 of the above algorithm is defined as follows

$$P\{h_*(c^i) + h_*'(c^i)(v^i - c^i)\}_j = 0, \qquad j \in N_1(c^i,\eta^i) \tag{5.129a}$$

$$P\{h_*(c^i) + h_*'(c^i)(v^i - c^i)\}_j = [h_*(c^i) + h_*'(c^i)(v^i - c^i)]_j, \quad j \notin N_1(c^i,\eta^i) \tag{5.129b}$$

where

$$N_1(c^i,\eta^i) = \{j : \ \eta_j = 0 \ \text{ and } \ [h_*(c^i) + h_*'(c^i)(v^i - c^i)]_j \le 0\} \tag{5.130}$$

and the relaxation coefficient k_η^i is defined as follows

$$k_\eta^i = \begin{cases} k_\eta & \text{if } \; N_2(c^i,\eta^i) = \emptyset \\ \min\{k_\eta, \bar{k}_\eta^i\} & \text{if } \; N_2(c^i,\eta^i) \ne \emptyset \end{cases} \tag{5.131}$$

where

$$\bar{k}_\eta^i = \min_{j \in N_2(c^i,\eta^i)} \left\{ \frac{-\eta_j^i}{[h_*(c^i) + h_*'(c^i)(v^i - c^i)]_j} \right\} \tag{5.132}$$

$$N_2(c^i,\eta^i) = \{j:\ \eta_j > 0 \ \text{ and } \ [h_*(c^i) + h_*^{'}(c^i)(v^i - c^i)]_j < 0\}. \quad (5.133)$$

The purpose of using the strategy (5.131) is to exclude the possibility of η_j^{i+1} to become negative, as applied (by Lin *et al.*, 1988) according to the active set method of mathematical programming (Luenberger 1984).

The main difference between the algorithms ISOPEY1 and ISOPEY2 is the way the inequality output constraint $\psi(c,y) \leq 0$ is treated, especially in modified model optimization problems. In the first algorithm it is converted to the model-based constraint $h(c,\alpha) = \psi(c,F(c,\alpha)) \leq 0$ and added (as a constraint only) to the modified model-based optimization problem, see (5.118). The second algorithm it more sophisticated, the derivative of the model-based constraint is included into the performance function only (not into the constraint set) in a specific way, as a product with the multiplier, see (5.125). Moreover, the multiplier vector η^i is adjusted according to the value of the linearized process-based output constraint $h_*(c^i) = \psi(c^i, F_*(c^i)) \leq 0$, see (5.129). It is possible and easy since no new information is needed to calculate the derivative $h_*^{'}(c^i)$ because

$$h_*^{'}(c^i) = \psi_c^{'}(c^i, F_*(c^i)) + \psi_y^{'}(c^i, F_*(c^i))F_*^{'}(c^i) \quad (5.134)$$

and approximation of the process derivative $F_*^{'}(c^i)$ is calculated earlier in Step 1 of the algorithm. Note that in both algorithms the multiplier $\lambda(c^i,\alpha^i,\eta^i)$ is given by the same formula (5.116).

The algorithms ISOPEY1 and ISOPEY2 have been presented in their basic form, as given by (Brdyś *et al.*, 1986) and by (Lin *et al.*, 1988) — the only difference being the convexifying term $\rho\|c^i - v\|^2$ added to the performance function of modified model-based optimization problems. As in every ISOPE algorithm, a key factor determining practical efficiency of both algorithms is the technique of evaluation of the plant output derivative $F_*^{'}(c^i)$ in Step 1. The techniques described in Chapter 4 are generally possible here, i.e.,

- Finite difference approximation based on additional set-point perturbations around the current set-point c^i at each iteration (basic algorithm, Section 4.1),
- Application of the technique used in the ISOPED algorithm, leading to dual algorithms (Section 4.4),
- Evaluation the derivatives from a linear dynamic model of the controlled plant identified in the vicinity of each set-point c^i, as it was

presented in Section 4.3.

The first possibility is not efficient and the last one of limited applicability, as discussed in Chapter 4. The second technique is usually most effective and can easily be applied to the considered case with output dependent constraints. Namely, further development of the algorithms ISOPEY1 and ISOPEY2 to the dual formulations is simple and straightforward. The conditioning constraint must be only added to the modified model-based optimization problems and estimation of the process mapping derivative performed as in the dual ISOPE algorithm, see Section 4.4 in Chapter 4. Details of this formulation can de easily derived, so it is left to the reader. However, in the simulation study presented below results of applications of both algorithms in their basic (ISOPEY1 and ISOPEY2) and dual (ISOPEDY1 and ISOPEDY2) versions will be given.

It should be realized that it can happen that the values of the plant outputs $y^i = F_*(c^i)$ can temporarily violate the true constraints

$$\psi(c^i, y^i) = \psi(c^i, F_*(c^i)) \leq 0 \tag{5.135}$$

during the run of any of the two discussed algorithms. The reason is that only an approximate model of these constraints,

$$h(v^i, \alpha^i) = \psi(v^i, F(v^i, \alpha^i)) \tag{5.136}$$

is known — and only these model-based constraints are guaranteed to be satisfied at every iteration of the ISOPEY1 algorithm, at each solution v^i of the MMOP1 problem (5.118). Of course, the true constraints $\psi(c^i, F_*(c^i)) \leq 0$ are satisfied when the algorithm has converged since then $c^i = v^i$, and always $F_*(c^i) = F(c^i, \alpha^i)$. In the case of the ISOPEY2 algorithm even fulfillment of the model-based constraints is not guaranteed during its run. However, it has been theoretically proved and tested that the ISOPED2 algorithm converges to a point satisfying the true constraints as well (Lin *et al.*, 1988).

It follows directly from the above discussion that the approach treating output dependent constraints only algorithmically is suitable in fact for the constraints of soft type, i.e., those that can be temporarily violated. It is a major difference when comparing with ISOPE algorithms applied in the structure with feedback controlled output constraints, where true constraints are continuously satisfied with the accuracy of feedback controllers. Therefore, where applying the algorithms discussed in this Section, ISOPEY1 and ISOPEY2, much wider constraint safety bounds must usually be foreseen at the design stage when constraint violation should be avoided.

5.2.4 *Illustrative simulation example*

Example 5.3

The simple optimizing control problem given in Example 5.1 will be considered again, to illustrate the approach and for comparison with the algorithm with feedback controlled output constraints treated there. Recall that the initial problem is three-dimensional, $c \in \mathbb{R}^3$, with one-dimensional plant output, $y \in \mathbb{R}$ and the output constraint $y \geq 2$.

Recall that the steady-state dependence of the plant output on the set-points c is given by the equation

$$y = F_*(c_1, c_2, c_3) = 2c_1^{0.5} + c_2^{0.4} + c_3 + 0.2c_1c_2 + c_1c_3 \tag{5.137}$$

the performance function (to minimize) is given by

$$Q(c, y) = Q(c) = c_1 + c_2 + 2c_3 + (c_1 - 0.5)^2 + 0.5(c_2 - 0.5)^2 \tag{5.138}$$

and the constraint sets are as follows:

$$C = \{c \in \mathbb{R}^3 : \ 0 \leq c_1 \leq 1.5, \ \ 0 \leq c_2 \leq 1.5, \ \ -1 \leq c_3 \leq 2\} \tag{5.139}$$

$$Y = \{y \in \mathbb{R} : \ y \geq 2\}. \tag{5.140}$$

The optimal point for the plant is $\hat{c}_* = [0.655 \ \ 0.445 \ \ -0.242]^T$, with corresponding value of the true performance function equal to $Q(\hat{c}_*, F_*(\hat{c}_*)) = 0.6418$ and the output constraint active. The model F of the mapping (5.74) will be taken as (see Example 5.1)

$$y = F(c_1, c_2, c_3) = 2.14c_1 + 0.84c_2 + 1.75c_3 + 0.22 + \alpha. \tag{5.141}$$

For the ISOPEY1 algorithm, the modified model-based optimization problem (MMOPY1) takes the form

$$\begin{aligned}
&\text{minimize}_v \{Q(v) - \lambda(c^i, \alpha^i, \eta^i)^T v + \rho \|c^i - v\|^2\} \\
&\text{subject to}: \ 0 \le v_1 \le 1.5 \\
&\qquad\qquad\quad 0 \le v_2 \le 1.5 \\
&\qquad\qquad\quad -1 \le v_3 \le 2 \\
&\qquad\qquad\quad -(2.14c_1 + 0.84c_2 + 1.75c_3 + 0.22 + \alpha) + 2 \le 0
\end{aligned} \tag{5.142}$$

where $v = [v_1 \ v_2 \ \ v_3]^T$,

$$\lambda(c^i, \alpha^i, \eta^i) = -[F_c^{'}(c^i, \alpha^i) - F_*^{'}(c^i)]^T \eta^i. \tag{5.143}$$

For the ISOPEY2 algorithm, the modified model-based optimization problem (MMOPY2) takes the form

$$\begin{aligned}
&\text{minimize}_v \{Q(v) - \lambda(c^i, \alpha^i, \eta^i)^T v + \rho \|c^i - v\|^2 - [2.14 \ 0.84 \ 1.75]\eta^i v \ \} \\
&\text{subject to}: \ 0 \le v_1 \le 1.5 \\
&\qquad\qquad\quad 0 \le v_2 \le 1.5 \\
&\qquad\qquad\quad -1 \le v_3 \le 2.
\end{aligned} \tag{5.144}$$

For dual algorithms formulations, ISOPEDY1 and ISOPEDY2, the conditioning constraint

$$d(c^i + k_c(v - c^i), c^i, c^{i-1}, c^{i-2}) > \delta \tag{5.145}$$

must be added to the modified model-based optimization problem constraint sets, see Section 4.4.

Results of control simulations for ISOPEY1 and ISOPEY2 algorithms are given in Figures 5.13 and 5.14, in terms of performance function trajectories and output constraint values, respectively. The parameters applied were $\rho = 2$, $k_\eta = 1$, $k_c = 0.8$ for the ISOPEDY1 and $k_c = 0.4$ for the ISOPEDY2 algorithm. For the derivative estimation set-point deviations $\triangle c_j^i = 0.05$ in each component of the set-point vector c^i were applied at each iteration. The solution accuracy was set $\epsilon = 0.02$, intentionally lower than corresponding to $\triangle c_j^i$, to test better the convergence behaviour of the algorithms. The trajectories are given in the figures for all set-points, those corresponding to main iteration points c^i together with those corresponding to set-point deviations needed for derivative estimation (each c^i is followed by three set-points being its deviations).

It can be seen that both algorithm versions 1 and 2 (and especially Algorithm 1) are more efficient in the dual formulations approaching then

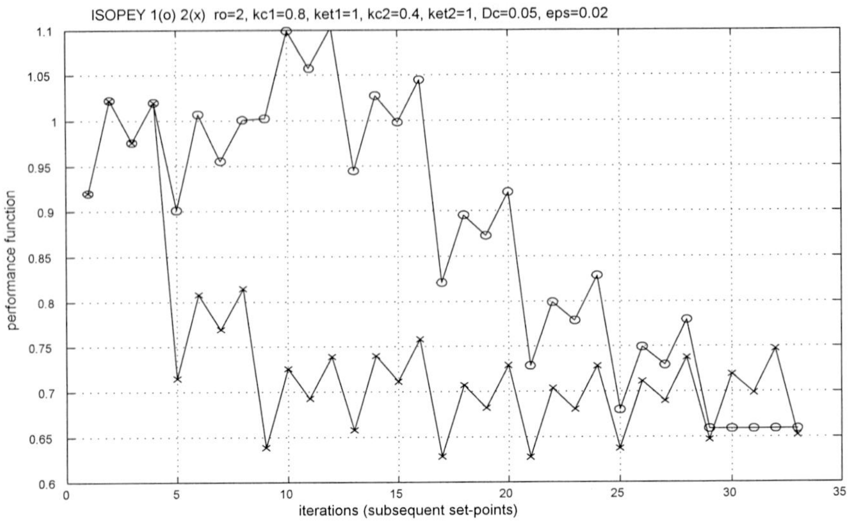

Fig. 5.13 Performance function trajectories of the ISOPEY1 and ISOPEY2 algorithms.

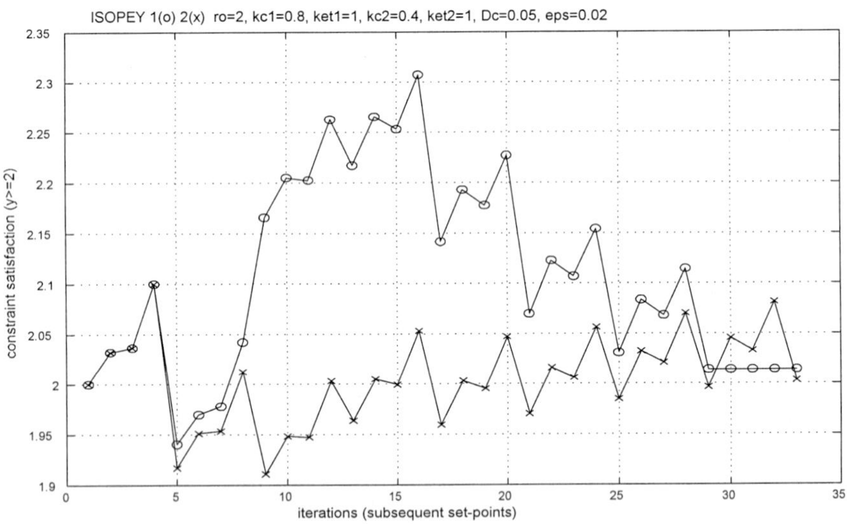

Fig. 5.14 Output constraint value (required ≥ 2) for the ISOPEY1 and ISOPEY2 algorithms.

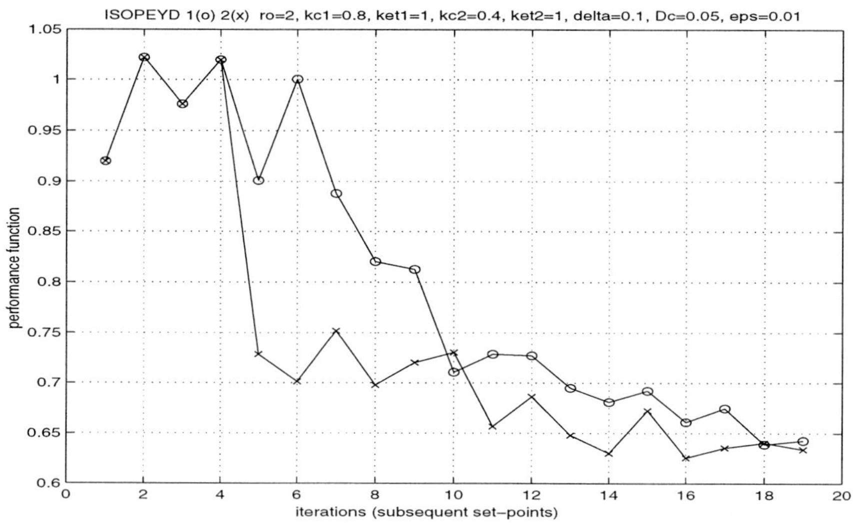

Fig. 5.15 Performance function trajectories of the ISOPEDY1 and ISOPEDY2 algorithms.

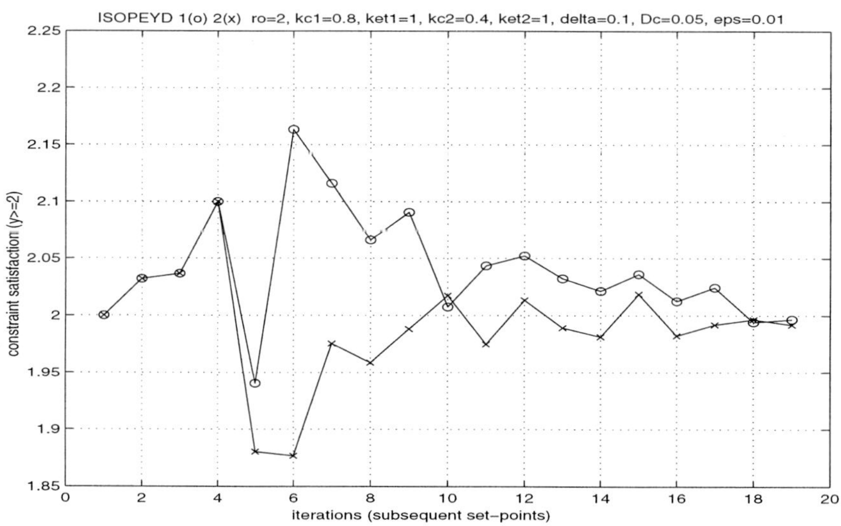

Fig. 5.16 Output constraint value (required ≥ 2) for the ISOPEDY1 and ISOPEDY2 algorithms.

vicinity of the true constrained optimum quicker, in a smaller number of set-point changes. Comparing Algorithms 1 and 2, both in basic and dual realizations, it can be concluded that version 2 was more effective in the case considered.

On the other hand, comparing the most efficient case taken from those presented above, the results for the ISOPEDY2 algorithm, with results obtained for the same example problem when applying the ISOPED algorithm for the structure with feedback controlled output constraints, presented in Fig. 5.3, one must conclude that the latter algorithm is significantly more effective approaching the true plant optimum about twice quicker, while keeping the output constraint satisfied during the iterations. Certainly, it is due to the application of an additional, set-point feedback controller and the following smaller dimensionality of the optimization problem. □

5.3 ISOPE as Optimization Algorithm for Nonlinear Models

As discussed in Chapter 2, if the steady-state plant model is sufficiently accurate and can be effectively identified on-line, then the steady-state optimizing control problem may be implemented as pure model-based nonlinear optimization problem.

Let us consider the basic *model optimization problem* (MOP) introduced in Chapter 2 (Section 2.1)

$$\begin{aligned} &\text{minimize } Q(c,y) \\ &\text{subject to: } \; y = F(c, w_e, \alpha) \\ &\qquad\qquad\quad\; g(c) \le 0 \\ &\qquad\qquad\quad\; \psi(y) \le 0 \end{aligned} \tag{5.146}$$

where w_e (uncontrolled inputs measurements or estimates) and α (model parameters) are constant parameters for the optimization procedure. The optimization problem (5.146) is usually nonconvex, nonlinear input-output controlled plant model $y = F(c, w_e, \alpha)$ being the main or sole source of nonlinearity. Moreover, this mapping is often given in implicit way, as discussed in Section 2.1. For solving model-based optimization problems numerical procedures of nonlinear programming must be applied, appropriately chosen and adjusted to properties of the task to be solved. Presentation and discussion of such procedures is out of the scope of this book, the reader is referred to widely known good reference books, see, e.g., (Fletcher, 1987; Bertsekas, 1995) and to commercially available program packages, e.g., procedures from *Optimization Toolbox* for MATLAB.

However, the ISOPE algorithms presented in this book can also be successfully used for solving nonlinear programming problems of structure (5.146), being an alternative for more standard procedures. The reason is the structure of the problem (5.146), where nonlinear constraint $y = F(c, w_e, \alpha)$ is the source of nonlinearity and computational difficulty. Moreover, if this constraint is given only in an implicit form $\varphi(y, c, w_e, \alpha) = 0$ (or even in more complex form (2.12), (2.13), as explained in Section 2.1) then calculation of every value of the output y means usually a difficult iterative process. In this situation numerical procedures for the optimization of (5.146) should be chosen mainly as those yielding the optimum in minimal number of iterations, i.e. minimal number of changes in c and calculations of y (certainly, there is no direct need to solve implicit equations when optimizing both with respect to c and y, but then dimensionality of the nonlinear problem grows). Let us note now that optimizing control algorithms of ISOPE type are constructed precisely in this way, they aim at achieving the optimum in possibly smallest number of set-point changes.

Taking the difficult nonlinear controlled plant model $F(c, w_e, \alpha)$ as a representation of the true controlled plant and constructing its *much simpler computationally approximation* $F_a(c, w_e, \alpha)$ (e.g., a linear mapping) treated as a "plant model for control purposes" one can directly apply the basic ISOPE (or dual ISOPE) algorithm for the problem (5.146). During the iterations of this algorithm only the modified model-based optimization problem (MMOP) will be optimized, using the much simpler mapping $F_a(c, w_e, \alpha)$ only. Of course, applying ISOPE not to on-line optimizing control but to pure model optimization eliminates practically problems connected with inaccuracies of on-line output measurements — values of the "plant" outputs $y = F_*(c, w)$ will be generated with the accuracy of numerical procedures used.

The reasoning given above can be easily repeated also for nonlinear model-based optimization problems for the structure with feedback controlled output constraints. *Particularly in this case* application of the ISOPE algorithms may lead to good results, since precise nonlinear model of the controlled plant together with models of constrained outputs stabilized by feedback controllers create together an implicit nonlinear model usually difficult for numerical calculations.

Chapter 6

Iterative Algorithms for Dynamic Optimizing Control

6.1 Optimizing Control of Batch Processes

6.1.1 *Control task formulation*

An objective of the dynamic optimizing control was formulated in Chapter 2. A key element of its strategies is the optimal control problem, presented by equations (2.6a), (2.4), (2.6c), (2.6d), (2.6e) and (2.6f) describing the performance function to be minimized, the process dynamics seen by the optimizing control layer and the constraints imposed on the process outputs and manipulated variables (set-points or controls), respectively. The process performance function is to be minimized over horizon $t \in [0, T]$ under uncertain disturbance input $w(t)$ represented by its prediction $w_e(t)$. Hence, the process input-output mapping is described in the state space form as

$$\dot{x}(t) = f_c(x(t), c(t), w(t)) \tag{6.1a}$$

$$y(t) = g_c(x(t)) \tag{6.1b}$$

$$\text{where} \quad x(t) \in \mathbb{R}^{n_x}, \; c(t) \in \mathbb{R}^{n_c}, \; y(t) \in \mathbb{R}^{n_y}, \; w(t) \in \mathbb{R}^{n_w}.$$

Only the outputs but not the states are measured. Note that as the input $w(t)$ is not exactly known there is an uncertainty in the mapping $f_c(\cdot)$. Also, in order to simplify technical derivations it was assumed that there is no direct link between the output $y(t)$ and the control input $c(t)$. Hence, the variable $c(t)$ is not an argument in the output mapping described by (6.1b). This assumption is met for most of the industrial processes. The above formulation of the dynamic optimizing control task is time continuous and the resulting optimized set-point trajectories are also time continuous. Hence, it would be a rather demanding task for the actuating system to force the process to follow these trajectories. Also, a numerical nonlinear dynamic optimization algorithm is to be involved in determining the optimized set-points. Hence, it is natural to limit the set of admissible time continuous control trajectories to piecewise constant functions and to discretize the system dynamics described by (6.1a) and (6.1b), accordingly. The resulting discrete time dynamics can be modeled as:

$$x(t+1) = f_*(x(t), c(t), w(t)) \quad \text{for} \quad t = 0, 1, 2, \ldots.., N-1 \tag{6.2a}$$

$$y(t) = g_*(x(t)) \quad \text{for} \quad t = 0, 1, 2, \ldots.., N \tag{6.2b}$$

where t denotes discrete time instants that determine time steps over which set-points are constant.

Denoting the state, control and output trajectories over the control horizon as

$$x = [x(1)^T, x(2)^T, \ldots.., x(N)^T]^T \in \mathbb{R}^{Nn_x} \tag{6.3a}$$

$$c = [c(0)^T, c(1)^T, \ldots.., c(N-1)^T]^T \in \mathbb{R}^{Nn_c} \tag{6.3b}$$

$$y = [y(0)^T, y(1)^T \ldots.., y(N-1)^T, y(N)^T]^T \in \mathbb{R}^{(N+1)n_y} \tag{6.3c}$$

$$w = [w(0)^T, w(1)^T, \ldots.., w(N-1)^T]^T \in \mathbb{R}^{Nn_w} \tag{6.3d}$$

the performance function reads:

$$Q(c, y) = \sum_{t=0}^{N-1} Q(c(t), y(t), t) + Q_N(y(N)). \tag{6.4}$$

The term $Q_N(y(N))$ forces process output to reach desired value at the end of the control horizon.

The initial condition $x(0)$ is assumed to be fixed. Hence, it was not included into definition (6.3a) of the variable and control dependent state trajectory. The dynamic optimization task can now be cast into a static optimization framework. Indeed, the prescribed vector c of control actions applied to the process over control horizon produces (together with the disturbance trajectory) the process state trajectory $x_*(c, w)$ according to the state equations (6.2a). The state trajectory $x_*(c, w)$ yields the process output trajectory $y_*(c, w)$ according to the output equation (6.2b). Hence, under fixed state initial condition $x(0)$ (and fixed disturbance scenario $w(\cdot)$) the control inputs c and the controlled outputs y are related by the process input-output mapping F_* as follows:

$$\begin{aligned} F_* \quad &: \quad \mathbb{R}^{Nn_c} \mapsto \mathbb{R}^{(N+1)n_y} \\ c \quad &\in \quad \mathbb{R}^{Nn_c} \longmapsto x_*(c,w) \in \mathbb{R}^{Nn_x} \longmapsto y_*(c) \in \mathbb{R}^{(N+1)n_y} \\ & \quad F_*(c) \triangleq y_*(c). \end{aligned} \tag{6.5}$$

It is pointed out that the mapping $F_*(\cdot)$ is defined for specific values of the state initial condition $x(0)$ and the disturbance scenario $w(\cdot)$ in the system over the control horizon (see (6.3d)). For the sake of a notational simplicity this dependence is not reflected in the mapping argument set. The *dynamic optimizing control problem (DOCP)* can now be formulated in a "static" manner as

$$\text{minimize}_{c,y}\ Q(c,y) \quad = \quad \sum_{t=0}^{N-1} Q(c(t), y(t), t) + Q_N(y(N)) \tag{6.6a}$$

subject to constraints :

$$y \quad = \quad F_*(c) \tag{6.6b}$$

$$c(t) \quad \in \quad C(t) \qquad t = 0, 1, 2, ...N-1 \tag{6.6c}$$

$$\|c(t) - c(t+1)\| \quad \leq \quad \Delta^{\max} \quad t = 0, 1, ...N-2 \tag{6.6d}$$

$$y(t) \quad \in \quad Y \qquad t = 1, ..., N. \tag{6.6e}$$

where $\|\ \|$ denotes a vector norm properly chosen to accommodate the rate constraints. The rate constraints (6.6d) although very important do not require special methodological treatment and therefore, will not be further pursued in order to simplify notation. The output constraints require special structural attention as discussed in Chapter 5 and they will be omitted. Hence, we shall further consider the DOCP in the form:

$$\begin{aligned} &\text{minimize}_{c,y}\ Q(c,y) \\ &\text{subject to: } y = F_*(c) \\ &\qquad\qquad\quad c \in C \subset \mathbb{R}^{Nn_c} \\ &\qquad\qquad\quad y \in \mathbb{R}^{(N+1)n_y} \end{aligned} \tag{6.7}$$

There is an uncertainty in the above formulation of the DOCP even if the real process mappings are exactly known, that is if the disturbance input is known over $\overline{0:T}$ and there are no modeling errors in the state and output equations (6.1a) and (6.1b), respectively. Indeed, even if very accurate simulator is used in order to evaluate values of the mappings $f_*(\cdot)$ and $g_*(\cdot)$ the discretization error is inevitable. However, the modeling and disturbance prediction errors are common in real life situations. Hence, the process input-output mapping $F_*(\cdot)$ is not exactly known due to the discretization error, disturbance prediction and modeling error.

Introducing parameters in the relationships (6.2a) and (6.2b) we can obtain a point-parametric model $F(c, \alpha)$ of the mapping $F_*(c)$ that would be suitable to apply ISOPE algorithm to the (6.7) formulation of the DOCP in order to determine optimal set-point trajectory for the real process.

However, a number of iterations are required that involve applications of the set-point trajectory at hand to the real process in order to iteratively reach an optimal set-point trajectory. As the mapping $F_*(\cdot)$ depends on the initial state of the process that is fixed this means starting several times from the same initial state of the process. Because real time can not be reversed such operation is possible only if the process operates in so called batch mode when at the beginning of a new batch the same initial condition of the process state is restored. The chemical or biological reactor operating in a batch mode serve examples of such processes where after a certain and fixed time period the production is completed, the reactor is loaded again and a new process batch starts from the same initial conditions. In this case the state initial condition is restored by employing a routine technological operation. Another situation when this happens is if the process itself is continuous but the controls applied are periodic functions of time. Then, the process state at the end of the period is the same as at the beginning of the period. The process batch is then defined as its operation over a period.

Moreover, the mapping $F_*(\cdot)$ depends on a disturbance scenario in the process over the control horizon. This would ruin the ISOPE convergence properties if the disturbance input trajectories significantly differ over different batches. Hence, it is assumed that the disturbance inputs are periodic functions of time over the batches. In practice, this means that the disturbance profiles slowly vary from batch to batch. In this case the process output response trajectories to the same control trajectories are the same over each batch. Finally, it is assumed that the output y defined by (6.3c) is available from the process measurements and the point parametric model $F(c, \alpha)$ of the mapping $F_*(c)$ is also available.

6.1.2 Optimizing control algorithm structure

We shall now derive a dynamic ISOPE optimizing control algorithm structure for batch processes. If $x(0) = x(N)$ then because $x(0)$ is fixed the term $Q_N(y(N))$ is constant and it does not depend on the control inputs. Hence, it can then be omitted in (6.6). Let us start with defining the point parametric model $F(c, \alpha)$ of the batch process input-output mapping $F_*(c)$. The batch process model mappings and parameters are introduced:

$$\begin{array}{ll} f(x(t), c(t), \alpha(t)) & \text{for } t = 0, 1, 2, ..., N-1 \\ g(x(t), \alpha(t)) & \text{for } t = 0, 1, ..., N \\ \alpha(t) \in \mathbb{R}^{n_\alpha} & \text{for } t = 0, 1, 2, ..., N \\ \multicolumn{2}{l}{\alpha \triangleq [\alpha(0)^T, \alpha(1)^T, ..., \alpha(N)]^T \in \mathbb{R}^{(N+1)n_\alpha}} \end{array} \tag{6.8}$$

in order to model the unknown real process state and output mappings $f_*(x(t), c(t), w(t))$ and $g_*(x(t), c(t)))$ (see (6.2a) and (6.2b), respectively). The vector variable $\alpha(t)$ represents the model parameters at time t while α is the parameter trajectory over a batch. The batch process model equations in a state space form read

$$x(t+1) = f(x(t), c(t), \alpha(t)) \quad \text{for } t = 0, 1, 2, ..., N-1 \tag{6.9}$$

$$y(t) = g(x(t), \alpha(t)) \quad \text{for } t = 0, 1, 2, ..., N. \tag{6.10}$$

The model is point parametric which means that for any $\overline{x}(0), \overline{w}, \overline{c}$ there is such parameter trajectory over a batch that the model state and output trajectories $x(\overline{c}, \overline{\alpha})$ and $y(\overline{c}, \overline{\alpha})$ produced by the model described by the equations (6.9) and (6.10) over a batch exactly match the corresponding real process trajectories $x_*(\overline{c}, \overline{w})$ and $y_*(\overline{c}, \overline{w})$ over a batch. Hence, the following hold: $x(\overline{c}, \overline{\alpha}) = x_*(\overline{c}, \overline{w})$ and $y(\overline{c}, \overline{\alpha}) = y_*(\overline{c}, \overline{w})$.

A point parametric model $F(c, \alpha)$ of $F_*(c)$ is then defined for fixed $x(0)$ and w similarly as in (6.5):

$$F : \mathbb{R}^{Nn_c} \times \mathbb{R}^{(N+1)n_\alpha} \mapsto \mathbb{R}^{(N+1)n_y}$$

$$(c, \alpha) \in \mathbb{R}^{Nn_c} \times \mathbb{R}^{(N+1)n_\alpha} \longmapsto (x(c, \alpha), \alpha) \in \mathbb{R}^{Nn_x} \times \mathbb{R}^{(N+1)n_\alpha} \longmapsto$$

$$\longmapsto y(c, \alpha) \in \mathbb{R}^{(N+1)n_y}$$

$$F(c, \alpha) \triangleq y(c, \alpha). \tag{6.11}$$

The dynamic optimizing control problem DOCP described by (6.6) can now be written in an equivalent form

$$\begin{aligned} &\text{minimize}_{c,\alpha}\ q(c,\alpha) \\ &\text{subject to: } F(c,\alpha) = F_*(c) \\ &\qquad\qquad\quad c \in C \subset \mathbb{R}^{Nn_c} \end{aligned} \tag{6.12}$$

where

$$q(c,\alpha) = Q(c, F(c,\alpha)). \tag{6.13}$$

Following the ISOPE development route from Chapter 4, additional variable ν is introduced as

$$\nu = [\nu(0)^T, \nu(1)^T, \dots\dots, \nu(N-1)^T]^T \in \mathbb{R}^{Nn_c} \tag{6.14}$$

in order to separate a parameter estimation problem and a modified model-based optimization problem. DOCP (6.12) can now be written as

$$\begin{aligned} &\text{minimize}_{c,\alpha}\ \{q(c,\alpha) + \sum_{t=0}^{t=N-1} \rho(t) \parallel c(t) - \nu(t) \parallel^2\} \\ &\text{subject to:} \quad F(c,\alpha) = F_*(c) \\ &\qquad\qquad\quad c \in C \subset \mathbb{R}^{Nn_c} \\ &\qquad\qquad\quad \nu = c. \end{aligned} \tag{6.15}$$

Note that different weights $\rho(t)$, $t \in \overline{0; N-1}$ are applied over a batch in order to obtain suitable conditioning of the minimized function through the convexifying term. Given the parameters α and set-points c, the *dynamic modified model-based optimization problem* (DMMOP) is defined as

$$\begin{aligned} &\text{minimize}_{\nu}\ \{q(\nu,\alpha) - \lambda(c,\alpha)^T\nu + \sum_{t=0}^{t=N-1} \rho(t) \parallel c(t) - \nu(t) \parallel^2\} \\ &\text{subject to: } \nu \in C \subset \mathbb{R}^{Nn_c}. \end{aligned} \tag{6.16}$$

The modifier trajectory over a batch $\lambda(c,\alpha) = [\lambda(0)(c,\alpha)^T, \lambda(1)(c,\alpha)^T, \dots, \lambda(N-1)(c,\alpha)^T]^T = \{\lambda(t)(c,\alpha)\}_{t=0}^{t=N-1} \in \mathbb{R}^{Nn_c}$, can be calculated from the formula derived in Chapter 4 (see (4.10)):

$$\lambda(c,\alpha) = [F_c'(c,\alpha) - F_*'(c)]^T Q_y'(c, F(c,\alpha))^T. \tag{6.17}$$

In the formula (6.17) there are derivatives of the mappings that involve a mapping $x(c,\alpha)$. Given c and α the mapping $x(c,\alpha)$ returns state trajectory over a batch. Efficient procedure for calculating derivative of this mapping based on the process state space model described by (6.9) and (6.10) will be derived in the next section. As the real process mappings $f_*(\cdot)$ and $g_*(\cdot)$ are not known this technique can not be applied to calculating $F_*^{'}(c)$. this derivative is calculated by parsimoniously applying perturbations of controls over a batch or using past control applications to real process as described in Chapter 4 for dual ISOPE algorithm.

A *dynamic parameter estimation problem* (DPEP) is defined as a problem of adapting the model parameters over a batch α at an operating point of the process $(c, y_*(c))$ so that the model response matches the process response. As the process outputs are measured over the batch this can be done by solving, with respect to α, the following set of equations:

$$F(c, \alpha) = F_*(c). \tag{6.18}$$

This is a very simple task if the parameters α are used in the model mappings (see (6.8)) in an additive manner. Otherwise, the DPEP equations need to be solved as a set of nonlinear dynamic equations by applying suitable numerical algorithm.

We are now in a position to formulate *Dynamic ISOPE (DISOPE) basic algorithm.*

Start. Given initial point c^0, relaxation coefficient k_c, $0 < k_c \leq 1$, convexifying coefficients $\rho(t), t \in \overline{0 : N-1}$ and solution accuracy $\varepsilon > 0$. Set $i := 0$.

Step 1. Apply c^i to the controlled plant and measure $y^i = F_*(c^i)$. Perform additional linearly independent perturbations around c^i and measure corresponding values of the plant outputs. Based on this measurements find a finite difference approximation of the plant output mapping derivative $F_*^{'}(c^i)$.

Step 2. The *dynamic parameter estimation problem* DPEP: using the obtained new measurements update parameters α so that the model outputs match the actual process outputs at c^i. This yields $\alpha^i = \hat{\alpha}(c^i)$ satisfying

$$y^i = F(c^i, \alpha^i) = F_*(c^i). \tag{6.19}$$

Step 3. For $c = c^i$ and $\alpha = \alpha^i$ and, therefore, $\lambda(c, \alpha) = \lambda(c^i, \alpha^i)$ solve the *dynamic modified model-based optimization problem* DMMOP

$$\begin{aligned}
&\text{minimize}_\nu \ \{q(\nu, \alpha^i) - \lambda(c^i, \alpha^i)^T \nu + \\
&\qquad\qquad + \sum_{t=0}^{t=N-1} \rho(t) \parallel c^i(t) - \nu(t) \parallel^2\} \\
&\text{subject to}: \ \nu \in C \subset \mathbb{R}^{Nn_c}.
\end{aligned} \tag{6.20}$$

Let $v^i = \hat{v}^i(c^i, \alpha^i)$ be the solution. If

$$\|c^i - v^i\| \leq \varepsilon \tag{6.21}$$

then terminate. Solution found.

Step 4. Set

$$c^{i+1} := c^i + k_c(v^i - c^i) \tag{6.22}$$

set $i := i + 1$ and continue from Step 1.

By following reasoning from Chapter 4, it can easily be seen that the algorithm is constructed in such a way that when it terminates, the necessary optimality conditions for the DOCP in a form described by (6.15) problem are satisfied.

The whole DISOPE algorithm can be regarded as of fixed-point type, since the set-points c are iterated in such a way as to fulfill the equation $\nu = c$, which in the algorithm realization takes the form

$$\hat{v}(c, \hat{\alpha}(c)) = c. \tag{6.23}$$

The iterative formula (6.22) is a simple adjustment rule for finding a fix point of (6.23), usually called iteration of a relaxation type and the parameter k_c called the relaxation coefficient. Notice that if $k_c = 1$ then this formula becomes a direct substitution rule

$$c^{i+1} = v^i. \tag{6.24}$$

Convergence of this iterative scheme is then guaranteed if $\hat{v}(c, \hat{\alpha}(c))$ is the contraction mapping (Kantorovich and Akilov, 1963).

6.1.3 *Implementation of DISOPE algorithm*

A dynamic optimizing control of a batch process problem has been cast in the subsection 6.1.1 into a dynamic optimizing control over one batch. The ISOPE has been applied in subsection 6.1.2 and the dynamic optimizing control algorithm structure has been derived. The ISOPE iterations are carried out in real time over the batches to converge to the process control set-point trajectory that is optimal for the real process over a batch. We shall call the resulting algorithm dynamic ISOPE (DISOPE). If the disturbance inputs profile over a batch change the DISOPE starts its iterations again and new optimal set-point trajectory is determined. Note that an iteration over a batch in a dynamic case corresponds to an instantaneous iteration in a steady state case. Hence, the *parameter estimation problem* and *modified model-based optimization problem* that need to be solved during one iteration of DISOPE are now dynamic problems. Formulation of these problems (see (6.18) and (6.16)) at present is explicit by using mappings F, F_*, q and λ that are explicit functions of control variables. The mappings q and λ are defined via the mappings F, F_*. However, the mappings F, F_* are defined by first solving the state equations and then by using the solutions $x(c, \alpha)$ and $x_*(c, w)$ in the output equations (see (6.11) and (6.5), respectively). The explicit solutions of the state equations are not known and therefore, the explicit analytical representations of the mappings $x(c, \alpha)$ and $x_*(c, w)$ are not known. Hence, the explicit analytical formulae describing the mappings F, F_* are not available. Regarding the mapping F_*, as the real process state and output mappings f_* and g_* are not known the values of F_* can be known only from the measurements and an approximation of the the derivative $F_*^{'}$ is calculated also form these measurements. As the measurements are noisy, special filtering is needed to process the raw measurements. As opposed to the mapping F_*, the values and derivatives of the mapping F can be accurately calculated by using the state and output models (see (6.8), (6.9) and (6.10)).

An *explicit architecture* of the DISOPE implies an expression for the modifier λ (see (6.17)) that constitutes a core of the DISOPE feedback mechanism from a real process allowing to reach a true process optimal set-point trajectory. However, the state variables can participate in the DPEP and DMMOP (see (6.18) and (6.16)) either as the *search variables* or as the *calculated variables.* In the former case the state variables explicitly participate in the search activities that are carried out in order to solve

DPEP and DMMOP tasks under constraints imposed by the state space model equations. In the latter case the state variable values are calculated from these equations, given control variable values. An implementation of DISOPE with state variables as the search variables is called *explicit*, as the state variables are seen by the searching algorithm. The DISOPE implementation with calculated state variables is called *implicit* as the sate variables are then not seen by the searching algorithm. We shall now present in detail these implementations.

6.1.3.1 *Implicit implementation of DISOPE algorithm*

We shall first consider DMMOP to be solved in the Step 3 of the algorithm (see (6.20)). Assuming a gradient type solver applied to solve this dynamic optimization task the performance function gradient with respect to ν needs to be calculated at each iteration of the solver. As the second and third terms of the performance function are explicitly expressed in terms of ν their gradients can be easily calculated. However, an analytical expression defining the mapping $q(\cdot, \alpha^i)$ is not available (see (6.13)). Utilizing the output equations (6.10) in the state space model we can express this mapping as (see (6.4))

$$\begin{aligned} q(\nu, \alpha^i) &= \sum_{t=0}^{N-1} Q(\nu(t), g(x(t), \nu(t), \alpha^i(t)), t) + Q_N(g(x(N), \alpha^i(N))) \\ x(t+1) &= f(x(t), \nu(t), \alpha^i(t)) \ \text{ for } \ t \in \overline{0; N-1}, \qquad x(0) \text{ given.} \end{aligned} \tag{6.25}$$

An efficient algorithm for calculating $q^{'}_{\nu}(\nu, \alpha^i)$ when treating state as the calculated variables is available, see e.g., (Sage, 1968; Lewis and Syrmos, 1995). Let us define Hamiltonian function at time instant $t \in \overline{0 : N-1}$:

$$\begin{aligned} H(x(t), \nu(t), \alpha^i(t), \mu(t), t) \triangleq & -Q(\nu(t), g(x(t), \nu(t), \alpha^i(t)), t) + \\ & +\mu(t)^T f(x(t), \nu(t), \alpha^i(t)) \end{aligned} \tag{6.26}$$

where $\mu(t) \in \mathbb{R}^{Nn_x}$.

Denote by $x(\nu, \alpha^i) = [x(1)(v, \alpha^i)^T, ..., x(t)(\nu, \alpha^i)^T, ..., x(N)(v, \alpha^i)^T]^T$ a state trajectory produced by the model state equations (6.9) with $c = \nu$ and $\alpha = \alpha^i$. The equations for the multipliers $\mu(t)$ are called *conjugate equations* and they read:

$$\mu(t-1) = H'_{x(t)}(x(t)(\nu,\alpha^i),\nu(t),\alpha^i,\mu(t),t)^T, \quad t\in\overline{1:N-1} \tag{6.27a}$$

$$\mu(N-1) = \\ -g'_{x(N)}(x(N)(\nu,\alpha^i),\alpha^i(N))^T Q'_{N,y(N)}(g(x(N)(\nu,\alpha^i),\alpha^i(N))^T. \tag{6.27b}$$

Finally, the derivative of the performance function $q(\cdot,\alpha^i)$ with respect to ν equals to:

$$q'_{\nu(t)}(\nu,\alpha^i) = -H'_{\nu(t)}(x(t)(\nu,\alpha^i),\nu(t),\alpha^i,\mu(t),t), \quad t\in\overline{1:N-1}$$
$$q'_{\nu}(\nu,\alpha^i) = [q'_{\nu(0)}(\nu,\alpha^i),..,q'_{\nu(t)}(\nu,\alpha^i),..,q'_{\nu(N-1)}(\nu,\alpha^i)]. \tag{6.28}$$

Hence, a procedure to calculate $q'_\nu(\cdot,\alpha^i)$ at a point ν starts with a calculation of the model state response trajectory $x(\nu,\alpha^i)$ by using (6.9). Next the multipliers over a batch are calculated backwards by using equations (6.27a) and starting from the "initial point" defined by (6.27b). Finally, the derivative is calculated by using formula (6.28) and by applying straightforward differentiations of explicit functions of $\nu(t)$. Notice that no iterations are required in order to calculate the derivative. With the performance function derivative available a suitable constrained nonlinear mathematical programming solver can be applied in order to solve DMMOP, see e.g., (Fletcher, 1987).

Regarding the expression (6.17) defining the modifier λ, the same technique as above can be applied to calculate the derivative $F'_c(c,\alpha)$.

Let us now consider DPEP to be solved in the **Step 2** of the algorithm (see (6.19)). As the parameters $\alpha(t)$, $t\in\overline{0:N}$ are independent it may seem that their estimation can be done independently by solving $N+1$ (for $t\in\overline{0;N}$) independent sets of equations. Clearly, it is not so as the state variables are involved in the definition of $F(\cdot,\cdot)$ and so the parameters are that appear in the state equations. Hence, a time structure in the state equations is conveyed into the time parameter structure of DPEP which is as follows:

$$F(t)(c^i,\alpha) = F_*(t)(c^i), \quad \text{for } t\in\overline{0:N} \tag{6.29}$$

where

$$F(c^i,\alpha) = [F(0)(c^i,\alpha)^T,...,F(N)(c^i,\alpha)^T]^T. \tag{6.30}$$

This means that in order to match the model and real process outputs at time instant t not only parameters $\alpha(t)$ at t are to be manipulated. It is a dynamic problem due to a time structure of the state equations (6.9) with parameters $\alpha(t)$ as the inputs. Hence, one of possible efficient solvers would be a dynamic optimization solver with suitably selected performance function to accelerate the convergence rate. For example, taking squares of the errors between the model and measured outputs and weighting the residua by the inverse variance measurement errors r_j, $j \in \mathbb{R}^{n_y}$ in order to differentiate between a quality of less and more noise corrupted measurements, yields well known least squares formulation:

$$\text{minimize}_\alpha \{\varphi(\alpha, c^i) \triangleq \frac{1}{2}\sum_{t=0}^{t=N}[F(t)(c^i,\alpha) - F_*(t)(c^i)]^T \times \\ \times R^{-1}[F(t)(c^i,\alpha) - F_*(t)(c^i)]\}$$

$$\text{subject to: } \alpha = [\alpha^T(0), ..., \alpha^T(N)]^T \in \mathbb{R}^{(N+1)n_\alpha} \subset A \qquad (6.31)$$

where $R = diag\{r_j\}_{j=1}^{j=n_y}$ and the set A accommodates a prior knowledge about the estimated parameters.

We shall now assume that the model output equation at $t = N$ is not parameterized. This will allow to apply the Hamiltonian function based technique presented above for the DMMOP to calculate a derivative of a function that is minimized in the least squares implicit formulation of DPEP (6.31). Hence, $g(x(N), \alpha(N)) = g(x(N))$ and as due to (6.10),(6.9)

$$F(N)(c^i, \alpha) = g(x(N)(c^i, \alpha(0), \alpha(1), ..., \alpha(N-1))) \qquad (6.32)$$

the model output at $t = N$ does not depend on the parameter $\alpha(N)$. As the state $x(N)$ at $t = N$ depends on $\alpha(0), \alpha(1), .., \alpha(N-1)$ it still should be possible for reasonably parameterized models to match the process output and model output responses to $c = c^i$ by manipulating the remaining parameter values.

The formulation (6.31) reads then as

$$\text{minimize}_{\tilde\alpha} \{\varphi(\tilde\alpha, c^i) \triangleq \frac{1}{2}\sum_{t=0}^{t=N-1}[F(t)(c^i,\tilde\alpha) - F_*(t)(c^i)]^T R^{-1}[F(t)(c^i,\tilde\alpha)+ \\ - F_*(t)(c^i)] + [g(x(N)) - F_*(N)(c^i)]^T R^{-1}[g(x(N)) - F_*(N)(c^i)]\}$$

$$\text{subject to: } \tilde\alpha = [\alpha^T(0), ..., \alpha^T(N-1)]^T \in \mathbb{R}^{Nn_\alpha} \subset \tilde{A}. \qquad (6.33)$$

The derivative $\varphi'(\tilde{\alpha}, c^i)$ can be calculated by using Hamiltonian functions for $t \in \overline{0; N-1}$:

$$H(x(t), c^i, \alpha(t), \mu(t), t) \triangleq -[F(t)(c^i, \alpha) - F_*(t)(c^i)]^T R^{-1} \times \\ \times [F(t)(c^i, \alpha) - F_*(t)(c^i)] + \mu(t)^T f(x(t), c^i(t), \alpha(t)) \quad (6.34)$$

with the conjugate equations

$$\begin{aligned} \mu(t-1) &= H'_{x(t)}(x(t)(c^i, \tilde{\alpha}), c^i(t), \alpha(t), \mu(t), t)^T, \quad t \in \overline{1 : N-1} \\ \mu(N-1) &= -g'_{x(N)}(x(N)(c^i, \tilde{\alpha}))^T R^{-1} [g(x(N)(c^i, \tilde{\alpha})) - F_*(N)(c^i)] \end{aligned} \quad (6.35)$$

as

$$\begin{aligned} \varphi'_{\alpha(t)}(\tilde{\alpha}, c^i) &= -H'_{\alpha(t)}(x(t)(c^i, \tilde{\alpha}), c^i, \alpha(t), \mu(t), t), \quad t \in \overline{0 : N-1} \\ \varphi'_{\alpha}(\tilde{\alpha}, c^i) &= [\varphi'_{\alpha(0)}(\tilde{\alpha}, c^i), .., \varphi'_{\alpha(t)}(\tilde{\alpha}, c^i), .., \varphi'_{\alpha(N-1)}(\tilde{\alpha}, c^i)]. \end{aligned} \quad (6.36)$$

Now, the damped Gauss-Newton solver, see (Dennis and Schnabel, 1983) for example, can be applied to solve the DPEP in its implicit form (6.29). The problem can be also approached by so called recursive least squares method that is suited for real time applications, see e.g., (Banyasz *et al.*, 1973; Ljung and Soderstrom, 1987). However, we would recommend non-recursive algorithm if computing time allows for it as the recursive algorithm convergence properties in a nonlinear case are much weaker than for the non-recursive one. A comprehensive coverage of model parameter estimation technologies can be found in (Walter and Pronzato, 1997), see also (Schweppe, 1973).

A time structure in DPEP can be removed by parameterizing only the output equations. The model will then read

$$x(t+1) = f(x(t), c(t)) \quad (6.37a)$$

$$y(t) = g(x(t), \alpha(t)) \quad \text{for } t \in \overline{0 : N}. \quad (6.37b)$$

The DPEP in this case can be solved as follows. First, the model state response to c^i is calculated by using (6.37a) to produce the model state trajectory $x(t)(c^i)$, $t \in \overline{0 : N}$. Next, the parameters $\alpha(t)$ are determined by solving $N+1$ independent sets of equations:

$$g(x(t)(c^i), c^i(t), \alpha(t)) = F_*(t)(c^i) \quad \text{for } t \in \overline{0 : N} \quad (6.38)$$

where $F_*(t)(c^i)$ is the real process output available from the measurements performed at t.

The task defined by (6.38) requires employing a solver of a set of non-linear algebraic equations and such solver can be found in, e.g., (Ortega and Rheinboldt, 1970). The DPEP can be further simplified if the output equations are additive in the parameters

$$y(t) = \widetilde{g}(x(t)) + \alpha(t) \qquad \text{for } t =\in \overline{0:N}. \tag{6.39}$$

A simple solution of the DPEP is then available as

$$\alpha(t) = F_*(t)(c^i) - \widetilde{g}(x(t)(c^i)) \qquad \text{for } t \in \overline{0:N} \tag{6.40}$$

The easiness of solving DPEP should not be a decisive factor in selecting the models for the DISOPE algorithm. Models with smaller modeling error and then leading to more complicated numerical tasks when solving DPEP may lead to smaller number of iterations of the set-points needed to reach the true process optimal solution. Clearly, not computing time but the transients forced in the process during the DISOPE iterations are of primary importance.

6.1.3.2 *Explicit implementation of DISOPE algorithm*

In this implementation the first term of the performance function to be minimized in Step 3 of the DISOPE algorithm is written as defined by (6.25). However, as opposed to the implicit implementation the state variables remain as the decision variables in the DMMOP described by (6.20). The model state equations will now not be used to calculate the state response to prescribed ν but will be introduced as constraints to the DMMOP. The explicit formulation of the DMMOP has then the following form:

$$\begin{aligned}
\text{minimize}_{\nu,x} \; \{ &\sum_{t=0}^{N-1} Q(\nu(t), g(x(t), \alpha^i(t)), t) + Q_N(g(x(N), \alpha^i(N))) + \\
&- \lambda(c^i, \alpha^i)^T \nu + \sum_{t=0}^{t=N-1} \rho(t) \parallel c^i(t) - \nu(t) \parallel^2 \} \\
\text{subject to} :& x(t+1) = f(x(t), \nu(t), \alpha^i(t)), \; t \in \overline{0; N-1}, \\
& x(0) \text{ given}, \quad \nu \in C \subset \mathbb{R}^{Nn_c}.
\end{aligned} \tag{6.41}$$

The solution is the optimal set-point prediction $\widehat{\nu}(c^i, \alpha^i)$ and the optimal state trajectory prediction $\widehat{x}(c^i, \alpha^i)$. It is a structural property of the *explicit structure* of the overall DISOPE algorithm that only $\widehat{\nu}(c^i, \alpha^i)$ is used in further iterations. Solving the DMMOP in its explicit form described by (6.41) can be done, as previously, by a nonlinear mathematical programming solver, see e.g., (Fletcher, 1987). The solvers that efficiently take into account the constraints in (6.41) due to the state equations can be found in (Lewis and Syrmos, 1995).

Comparing the dimension of the search spaces of the implicit and explicit formulations, it can be easily noticed that it is greater in the explicit case as in this case the state variables are also the decision variables. This increase of the DMMOP dimension is clearly a drawback of the explicit formulation. However, the constraints due to state equations in the explicit formulation are sparse. It is because a value of the state vector at any time instant enters only the adjacent state equations. There are solvers available such as TOMLAB, for example, that can very efficiently take advantage of the sparsity during the calculations. On the other hand, a drawback of the implicit formulation is that new, with regard to the explicit formulation, local minima can be introduced. This is nicely illustrated by an example of a dynamic optimization by (Brdyś and Ruszczynski, 1985). Hence, it may happen to be easier to find a global optimum using an explicit formulation and consequently save on a number of set-point iterations. This is particularly valid if the process dynamics is highly nonlinear.

Applying similar reasoning, an explicit approach to DPEP to be solved in **Step 2** of the DISOPE algorithm leads to a task of solving the following set of equations:

$$g(x(t), c^i(t), \alpha(t)) = F_*(t)(c^i) \quad \text{for} \quad t \in \overline{0:N} \tag{6.42a}$$

$$x(t+1) = f(x(t), c^i(t), \alpha(t)) \quad \text{for} \quad t \in \overline{0:N-1}. \tag{6.42b}$$

The above set of equations is solved simultaneously with respect to states $x(t)$ and parameters $\alpha(t)$. It is a dynamic problem due to time structure of the state equations (6.42b). Hence, one of possible efficient solvers would be a dynamic optimization solver with suitably selected performance function to accelerate the convergence rate and to filter out the measurement errors. Following a least squares approach applied in subsection 6.1.3.1 yields:

$$
\begin{aligned}
\text{minimize}_{\alpha,x} \ & \sum_{t=0}^{t=N}[g(x(t),\alpha(t)) - F_*(t)(c^i)]^T R^{-1}[F(t)(c^i,\alpha) - F_*(t)(c^i)] \\
\text{subject to: } & x(t+1) = f(x(t), c^i(t), \alpha(t)), \ t \in \overline{0:N-1}, \ x(0) \text{ given} \\
& \alpha = [\alpha^T(0), \ldots\ldots, \alpha^T(N)]^T \in \mathbb{R}^{Nn_\alpha} \subset A \\
& x = [x(1)^T, x(2)^T, \ldots\ldots, x(N)^T]^T \in \mathbb{R}^{Nn_x}.
\end{aligned} \tag{6.43}
$$

As in the implicit case structuring of the state and output equation models allows more efficient formulation of the explicit DPEP.

Uncertainty in the output measurements has been treated by simple weighting of the residues in the least squares performance index. This can be insufficient for compensating impact of the measurement noise on the parameter estimates. The consequences are that the model output generated by the state space model of the process with the parameter estimates will differ from the true and unknown process output. In other words, the equality constraint condition in (6.12) will be satisfied for the measured outputs but not for the true process output values. As this condition is vital for the ISOPE algorithm optimality, the resulting set-points provided by the optimizing control algorithm once the convergence has occurred will only be suboptimal. In an extreme case the algorithm convergence may be lost and the closed-loop optimizing control system my become unstable. In order to guarantee the stability, to assess the loss on optimality and/or to reduce it, a more sophisticated approach to the parameter estimation problem is needed. As the models used within the ISOPE technology are not, as usual, parametric but point-parametric the problem is very new. It has recently received much attention and at present, parameter estimation in point-parametric models for monitoring and control purposes is a rapidly growing area in the system identification, see e.g., (Brdyś and Chang, 2002; Chang *et al.*, 2003; Chang *et al.*, 2004).

6.2 ISOPE for Traveling Load Furnace Control: A Case Study

6.2.1 *Introduction*

The pilot-scale traveling load furnace in the Control Engineering Centre of City University is shown in Fig. 6.1. An implicit implementation of the

ISOPE dynamic optimizing control algorithm (DISOPE), is now applied to the furnace and this tests the main features of the algorithm (Stevenson *et al.*, 1985; Stevenson, 1985).

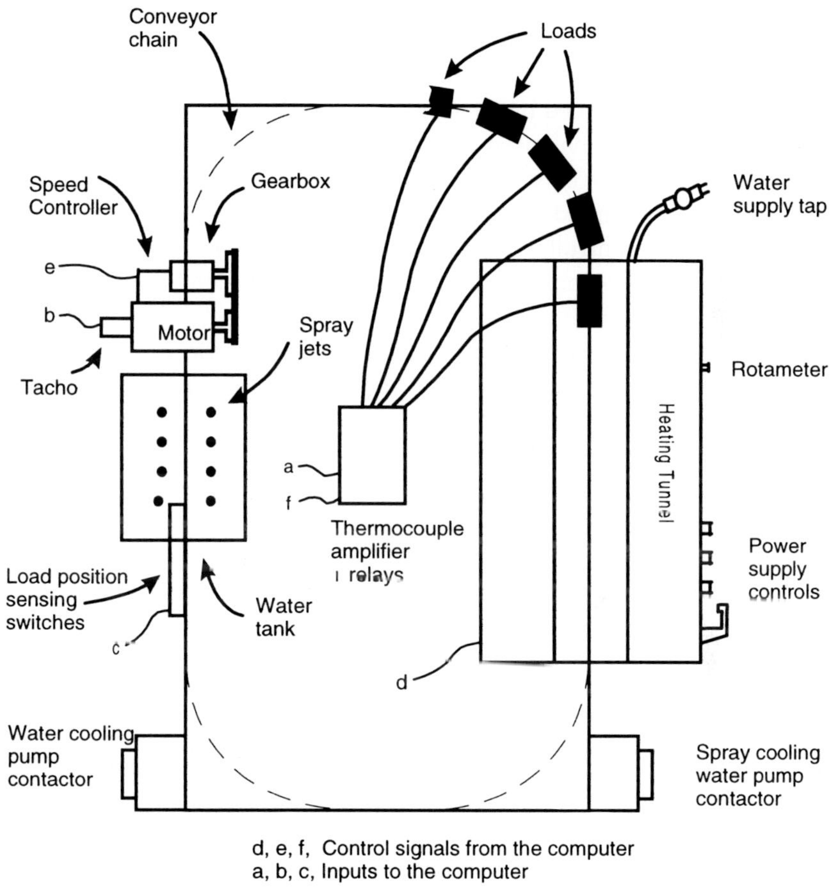

Fig. 6.1 A diagram of the furnace (cooling tunnel covers not shown).

6.2.2 Description of the process, measurement instrumentation and control architecture and hardware

Traveling load furnaces are widely used in industry, for example in steel reheating, baking and drying. The pilot-scale furnace has a heating tunnel composed of eight separately controllable radiant heating zones through which are propelled the loads to be heated. There are thirty aluminium loads which measure $8 \times 8 \times 4$ inches. They have been painted with black heat-resistant paint to enable them to absorb more of the incident radiation. Each load is attached by means of a hanger to a conveyor chain which is driven by a motor the speed of which is controlled using a phase controller. The phase controller derives its set-point from a digital to analogue converter (DAC) mounted in the furnace interface rack with the DACs used for the zone power controllers. A measurement of the motor speed, thus the conveyor speed, is available from a tachometer. The conveyor is normally operated at a constant speed.

Heating in each zone is accomplished by six electric radiant heaters, wired in parallel, three on each side of the heating tunnel, which are rated at 1kW each. The power applied in each zone is regulated using a phase controller which is situated in the main furnace casing near the furnace exit. Set-points for the phase controllers are obtained from a local control computer via DACs in the interface rack.

The DACs allow a digital input in the range 0 to 255, and are multiplexed on to a common data line. Thus to set a zone power or the motor speed the data must first be applied and then the multiplexer must be instructed to apply a strobe pulse to the appropriate DAC.

Load temperatures are measured by thermocouples attached to each load. Each thermocouple is connected to an interface box which is located centrally within the conveyor loop. This box contains filters and amplifiers for the thermocouple electromagnetic force, a cold junction compensation circuit (a means of measuring the ambient temperature), and reed relays which are used to select the thermocouple which is to be connected to the amplifier. The analogue to digital converter (ADC) in the local computer requires input signals in the range 0 to 10 volts, hence the thermocouple is amplified and the tachometer output is attenuated before being applied to the converter.

In total there are thirty loads, of which eight are being heated at any time. Once the loads have left the heating tunnel they are propelled through a water spray-cooling tunnel before being returned for heating again. Load position within the heating tunnel is determined by six microswitches which are attached to the conveyor frame and are operated sequentially by bolts attached to the load supports as the loads move along through a zone.

The local control computer uses the foreground/background monitor (the single job monitor is unsuitable because of the need to queue interrupt requests). This computer forms part of the lower level of the distributed hierarchical computer network in the control laboratory. It is linked to the higher level computer using parallel optically isolated link. The higher level computer has a multi-user time shared operating system which provides an operating environment allowing four time simultaneous real time users. Peripherals available to this computer include intelligent colour graphics terminal which may be used to display measurements from the furnace and controls applied to it on a minic diagram in real time. A graphical terminal may also be used to display the current furnace temperature profile. Results obtained during an experiment may be transferred to a Prime minicomputer for analysis and plotting.

In order to test the system controllability and also to design an infrastructure for data gathering for the modeling purposes a decentralized PID based controllers were designed and implemented by (Sheena, 1977). There are eight loads in the furnace. A temperature to be reached by a selected slab at each zone is prescribed. The PID control objective is to adjust the zone powers so that the prescribed temperature profile along the furnace is followed by the slab as it travels through the furnace. A typical result obtained by (Sheena, 1977) is given in Fig. 6.2. The power is scaled 0 to 255 which represents an actual power range of 0 to 6 kW. The other notation denotes the following: S_p – motor speed (255 represents a full speed), T_s – sampling interval in seconds and the PID controller parameters K – proportional gain, T_i – integral action time in seconds, and T_d – derivative action time in seconds. The zone set-points are represented by dashed lines.

It can be seen that successful temperature profile tracking can be achieved by the traveling slab. This PID decentralized controller with a separated PID control loop for each of the heating zones would be used to maintain a prescribed load temperature at the furnace exit and this is a major control objective. However, there are variety of temperature profiles in the zones, following which can be achieved. Hence, the least energy cost profile is wanted. This gives rise to an optimizing control of the furnace.

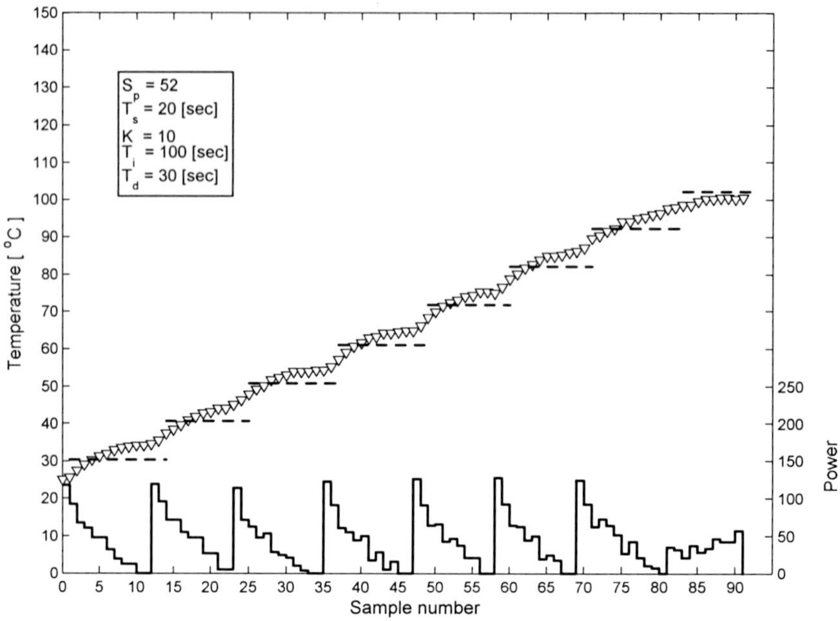

Fig. 6.2 PID control result.

The DISOPE is to be applied in order to derive the optimizing control algorithm. We shall start with defining a batch for the furnace.

6.2.3 *Defining the batch for the pilot-scale traveling load furnace*

The actual furnace zones are now linked to simulate a steel reheat furnace as is illustrated in Fig. 6.3. The eight heating zones are integrated into the three zones that are the soak, heat and preheat zones. The control inputs are three power set-points for these three zones: c_1, c_2 and c_3.

Fig. 6.3 shows the position of the eight loads defining a batch within the furnace at the start of the batch. The time horizon defining the batch is taken for the last load in the batch (at the furnace entrance) to travel through the furnace. Before the control algorithm could be developed it was necessary to verify that this batch satisfied the batch conditions required for batch type control that, for each batch, given the same initial

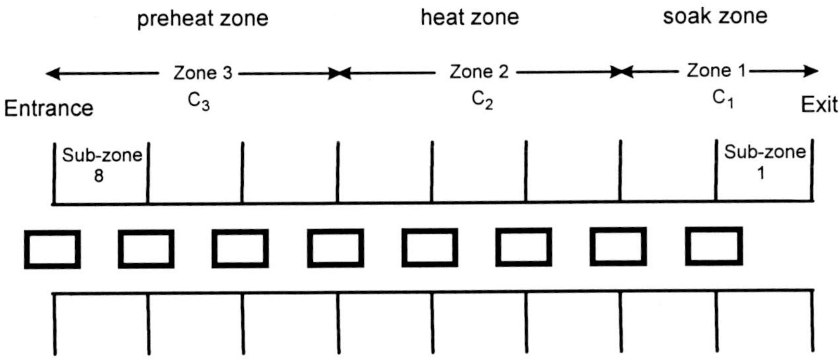

Fig. 6.3 Load locations at the start of a batch.

load temperature and set-points a nearly identical performance could be obtained.

An experiment was performed to verify this as follows. Firstly, a time-varying control was applied for two batches to ensure that the initial load temperature for the second batch had stabilized to the value determined by this control. A measurement of the performance defined as the sum of the temperatures of the load at the exit of the furnace during the batch squared was calculated for the second batch. Secondly, a different time varying control was applied for the next two batches and the performance was calculated for the second of these two batches as before. The first control was then reapplied for the next two batches, then the second control for two batches, *etc.* A comparison was then made between the performances calculated for the same control. It was found that, if the water spray cooling was not used, the temperature of the loads entering the furnace was not constant, i.e., they had not cooled sufficiently whilst moving from the exit of the furnace back to its entrance giving rise to unacceptable change in performance. When water cooling was used the maximum variation in the temperature between batches subjected to the same control was found to be 4% which was thought to be acceptable. The batch time horizon or control horizon depends on the traveling speed that is constant over time and sufficiently long to allow the iterative optimizing control algorithm to reach an optimal solution and to benefit from using it. If a disturbance profile over a batch changes then the algorithm will adjust the controls to find a new optimal sequence of controls over a batch (new optimal batch).

6.2.4 *Mathematical model of the batch and the parameter estimation problem*

As the slabs along the furnace form a distributed in space system, a temperature profile along the furnace over a batch is described by a highly nonlinear partial differential equation. With the measurements available it was found to be impossible to identify such model dynamics. As the ISOPE technique does not require perfect model of a process an approximation of the temperature profile by the temperatures of individual traveling loads can be attempted. There are then eight state variables being the load temperatures and three control inputs. A traveling load j temperature dynamics is described by a linear discrete time equation

$$
\begin{aligned}
T_j(t+1) &= \Phi(t)x(t) + \Gamma(t)c(t) \qquad (6.44)\\
\text{where} \quad x(t) &= [T_1(t), T_2(t), ...T_8(t)]^T \\
c(t) &= [c_1(t), c_2(t), c_3(t)]^T, \quad j \in \overline{1:8}
\end{aligned}
$$

where t denotes sampling time instant and the matrices $\Phi(t), \Gamma(t)$ are composed of the time-varying parameters to be estimated from the temperature and zone power measurements; $c(t)$ denotes the power supplied at t.

A temperature of the load at the furnace exit at instant t is the system output temperature $T_e(t)$ at the instant t. Hence, the model output equation can be written by identifying at t the load at the exit and then allocating its temperature to the output variable $T_e(t)$. It all sounds reasonable. Let us then investigate if the model (6.44) would be suitable to act as parametric model of the furnace. This depends on how good the linear model structure is to approximate the measurements by manipulating the model time-varying parameters that are not input dependent. There is considerable interaction between adjacent heating zones, i.e., a set-point applied to a zone can affect the temperature of the loads in the next few adjacent zones significantly, see (Bailey, 1980). Hence, the matrices $\Phi(t), \Gamma(t)$ need to be considered full. Hence, there are $8 \times 8 + 3 \times 8 = 84$ parameters to be estimated at a given sampling instant. In the light of 11 instantaneous measurements available, it means that there is no sensible way of estimating the parameter values on a routine basis. The estimation problem is heavily overparametrized. Hence, although by applying a method of trial and error sensible match of the measurement data and the model response can be achieved the model is not suitable for any control application. Several nonlinear models were tried such as Hammerstein model for example, see e.g., (Banyasz *et al.*, 1973) in order to obtain more parsimonious parameterization of the model which improved the situation but still did not satisfactorily solve the problem.

As the ISOPE technology needs only a point-parametric model then with the model parameter being input dependent the number of parameters can be vastly reduced. For example, with the model structure given by (6.37a) and (6.39) there are only eight time-varying parameters needed in order to obtain the required point-parametric model. However, we will utilize some physical and experimental knowledge in order to further reduce the parameter estimation burden when solving DPEP. Such an approach is known as a grey-box modeling and and it was successfully applied to numerous case-studies including recent application to modeling of a biological reactor in a wastewater system for optimizing robust model predictive control (Rutkowski *et al.*, 2004; Kim *et al.*, 2001). Earlier research by (Caffin, 1972) established that a model of the furnace would probably contain nonlinear terms in the load temperature. Hence, the following nonlinear models of a particular load j temperature throughout its journey through the furnace were tried:

$$T_j(t+1) = a_j T_j(t) + b_j T_j^2(t) + d_j \widetilde{c}_j(t) \tag{6.45a}$$

$$T_j(t+1) = a_j T_j(t) + b_j T_j^2(t) + d_j T_j^3(t) + e_j \widetilde{c}_j(t) \tag{6.45b}$$

where a_j, b_j, d_j, e_j are constant parameters and $\widetilde{c}_j(t)$ is a power applied to the load j at t.

The power computer units range is $0 : 255$. Notice that $\widetilde{c}_j(k)$ is determined by the load position at t and the power applied at t at this position. It should be pointed out that although the model parameters are constant in time they are input dependent.

The models were tested and data collected during a furnace run under PID control described in Subsection 6.2.2. A least squares method described in Subsection 6.1.3.1 was applied. As the parameters are constant and the state is measured, a purely static LS formulation was derived and solved with respect to the parameters by employing a simple linear solver of a set of resulting linear algebraic equations to produce:

$$\begin{aligned} a_j &= 1.014 \\ b_j &= 0.738 \times 10^{-2} \\ d_j &= -0.1055 \times 10^{-3}. \end{aligned} \tag{6.46}$$

From the experimental results the choice of model described by (6.45a) was found most appropriate. A result obtained using this model is given graphically in Fig. 6.4.

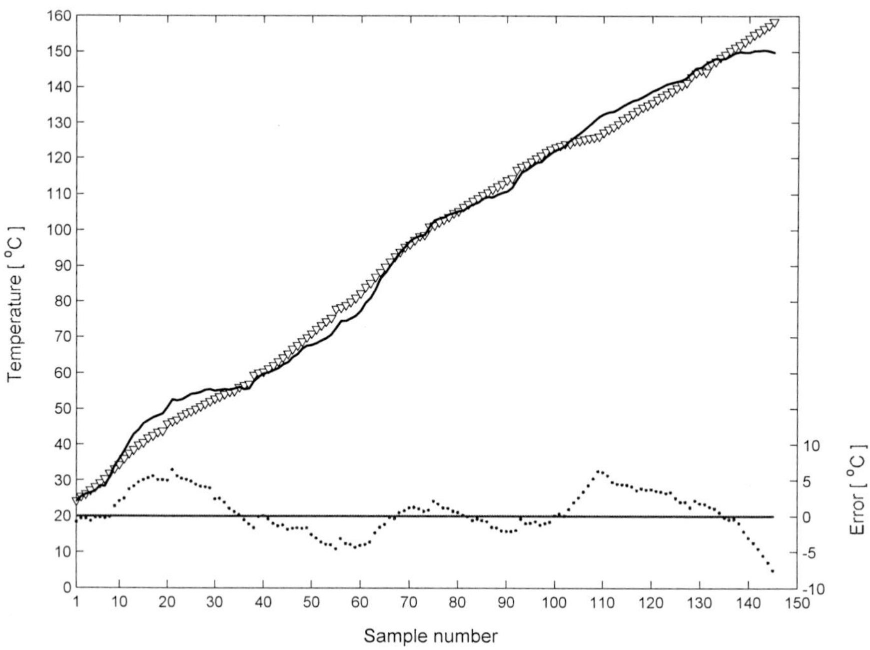

Fig. 6.4 Graph of actual and predicted load temperature.

In this graph the model output is depicted using symbols and the actual load temperature by the solid line. It can be seen that a good fit is obtained during the early part of the load progress through the furnace. As the furnace exit is approached significant extra non-linearities caused by the furnace end doors become significant. If the LS weighting factors were introduced into the estimation to compensate for this error then a better performance may be obtained at the end of the furnace but only at the expense of a poorer fit at the start.

Another method of overcoming this problem is to make the parameters time-varying. That is to obtain a separate set of parameters for the last two zones. However, the model complexity then increases. The final approach is to use a scaling factor to correct the model output at the furnace exit.

To obtain a model of the temperature of the load at the exit of the furnace, that is the process output variable T_e model, it is necessary to suitably combine the temperature models for each load in the batch (see (6.45a)). Since the resulting output model will be used for optimization it is important to ensure that it and its derivatives with respect to the variables involved are continuous. As the load moves form one zone to the next, as it travels along the heating tunnel, the power applied to it is a combination of the power applied in the zones it is just leaving and that applied in the zone it is entering. This may be reflected in the model by fuzzification of the load position and defining the power applied to the load as a weighted some of the powers applied to the zones with the weights being the grades of memberships of the load positions at each of the zones (Yen and Langari, 1999). This yields:

$$\widetilde{c}_j(t) = \psi_1(x)c_1(t) + \psi_2(x)c_2(t) + \psi_3(x)c_3(t) \tag{6.47}$$

where $\psi_i(x)$, $i \in \overline{1:3}$ are the load position membership functions related to the load positions in the soak, preheat and heat zones; x denotes the load distance at t in the furnace.

Each sub-zone is of 35 cm long. Hence the preheat zone, heat zone and soak zone are of 105 cm, 105 cm and 70 cm long, respectively. The membership functions are defined as:

$$\begin{aligned}
\psi_1(x) &= \frac{1}{1+\exp(K_\psi(210-x))} \\
\psi_2(x) &= \frac{1}{1+\exp(K_\psi(x-210))+\exp(-105K_\psi)+\exp(K_\psi(105-x))} \\
\psi_3(x) &= \frac{1}{1+\exp(K_\psi(x-105))}
\end{aligned} \tag{6.48}$$

where K_ψ was chosen as 40/175.

The functions chosen are illustrated in Fig. 6.5. They overlap and are continuously differentiable thus ensuring a smooth transition.

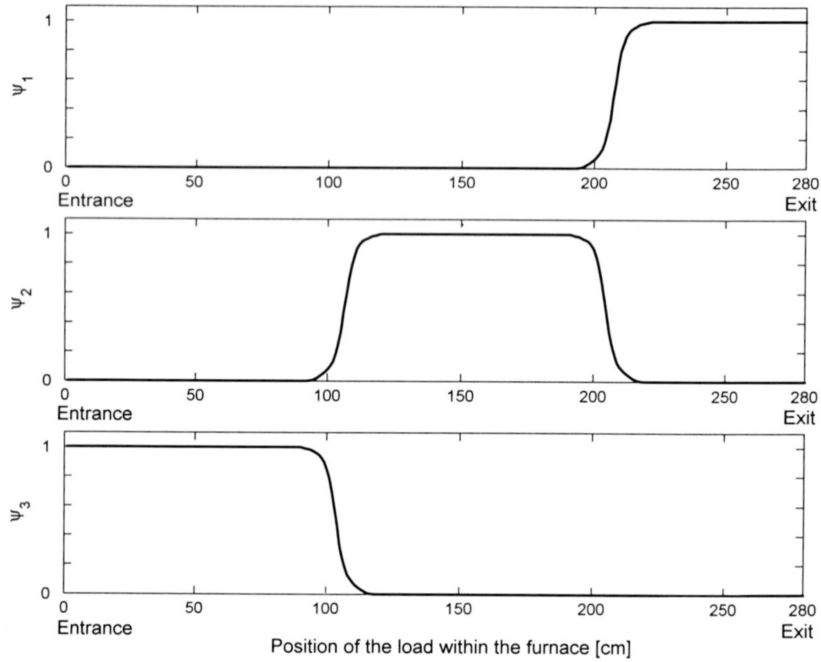

Fig. 6.5 Load position membership functions for smoothing the power applied to a load travelling throughout the furnace.

A similar approach is also required to switch between loads as one leaves the furnace and the next enters the exit zone in order to derive a continuously differentiable output model. Based on the sub-zone length and the load length the following load position membership functions are defined:

$$\begin{aligned}
\varsigma_1(\widetilde{x}) &= \frac{1}{1+\exp(K_\varsigma(245-x_p))} \\
\varsigma_2(\widetilde{x}) &= \frac{1}{1+\exp(K_\varsigma(210-x_p))+\exp(K_\varsigma(x_p-245))+\exp(-35K_\varsigma)} \\
\varsigma_3(\widetilde{x}) &= \frac{1}{1+\exp(K_\varsigma(175-x_p))+\exp(K_\varsigma(x_p-210))+\exp(-35K_\varsigma)} \\
\varsigma_4(\widetilde{x}) &= \frac{1}{1+\exp(K_\varsigma(140-x_p))+\exp(K_\varsigma(x_p-175))+\exp(-35K_\varsigma)} \\
\varsigma_5(\widetilde{x}) &= \frac{1}{1+\exp(K_\varsigma(105-x_p))+\exp(K_\varsigma(x_p-105))+\exp(-35K_\varsigma)}
\end{aligned}$$

$$
\begin{aligned}
\varsigma_6(\widetilde{x}) &= \frac{1}{1+\exp(K_\varsigma(70-x_p))+\exp(K_\varsigma(x_p-105))+\exp(-35K_\varsigma)} \\
\varsigma_7(\widetilde{x}) &= \frac{1}{1+\exp(K_\varsigma(35-x_p))+\exp(K_\varsigma(x_p-70))+\exp(-35K_\varsigma)} \\
\varsigma_8(\widetilde{x}) &= \frac{1}{1+\exp(K_\varsigma(x_p-35))}
\end{aligned}
\tag{6.49}
$$

where

$$
\begin{aligned}
&x_p = 280 - \widetilde{x} \\
&\widetilde{x} \;-\; \text{position of the last load in the batch in } cm \\
&K_\varsigma = 60/175
\end{aligned}
\tag{6.50}
$$

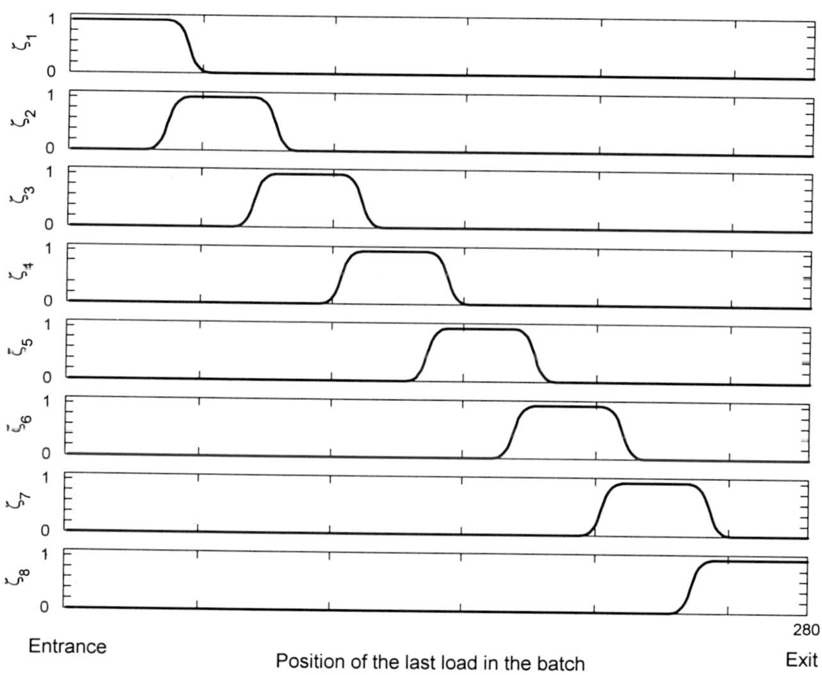

Fig. 6.6 Load position membership functions for smoothing the output temperature at the furnace exit.

The functions chosen are illustrated in Fig. 6.6. They overlap and are continuously differentiable thus ensuring a smooth output model.

The output model reads:

$$T_e(t) = \sum_{j=1}^{i=8} \varsigma_j(\widetilde{x}) T_j(t) \tag{6.51}$$

where $T_j(t)$, $j \in \overline{1:8}$ is described by (6.45a) and $\widetilde{c}(t)$ is determined by (6.47) with the smoothing functions $\psi_i(x)$, $i \in \overline{1:3}$ described by (6.48).

Since the functions provide a gradual changeover in the load used to determine T_e it is necessary to calculate the temperature of a load for a few samples after it has technically left the furnace (in reality it is going through the exit doors).

Although both sets of the membership functions are functions of distance, it is possible, since the loads are propelled through the furnace at a constant speed, to replace the distances x in (6.48) and $\widetilde{x}$ in (6.49) with time and hence sampling instant. Once it has been done the membership functions become the functions of the sampling instant as

$$\widetilde{\psi}_{j,i}(k) \quad \text{for} \quad j \in \overline{1:8} \quad \text{and} \quad i \in \overline{1:3} \tag{6.52a}$$

$$\widetilde{\varsigma}_j(k) \quad \text{for} \quad j \in \overline{1:8}. \tag{6.52b}$$

The complete furnace model becomes:

$$T(t+1) = AT(t) + BT^2(t) + D\widetilde{\psi}(t)c(t) \tag{6.53a}$$

$$T_e(t) = \widetilde{\varsigma}(t)T(t) \tag{6.53b}$$

where

$$\begin{aligned} T(t) &= [T_1(t), \dots\dots, T_8(t)]^T \\ T^2(t) &= [T_1^2(t), \dots\dots, T_8^2(t)]^T \\ c(t) &= [c_1(t), c_2(t), c_3(t)]^T. \end{aligned} \tag{6.54}$$

The equations (6.53a) represent the furnace state equations while the equation (6.53b) is the furnace output equation. The two equations together represent the furnace state space model.

The diagonal matrices A, B and D are composed of the furnace model constant parameters (see (6.45a) as

$$\begin{aligned} A &= diag\{a_j\}_{j\in\overline{1:8}} \\ B &= diag\{b_j\}_{j\in\overline{1:8}} \\ D &= diag\{d_j\}_{j\in\overline{1:8}}. \end{aligned} \tag{6.55}$$

The triple (a_j, b_j, d_j), $j \in \overline{1:8}$, constitutes the parameter set for the load j.

The control input, state, output and parameter space dimensions are (see (6.3a), (6.3b), (6.3c) and (6.8)): $n_c = 3$, $n_x = 8$, $n_y = 1$ and $n_\alpha = 24$, respectively. However, the overall dimensions of the control input, state, output and parameter vectors: Nn_c, Nn_x, $(N+1)n_y$ and $(N+1)n_\alpha$, respectively, depend on the load traveling speed and sampling rate that determine a number of samples $N+1$ over a batch. Typically, $N+1 = 90$ giving 267, 712, 90 and 24 corresponding signals over a batch (the parameters are constant over a batch).

The matrices $\widetilde{\psi}(t)$ and $\widetilde{\varsigma}(t)$ read

$$\widetilde{\psi}(t) = \begin{bmatrix} \widetilde{\psi}_{1,1}(t) & \widetilde{\psi}_{1,2}(t) & \widetilde{\psi}_{1,3}(t) \\ \widetilde{\psi}_{2,1}(t) & \widetilde{\psi}_{2,2}(t) & \widetilde{\psi}_{2,1}(t) \\ \widetilde{\psi}_{3,1}(t) & \widetilde{\psi}_{3,2}(t) & \widetilde{\psi}_{3,1}(t) \\ \widetilde{\psi}_{4,1}(t) & \widetilde{\psi}_{4,2}(t) & \widetilde{\psi}_{4,1}(t) \\ \widetilde{\psi}_{5,1}(t) & \widetilde{\psi}_{5,2}(t) & \widetilde{\psi}_{5,1}(t) \\ \widetilde{\psi}_{6,1}(t) & \widetilde{\psi}_{6,2}(t) & \widetilde{\psi}_{6,1}(t) \\ \widetilde{\psi}_{7,1}(t) & \widetilde{\psi}_{7,2}(t) & \widetilde{\psi}_{7,1}(t) \\ \widetilde{\psi}_{8,1}(t) & \widetilde{\psi}_{8,2}(t) & \widetilde{\psi}_{8,1}(t) \end{bmatrix} \tag{6.56}$$

$$\widetilde{\varsigma}(t) = \begin{bmatrix} \widetilde{\varsigma}_1(t) & \widetilde{\varsigma}_2(t) & \widetilde{\varsigma}_3(t) & \widetilde{\varsigma}_4(t) & \widetilde{\varsigma}_5(t) & \widetilde{\varsigma}_6(t) & \widetilde{\varsigma}_7(t) & \widetilde{\varsigma}_8(t) \end{bmatrix}. \tag{6.57}$$

Notice that raw j of the matrix $\widetilde{\psi}(t)$ contains the load j membership functions at sampling instant t.

An important feature of constructing the model in this manner is that the elements of the parameter matrices for a particular load in the batch can be determined as soon as the load has left the furnace, since there is no coupling between the state equations through the state variables. The couplings that exist in reality are accommodated into the parameters. For the load which is nearest to the furnace exit at the start of the batch the parameters will be obtained using approximately one eighth of the measurements used for the load at the furnace entrance. Since the parameter estimation algorithm gives an equal weight to all the data points the resulting error in temperature prediction, mentioned already, is less accurate for the first few loads to leave the furnace and it is only necessary to correct the model for the last two loads. This is most easily achieved by scaling the appropriate membership functions. A typical graph of model (dashed line) and actual output (solid line) temperatures is shown in Fig. 6.7. The model outputs for the eight loads in the batch are illustrated in Fig. 6.8.

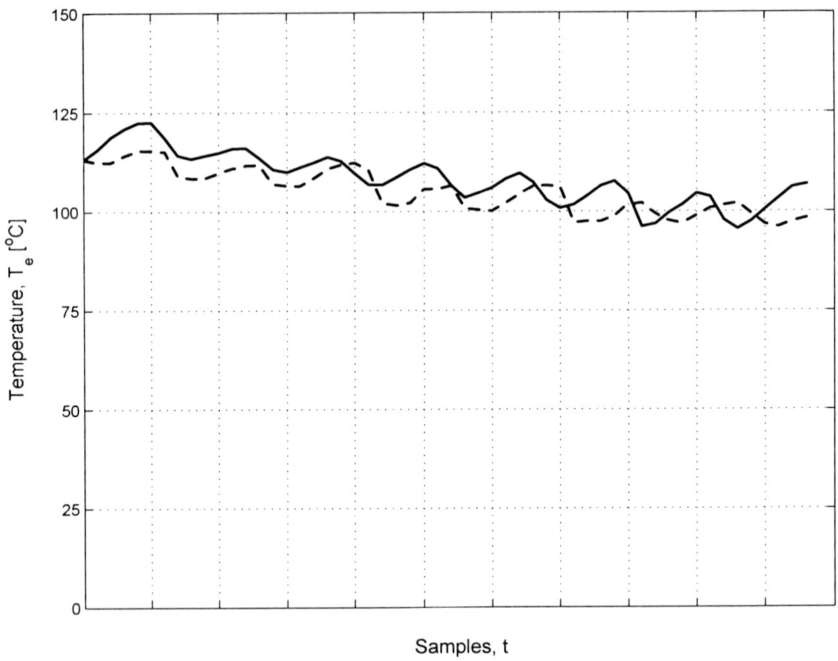

Fig. 6.7 Typical model and actual output temperatures for a batch.

It can be seen that a reasonable agreement between the predicted and and actual temperature is obtained. The apparent shift between the model and actual outputs is caused by the uneven spacing between the loads which is ignored in the model, where the loads are assumed to be spaced an equal distance apart.

Let us now consider DPEP (see (6.19)). As it has been pointed out already the DPEP breaks down into eight independent parameter estima-

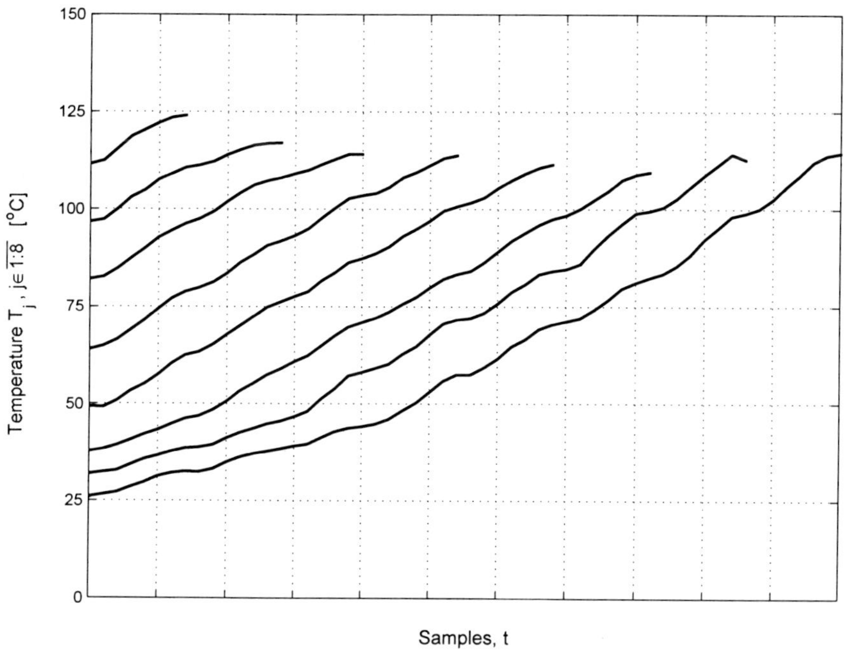

Fig. 6.8 The temperatures of the eight loads in the batch.

tion problems associated with individual loads. Solving such problem for the load j in the batch yields the parameters a_j, b_j, d_j (see (6.45a)). The problem is solved immediately after the load j has left the furnace as all the measurement data are then available, since there are no interactions in the state equations model through the state variables. Hence, the overall DPEP is solved at once but in real time by sequentially solving the individual load parameter estimation tasks for the loads in the batch that have just left the furnace. As all the state variables are measured each load based DPEP is not a dynamic but a static problem that is efficiently solved by standard LS technique employing standard linear optimization solver, since the model is linearly parameterized. Summarizing, a very handy point-parametric mathematical model of the furnace has been derived for DISOPE purposes that allows for an application of extremely efficient parameter estimation algorithm.

Clearly, any open-loop optimizing control technique would require much more complicated parametric model of the furnace, as there would be no possibility for the control algorithm to adapt the model during controlling the system. In the light of physical complexity of the furnace process and limitations in the measurements available it is unlikely that an accurate enough model would be developed in order to successfully apply an open-loop optimizing control to the furnace.

6.2.5 *Performance function*

To apply any optimizing control algorithm a performance function must first be selected. A basic operational objective of the furnace system is to heat the incoming loads to the furnace so that they can reach a prescribed temperature at the furnace exit. These is achieved by proper heating at the three zones when the loads travel throughout the furnace. However, it is essential that the associated energy cost is minimized and therefore, the furnace operates in an economically optimal manner. A total energy cost over a batch is a sum of the energy costs due to heating at the soak, heat and preheat zones. These costs are proportional to the power used when heating in these zones over a batch. A deviation of the temperature of the load at the furnace exit from its prescribed set-point over a batch is inevitable and it depends on the power used. However, reducing this deviation to a very small may vastly increase the energy cost. Hence, the energy cost needs to be compromised with a desired exit temperature tracking accuracy. A natural choice of the performance function is then as (see (6.4)):

$$Q(c, T_e) = \sum_{k=0}^{N-1} \{\gamma_1 (T_e(t) - T^{ref})^2 + \sum_{i=1}^{i=3} c_i(t)\} + \gamma_2 (T_e(N) - T^{ref})^2 \tag{6.58}$$

where a number of samples over a batch equals $N + 1$, T^{ref} is a desired exit temperature set-point $[{}^0C]$, γ_1 and γ_2 are the factors weighting the deviation of the temperature of the load at the furnace exit from its set-point, c_i is a power set-point at zone i.

The weighting factors are suitably selected so that a required compromise between the desired temperature set-point tracking accuracy and the associated energy cost can be achieved.

With the furnace model described by (6.53a), (6.53b), (6.54)) and with the performance function defined by (6.58), solving DMMOP (see (6.20)) in the **Step 3** of the ith iteration of the DISOPE algorithm can be approached, given controls c^i over a batch, $c^i(t) = [c_1^i(t), c_2^i(t), c_3^i(t)]$, $t \in \overline{0:N-1}$ and parameters $\alpha^i = (a_j^i, b_j^i, d_j^i)$, $j \in \overline{1:8}$.

6.2.6 *Dynamic modified model-based optimization problem*

There are constraints on the control variables due to limited capacity of the heaters. The set $C(t)$ in (6.6c) is defined by the inequalities:

$$0 \leq c_i(t) \leq 255 \quad \text{for } t \in \overline{0:N-1} \quad \text{and } i \in \overline{1:3}. \tag{6.59}$$

There are natural constraints on the state variables and the output variable defining set Y in (6.6e):

$$0 \leq T_j(t) \leq T^{\max} \quad \text{for } t \in \overline{0:N} \quad \text{and } j \in \overline{1:8} \tag{6.60}$$

$$0 \leq T_e(t) \leq T_e^{\max} \quad \text{for } t \in \overline{0:N}. \tag{6.61}$$

The model response always yields the load and exit temperatures non-negative, hence the lower bound constraints (6.60) and (6.61) can be omitted in a formulation of (6.20) while the upper bound constraints are not active. It means that the constraints (6.60) and (6.61) are always satisfied and they can be omitted. Hence, only the control constraints (6.59) need to be considered.

Regarding the dimension of the control space, it is rather high. For typical traveling speed and sampling rate, ninety control signals for each zone over a batch are obtained giving in total two hundred seventy control signals over a batch. Hence, $Nn_c = 270$ (see (6.3b)).

Let us now consider an implicit approach to solving the DMMOP (6.20). A key issue is to calculate the derivative $q'_\nu(\nu, \alpha^i)$ (see (6.25)), where

$$q(\nu, \alpha^i) = \sum_{t=0}^{N-1} \{\gamma_1[\varsigma(t)T(t) - T^{ref}]^2 + \sum_{i=1}^{i=3} \nu_i(t)\} + \gamma_2[\varsigma(N)T(N) - T^{ref}]^2. \tag{6.62}$$

The Hamiltonian function for the furnace reads (see (6.58), (6.53a), (6.53b) and (6.26)):

$$H(T(t), \nu(t), \alpha^i(t), \mu(t), t) = -\gamma_1[\varsigma(t)T(t) - T^{ref}]^2 - \sum_{i=1}^{i=3} \nu_i(t) + \\ +\mu(t)^T[A^iT(t) + B^iT^2(t) + D^i\widetilde{\psi}(t)\nu(t)]. \tag{6.63}$$

Applying (6.27a) and (6.27b) gives the conjugate equations and the boundary conditions as:

$$\begin{aligned}\mu(t-1) &= H'_{T(t)}(T(t),\nu(t),\alpha^i(t),\mu(t),t) \\ &= -2\gamma_1[\widetilde{\varsigma}^i(t)T(t)-T^{ref}]\widetilde{\varsigma}^i(t)+\mu(t)^T[A^i+diag\{2b^i_j\}_{i\in\overline{1:8}} \\ &\qquad \text{for } t\in\overline{1:N-1} \qquad (6.64a)\\ \mu(N-1) &= -2\gamma_2[\widetilde{\varsigma}^i(N)T(N)-T^{ref}]\widetilde{\varsigma}^i(N) \qquad (6.64b)\end{aligned}$$

where $T(t)$ is a vector of the load temperatures over a batch calculated from the model (6.53a) for $c(t)=c^i(t)$ and $A=A^i, B=B^i, D=D^i$. The conjugate equations (6.64) are easily solved by a backward enumeration starting from the final value $\mu(N-1)=-2\gamma_2[\widetilde{\varsigma}^i(N)T(N)-T^{ref}]\widetilde{\varsigma}^i(N)$.

Finally, applying the formula (6.28) (see also (6.62)), the derivative of the mapping $q(\cdot,\alpha^i)$ with respect to ν reads:

$$\begin{aligned}q'_{\nu(t)}(\nu,\alpha^i) &= -H'_{v(t)}(T(t),\nu(t),\alpha^i(t),\mu(t),t)\\ &= [1,1,1]-\mu(t)^TD^i\widetilde{\psi}(t) \quad \text{for } t\in\overline{0:N-1} \qquad (6.65)\end{aligned}$$

and

$$\begin{aligned}q'_{\nu}(\nu,\alpha^i) &= [q'_{\nu(0)}(\nu,\alpha^i),..,q'_{\nu(t)}(\nu,\alpha^i),..,q'_{\nu(N-1)}(\nu,\alpha^i)]\\ &= \{[1,1,1]-\mu(0)^TD^i\widetilde{\psi}(0),...,[1,1,1]-\mu(t)^TD^i\widetilde{\psi}(t),...\\ &\qquad ...,[1,1,1]-\mu(N-1)^TD^i\widetilde{\psi}(N-1)\}. \qquad (6.66)\end{aligned}$$

With the multipliers now known the performance derivative can easily be calculated by applying formula (6.65).

6.2.7 *Reducing number of control signals*

A major problem with an implementation of the DISOPE optimizing control algorithm is dimension of the control vector over a batch. Even with the derivative of the performance function in the DMMOP efficiently computed as above, the optimization problem is nonlinear and its numerical solution requires significant computing time. However, this is not a critical factor in the algorithm implementation. A large number of control signals makes it difficult to calculate the modifier λ. Even with an efficient use of the output measurements from the previous iterations of the control algorithm according to the ISOPED technique (see Chapter 4) more than two hundred independent control perturbations are required for a typical batch to be applied to the furnace in order to refresh an existing information and calculate sufficiently accurate approximation of the furnace input-output mapping $F_*(\cdot)$ derivative $F_*^{'}(\cdot)$ at a current set-point value c^i (see (6.17)). Bearing in mind that in order to perform a control perturbation separate batch must be run, calculating λ becomes real time consuming and it may upset the furnace. Hence, for the practical purposes it is essential to reduce the number of control signals.

6.2.7.1 *Periodic control functions*

A first set towards reducing the number of control signals required can be made by making the control function *periodic* in time with a period equal to the time taken for a load to travel through one actual furnace zone. This is sensible since it insures that every new load that enters the furnace receives the same control actions as its predecessors in the batch. Hence, only one eighth of the total number of controls need to be calculated. Let us now formalize the above reduction of controls.

According to (6.3b) a vector of the control signals (controls) over a batch is expressed as:

$$c = [c(0)^T, c(1)^T, ..., c(N-1)^T]^T \tag{6.67}$$

Let us denote the control period n_P. A control vector over the period c_P reads:

$$c_P = [c(0)^T, c(1)^T, ..., c(n_P)^T]^T. \tag{6.68}$$

The furnace controls at a sampling time instant t are composed of the soak, heat and preheat zone controls at t (see (6.54)) as:

$$c(t) = [c_1(t), c_2(t), c_3(t)]^T. \tag{6.69}$$

Hence, a control vector over the period c_P reads:

$$c_P = [c_1(0), c_2(0), c_3(0), .., c_1(t), c_2(t), c_3(t), ..., c_1(n_P), c_2(n_P), c_3(n_P)]^T. \tag{6.70}$$

The controls over a batch can now be expressed as composed of the controls over a period c_P and the remaining controls c_{BR} over a batch as:

$$c = \begin{bmatrix} c_P \\ c_{BR} \end{bmatrix} \tag{6.71}$$

where

$$c_{BR} = Mc_P. \tag{6.72}$$

The matrix M can be easily determined once the period has been prescribed. It is a transformation matrix of the controls over a period into the remaining controls over a batch. The independent controls are those over the period and they are determined by (6.70).

The performance function $q(\cdot, \alpha^i)$ (see (6.62)) now becomes the function $\widetilde{q}(\cdot, \alpha^i)$ of the controls only over the period and due to (6.71) and (6.72)

$$q(c, \alpha^i) = q(c_P, c_{BR}, \alpha^i) \tag{6.73}$$

$$\widetilde{q}(c_P, \alpha^i) \triangleq q(c_P, Mc_P, \alpha^i). \tag{6.74}$$

A formula for the function $\widetilde{q}(\cdot, \alpha^i)$ derivative $\widetilde{q}'_{\nu_P}(\cdot, \alpha^i)$ with respect to ν_P is needed in order to solve the DMMOP in its implicit formulation in **Step 3** of the DISOPE algorithm.

Notice, that due to (6.73), the following holds:

$$\widetilde{q}'_{\nu_P}(\nu_P, \alpha^i) = q'_{\nu_P}(\nu_P, M\nu_P, \alpha^i) + q'_{\nu_{BR}}(\nu_P, M\nu_P, \alpha^i)M. \tag{6.75}$$

The performance function derivatives in the reduced control space can now be calculated as follows. First, the derivatives $q'_{\nu(t)}(\nu(t), \alpha^i)$ are calculated with respect to all controls $\nu(t)$, $t \in \overline{0 : N-1}$ over the batch, by applying the conjugate equations (6.64a) and (6.64b) and then by using the Hamiltonian expressions (6.65). Next, the obtained derivatives are combined according to the formula (6.75). In other words,the derivatives with respect to the reduced controls are calculated by taking the components of the total derivative vector (see (6.66)) that is obtained by using the Hamiltonian (6.63), due to the controls for the first period and adding the derivatives due to the controls for the next seven periods multiplied by the control transformation matrix M.

An experiment was performed where a periodic control was applied to the furnace and measurements were taken over a batch. By inspecting the results it was possible to determine the approximate average exit temperature during the batch. This was used as the reference temperature T_e^{ref} in the performance function (6.58) and the DISOPE was applied in the control periodic mode. The results are illustrated in Fig. 6.9 and Fig. 6.10. It can be seen that good tracking of the the furnace exit set-point temperature was achieved.

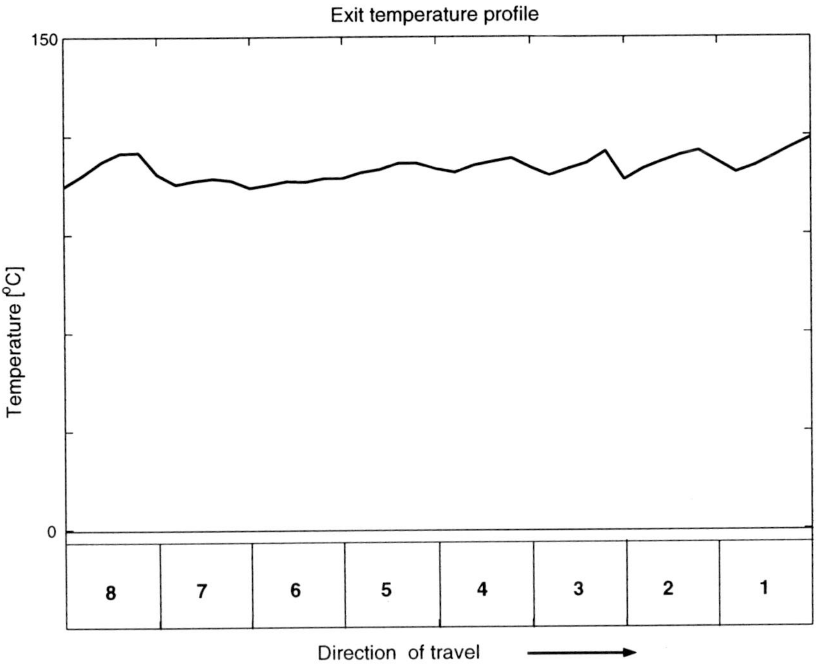

Fig. 6.9 Temperature profile at the furnace exit during DISOPE optimizing control with periodic control functions

6.2.7.2 *Parameterized control functions*

The shape of the optimal control functions obtained during DISOPE optimizing control with periodic control functions (see Fig. 6.10) suggests that it would be sensible to further reduce control space dimension by approximating the optimal control functions by piecewise linear functions. Accordingly, the admissible control functions were chosen as shown in the Fig. 6.11 for the soak zone, Fig. 6.12 for the heat zone and Fig. 6.13 for the preheat zone.

The parameters c_1, c_3 and c_2 in Fig. 6.11 are the control amplitudes at the start, middle and end of one period, respectively. They are subject to the constraints that they must lie within the acceptable range $0 \leq c_i \leq 255$, $i = 1, 2, 3$ and that c_1, c_2 must be not smaller than c_3, that is $c_3 \leq c_1$ and $c_3 \leq c_2$. The parameters c_4, c_5, c_6 and c_7 are amplitudes, as before, which

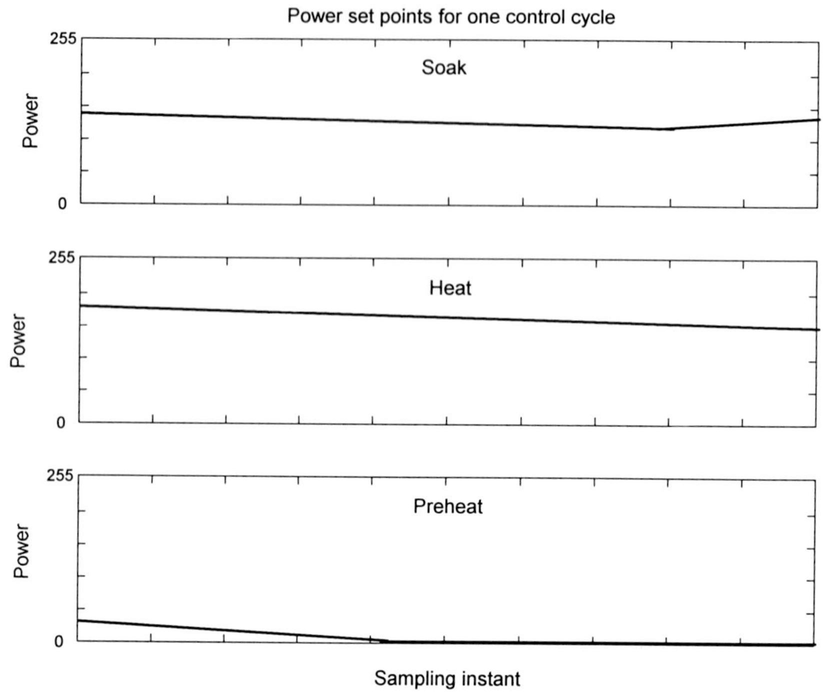

Fig. 6.10 Power profiles in the furnace zones during DISOPE optimizing control with periodic control functions.

are subject to the same range constraints as others.

It is interesting to note that the slope of the control in the soak zone is different from that in other zones. A possible interpretation for this is that towards the end of the control period, as the load is about to leave the furnace, it is subject to the cooling effect of the furnace end doors. Consequently, the power in the soak zone is increased to compensate.

Since the point at which the change in slope of the control functions for the soak zone takes place is fixed it is possible to relate the control parameters $c_i, i \in \overline{1:7}$ to the controls over one period c_P by a linear transformation matrix. Consequently, the derivatives of the performance function with respect to the control parameters may be obtained by multiplying those corresponding to the c_P by the transformation matrix. The derivative with respect to c_P is calculated from (6.75) via Hamiltonian as described

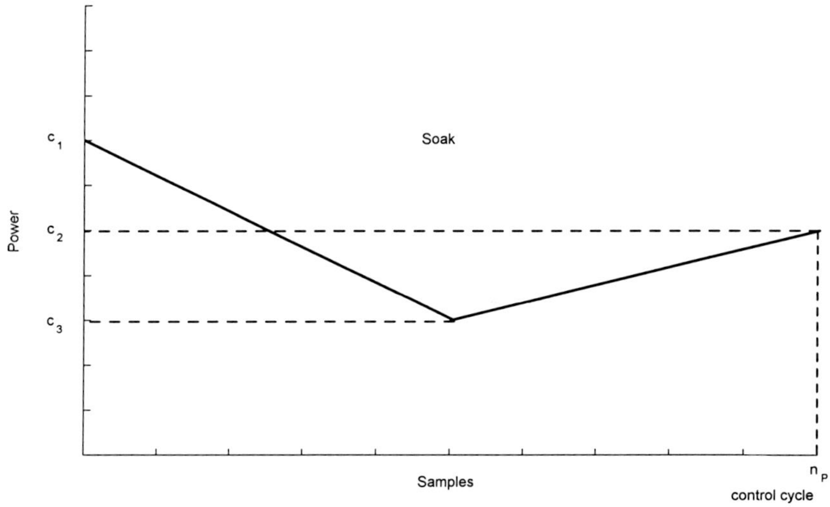

Fig. 6.11 Piecewise linear control function chosen in the soak zone for the DISOPE optimizing control of the furnace.

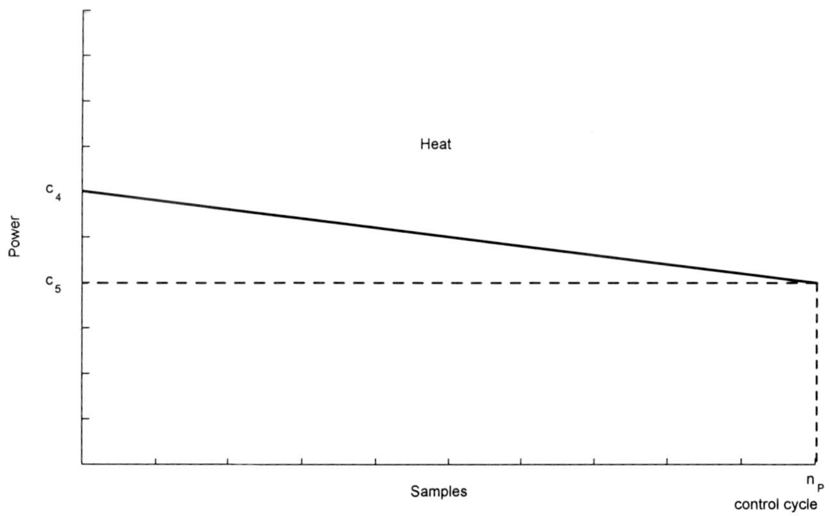

Fig. 6.12 Piecewise linear control function chosen in the heat zone for the DISOPE optimizing control of the furnace.

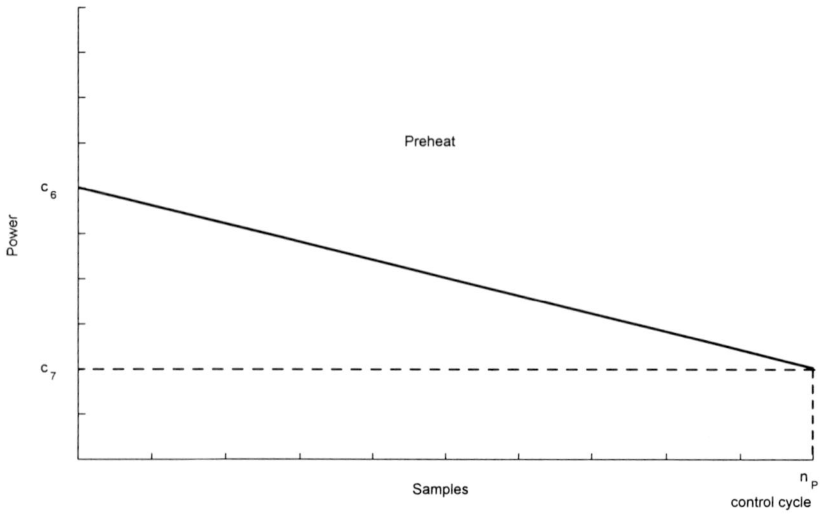

Fig. 6.13 Piecewise linear control function chosen in the preheat zone for the DISOPE optimizing control of the furnace.

in Section 6.2.7.1.

The decision variables in DMMOP are the control function parameters $c_i, i \in \overline{1:7}$. As described before, they are subject to linear constraints. Hence, the very efficient algorithm of (Rosen, 1960) with projection of gradient on a set of active constraints was applied to solve the DMMOP within the DISOPE optimizing control algorithm. In order to appreciate the importance of of the convexifying term in DMOPM (see (6.15)) the modified model- based optimization problem without this term was considered. The problem is highly nonlinear and the global minimum may not be found by one run of the optimization solver. The problem scaling although extensively tried did not improve the optimization algorithm. It was therefore necessary to choose different starting points for the Rosen's solver so that the global optimum could be assigned as that solution for which several initial starting points had given the same results.

The DISOPE optimizing control algorithm with control functions parameterized as above was applied to the furnace. The key parameters applied during the experiments were as follows:

- motor speed - 70, so that one batch was completed in approximately

30 minutes,

- temperature set-point - $T_e^{ref} = 120\ {}^0\mathrm{C}$,
- iterative loop gain k_c (see (6.22)) - 0.2,
- initial vector of control parameters - $[80, 80, 30, 60, 30, 10, 10]^T$.

The results are shown in Fig. 6.14. For ease of comparison three sets of results are given on the graph.

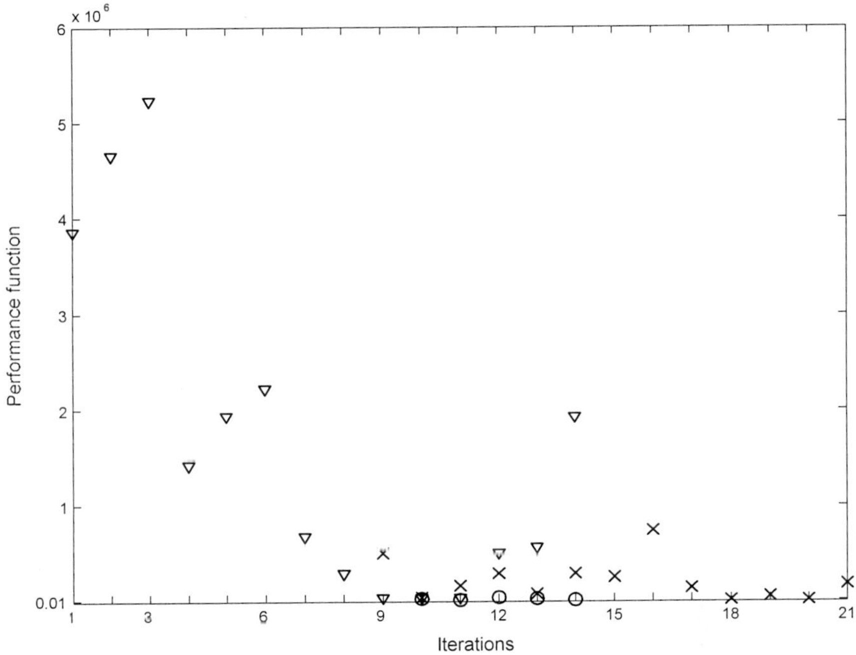

Fig. 6.14 Performance function of DISOPE optimizing control with parameterized control functions.

Firstly, consider the results shown using triangular symbols. Initially, the algorithm reduces the power applied and hence, because the start-up controls are insufficient in magnitude to heat the loads to the desired temperature, an increase in performance is observed. Since, at the start of the algorithm iterations, the real process derivatives have just been obtained using perturbations, it is unlikely that the deterioration is due to poor derivative information. A more plausible reason is that the modified optimization problem is not convex in this region causing the Rosen algorithm to find the wrong solution. After three iterations the values of the multipliers λ^i are such that the optimization may proceed in a favorable direction. Another worsening in the performance then occurs. At this point a considerable change has been made in the values of the controls, hence, since the real process derivatives are calculated using the changes in controls, these derivatives may be becoming inaccurate. After another two iterations the performance once more decreases, to a minimum, whereafter an increase in performance is observed. The increase in performance at this point may be reduced, or possibly eliminated, by performing additional perturbations to refresh the derivative information.

In the second set of results, shown on the graph using crosses, the algorithm was re-started at the end of the eight algorithm iterations and three additional perturbations were performed. After the final perturbation had been applied the optimization was recommenced. After two iterations the algorithm once more started to produce a deterioration in the performance that is not plotted. Clearly additional perturbations applied before the minimum point of the original run have little effect. Two additional perturbations were applied at iteration ten (the minimum point of the original run). These perturbations resulted in a significant improvement in algorithm performance with the previously observed deterioration greatly reduced. At iteration twenty the resulting control produces a performance less than obtained without the additional perturbations. The exit temperature produced by this optimal control, for a batch, is shown in Fig. 6.15.

A further experiment was performed where the original run was continued from iteration ten (no perturbation at iteration eight). These results are shown in Fig. 6.14 using circles. The deterioration in performance after iteration ten has been greatly reduced.

Let us now examine the optimal control. The control parameters are:

$$c^{opt} = [145, 94, 30, 140, 33, 188, 48]. \tag{6.76}$$

Note that there is significant difference between the start-up and optimal parameters.

The corresponding control functions are shown in Fig. 6.16.

It can be seen from Fig. 6.16 that the heating effect is distributed over all the furnace zones rather than being concentrated in one or two zones. This result is interesting because the optimizing control algorithm was free to

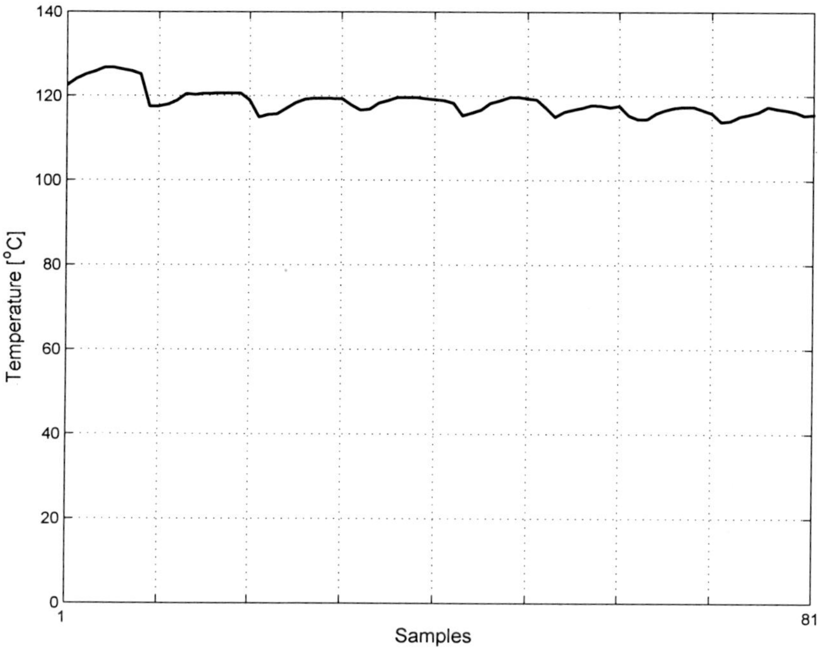

Fig. 6.15 Temperature profile at the furnace exit during DISOPE optimizing control with parameterized control functions suing the optimal batch control parameters.

distribute the power as it liked. Clearly, the most efficient furnace operation is obtained if the loads are heated gradually throughout their travel along the furnace. This is intuitively reasonable since if a higher power is applied only over two zones of the furnace, i.e., the preheat zone is not used, then the power losses will be greater and the heating will therefore be less efficient resulting in an increase in the performance.

An important practical observation can be made based on the obtained results with regard to the modifier λ. It turned out that the number of perturbations used in order to calculate the furnace output derivatives required to calculate the modifier was much smaller than that predicted by the theory. If the derivative approximations become poor during the iterations then the control algorithm may tend to refresh them by producing control actions which may cause a deterioration in performance. This may be seen a *learning phase before the correct control action can,* once more, be

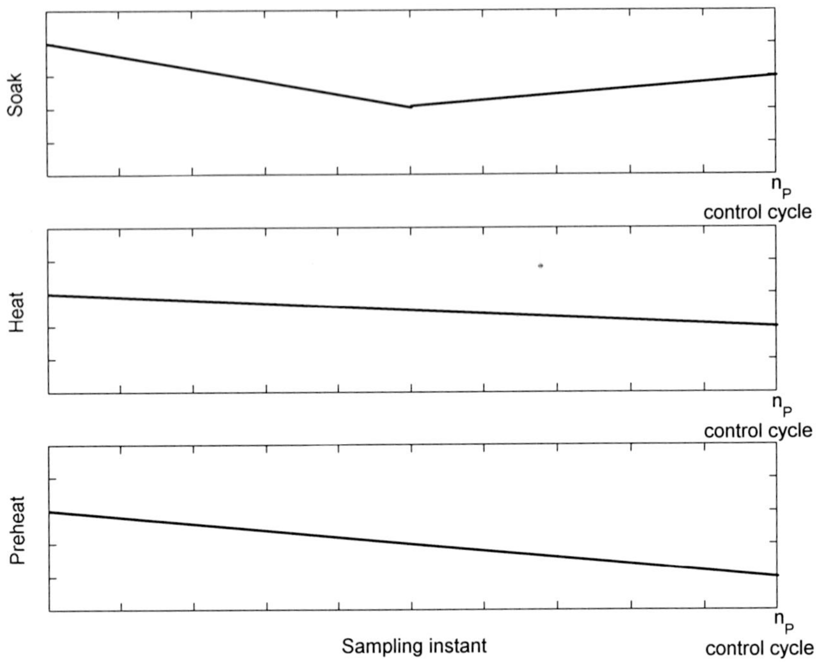

Fig. 6.16 The shape of the optimal control functions found for an optimal batch by DISOPE optimizing control algorithm applied to the furnace exit temperature tracking.

taken. Additional perturbations may also be performed manually based on monitoring the performance and other process variables. This leads to *soft intelligent control* with an operator in a loop or an intelligent automated supervisor.

Chapter 7

Optimizing Control of Interconnected Systems

7.1 The Interconnected System Description and Modeling

An arrangement of several processes where outputs of certain processes are connected with inputs to another ones is usually called *an interconnected system*, or *a complex system* — a short description *a system* will be usually used in this chapter. System elements are called *subsystems*, each of them being generally a nonlinear process with possibly many inputs and outputs. Optimizing control algorithms for interconnected systems will be presented in this chapter. Therefore, a system consisting of interconnected *regulatory controlled subsystems* will be considered, i.e., each subsystem together with its direct feedback controllers (with its actuating controller, see Sections 1.3 and 2.1). The direct (actuating) controllers constitute the first layer of an overall system multilayer control structure. Assuming that each subprocess is regulated locally by its feedback controllers means that a *decentralized control structure* of the regulatory control layer of the interconnected system is assumed, the structure commonly used in practice. The schematic description of a considered controlled subsystem is given in Fig. 7.1, where
– $c_i \in \mathbb{R}^{n_{c_i}}$ are the it-h subsystem steady-state controls (regulatory controller set-points),
– $u_i \in \mathbb{R}^{n_{u_i}}$ are the it-h subsystem interconnection (interaction) inputs,
– $y_i \in \mathbb{R}^{m_i}$ are the it-h subsystem interconnection outputs,
– $y_i^o \in \mathbb{R}^{m_i^o}$ are the it-h subsystem external outputs.
The latter are the i-th part of the overal system outputs to the process environment, while the interconnection outputs are connected to inputs of another subsystems, as shown schematically for an example system struc-

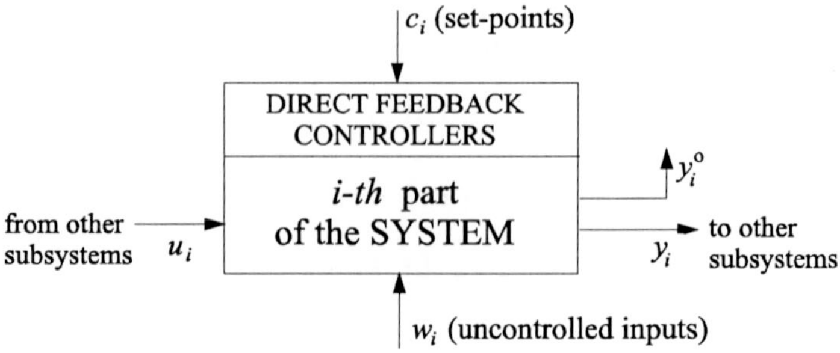

Fig. 7.1 The i-th subsystem (i-th part of the controlled system).

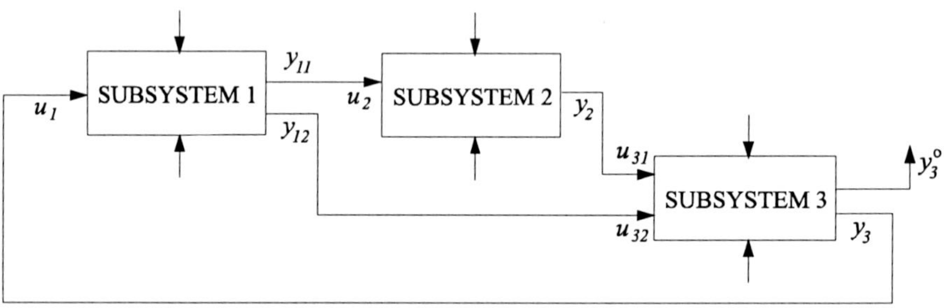

Fig. 7.2 A schematic drawing of an example interconnected system.

ture in Figure 7.2.

The interconnections create *the structure* of the system, which can be formally described by an input-output *interconnection matrix* H, called also a *structure matrix*. Each row of this matrix is associated with a single input of a subsystem. It will be assumed that each interconnection output from a subsystem is connected to only one input of another subsystem (and vice versa). Therefore, the interconnection matrix can be assumed to consist of zeros and ones only, with precisely one non-zero element in each row and each column. This way of describing the system structure is quite common, see e.g., (Findeisen *et al.*, 1980; Findeisen, 1997). It is general, since the case of one output stream splitting into two (or more) sub-streams feeding different subsystems can be easily described using the assumed convention, by incorporating splitting element into the subsystem generating this stream. As an example, the interconnection matrix H corresponding to the system structure presented in Fig. 7.2 is as follows

$$H = \begin{bmatrix} 0 & 0 & 0 & 1 \\ 1 & 0 & 0 & 0 \\ 0 & 0 & 1 & 0 \\ 0 & 1 & 0 & 0 \end{bmatrix} \tag{7.1}$$

where $u = Hy$, $u = [u_1 \; u_2 \; u_{31} \; u_{32}]^T \in \mathbb{R}^4$, $y = [y_{11} \; y_{12} \; y_2 \; y_3]^T \in \mathbb{R}^4$. Having introduced the matrix H, a complex system consisting of N interconnected subsystems can be presented in a way depicted in Figure 7.3.

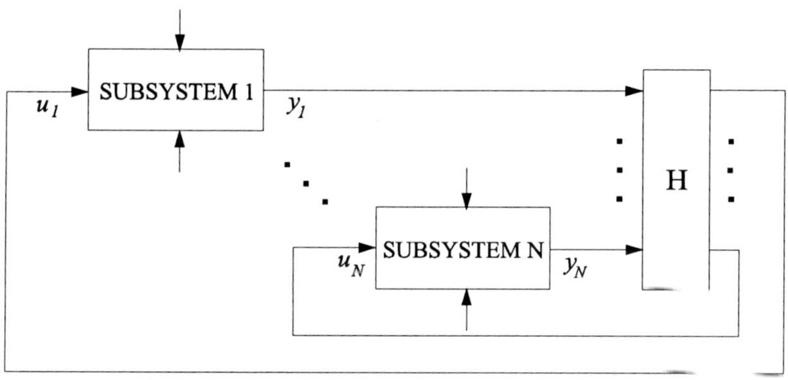

Fig. 7.3 The interconnected system presented with a structure matrix H.

The control of a single plant is usually such a complicated task that a multilayer control structure is applied, see e.g., (Findeisen *et al.*, 1980; Findeisen, 1997; Tatjewski, 2002) and Chapter 1 of this book. That is even more the case for an interconnected system. The task of each layer is difficult since the controlled system is now partitioned and often distributed in space, each subsystem being, generally, a multi-input multi-output nonlinear plant. Therefore, design of controllers or control algorithms for each layer should take into account this features, which results in decentralized and multilevel control structures. The decentralized structure of the direct feedback control layer is usually desired and implemented in industrial practice, while a multilevel hierarchical structure of the optimizing control layer may be a sound solution. Such an example multilevel control structure is depicted in Figure 7.4, where the optimizing control task is divided into local tasks associated with corresponding subsystems and being coordinated. In industrial implementations the task of each subsystem direct feedback control is usually performed by a local distributed control system, which can also perform local optimization tasks at computer workstations at a higher level of its own hierarchy.

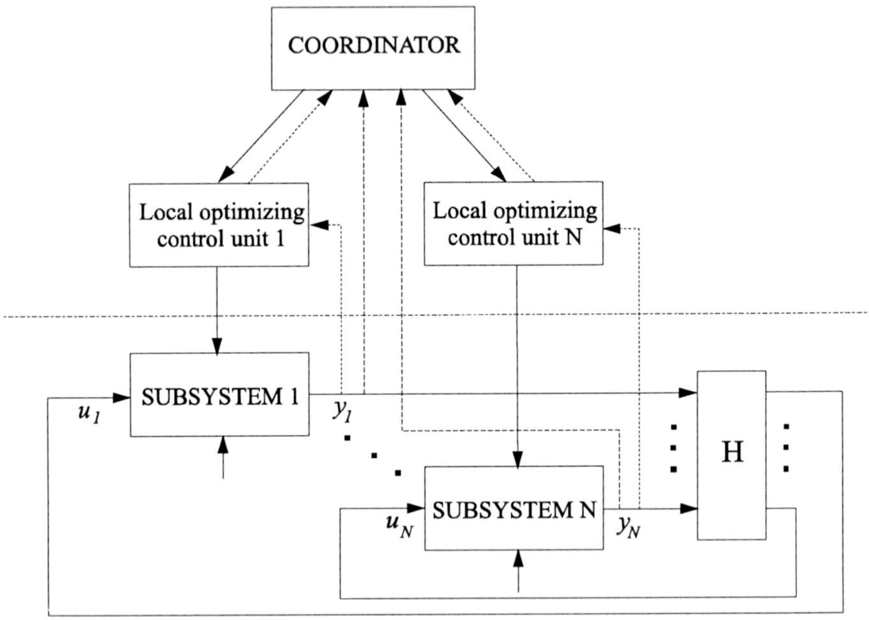

Fig. 7.4 Multilevel optimizing control of an interconnected system. Dotted lines show possible feedback paths (see (Findeisen *et al.*, 1980)).

Let us remind the reader of the important reasons for using decentralized or multilevel (hierarchical) rather then centralized control structures:

- The need to increase the overall system robustness (reliability) so that the system still operates when one of the control units fails or an information link breaks down.
- The possibility of decreasing the system sensitivity to disturbances if local control units can respond faster and more adequately than

one central decision unit.
- The developments in computer technology enabling and stimulating applications of powerful distributed computer control systems.

We will be concerned with methods for optimization of steady-states of the interconnected system under uncertainty in this chapter, i.e., methods applicable for finding optimal (or suboptimal) set-points for the system, for given values of uncontrolled system inputs (measured, estimated or unknown but constant for longer time periods). Therefore, steady-state mappings describing relations between system inputs and outputs will be of interest. Starting from a subsystem, let us denote steady-state mappings between its inputs and outputs by F_{*i} and F^o_{*i},

$$y_i = F_{*i}(c_i, u_i) \tag{7.2a}$$
$$y_i^o = F^o_{*i}(c_i, u_i), \qquad i = 1, ..., N \tag{7.2b}$$

where explicit dependence on the uncontrolled inputs (disturbances) has been omitted, to simplify the notation. Due to uncertainty in modeling of every subsystem itself and in the disturbances, the known steady-state description of every controlled subsystems is its steady-state model only, formulated in the following form

$$y_i = F_i(c_i, u_i, \alpha_i) \tag{7.3a}$$
$$y_i^o = F_i^o(c_i, u_i, \alpha_i), \qquad i = 1, ..., N \tag{7.3b}$$

where $\alpha_i \in A_i \subset \mathbb{R}^{s_i}$ are the i-th subsystem model parameters.

The interconnections between subsystems are described by the interconnection matrix H,

$$u_i = H_i y = \sum_{j=1}^{N} H_{ij} y_j, \qquad i = 1, \ldots, N \tag{7.4}$$

where

$$H = \begin{bmatrix} H_1 \\ H_2 \\ \vdots \\ H_N \end{bmatrix} = \begin{bmatrix} H_{11} & H_{12} & \cdots & H_{1N} \\ H_{21} & H_{22} & \cdots & H_{2N} \\ \vdots & \vdots & \cdots & \vdots \\ H_{N1} & H_{N2} & \cdots & H_{NN} \end{bmatrix} \tag{7.5}$$

and each submatrix H_{ij} describes interconnections between inputs to the i-th subsystem and outputs from the j-th subsystem.

We assume that the constraints restricting subsystem inputs have the following general form

$$(c_i, u_i) \in CU_i = \{(c_i, u_i) \in \mathbb{R}^{n_{c_i}} \times \mathbb{R}^{n_{u_i}} : g_i(c_i, u_i) \leq 0\} \quad (7.6)$$

where $g_i : \mathbb{R}^{n_{c_i}} \times \mathbb{R}^{n_{u_i}} \mapsto \mathbb{R}^{r_i}$ is a (vector) constraint function.

Denoting the overall system variables by

$$\begin{aligned} c &= (c_1, \ldots, c_N) \in \mathbb{R}^{n_c} \\ u &= (u_1, \ldots, u_N) \in \mathbb{R}^{n_u} \\ y &= (y_1, \ldots, y_N) \in \mathbb{R}^{m} \\ y^o &= (y_1^o, \ldots, y_N^o) \in \mathbb{R}^{m^o} \\ \alpha &= (\alpha_1, \ldots\alpha_N) \in A \subset \mathbb{R}^{s} \end{aligned} \quad (7.7)$$

where $s = s_1 + \ldots + s_N$ and

$$A = A_1 \times A_2 \times \ldots \times A_N \quad (7.8)$$

and the overall system model mappings by

$$F = [F_1^T, \ldots, F_N^T]^T \quad (7.9a)$$

$$F^o = [(F_1^o)^T, \ldots, (F_N^o)^T]^T \quad (7.9b)$$

$$g = [g_1^T, \ldots, g_N^T]^T \quad (7.9c)$$

we get the description of the whole system in the form

$$y = F_*(c, u) \quad (7.10a)$$

$$y^o = F_*^o(c, u) \quad (7.10b)$$

$$u = Hy \quad (7.10c)$$

while description of its model as

$$y = F(c, u, \alpha) \quad (7.11a)$$

$$y^o = F^o(c, u, \alpha) \quad (7.11b)$$

$$u = Hy \quad (7.11c)$$

with the constraints

$$(c, u) \in CU = \{(c, u) \in \mathbb{R}^{n_c + n_u} : g(c, u) \leq 0\} \quad (7.12)$$

where, obviously,

$$CU = CU_1 \times \cdots \times CU_N. \tag{7.13}$$

Interconnection inputs and outputs represent in fact the same physical variables, see the interconnection matrix H. Therefore, it will be often convenient to have the system description with only one set of these variables. Combining (7.10a) with (7.10c) leads to the equation

$$y = F_*(c, Hy) \tag{7.14}$$

defining implicitly the following *system explicit interconnection output mapping* $K_* : \mathbb{R}^{n_c} \longmapsto \mathbb{R}^m$,

$$y = K_*(c). \tag{7.15}$$

Dealing analogously with (7.11a) and (7.11c) gives the equation

$$y = F(c, Hy, \alpha) \tag{7.16}$$

defining implicitly the following *system explicit interconnection output mapping model* $K : \mathbb{R}^{n_c} \times \mathbb{R}^{n_s} \longmapsto \mathbb{R}^m$,

$$y = K(c, \alpha) \tag{7.17}$$

where $\alpha = (\alpha_1, \ldots, \alpha_N) \in \mathbb{R}^s$.

When looking at the interconnected system as at a single plant, the interconnection inputs and outputs can be treated as its internal variables, not essential for the plant environment. For this point of view, these variables can be eliminated from the plant and its model descriptions leading to their compact forms. For the overall plant, (7.10b) with (7.10c) gives the implicit equation $y = F_*^o(c, Hy)$ resulting in the *system explicit external output mapping* $K_*^o : \mathbb{R}^{n_c} \longmapsto \mathbb{R}^{m^o}$,

$$y^o = K_*^o(c) \tag{7.18}$$

while combining (7.11b) with (7.11c) we get the relation

$$y^o = F^o(c, Hy, \alpha) \tag{7.19}$$

defining implicitly the following *system explicit external output mapping model* $K^o : \mathbb{R}^{n_c} \times \mathbb{R}^{n_s} \longmapsto \mathbb{R}^{m^o}$,

$$y^o = K^o(c, \alpha). \tag{7.20}$$

Adding a performance function to the presented model, i.e., to the system output models (being equality constraints) and to the inequality constraints, we arrive at the complex model optimization problem. When the models are sufficiently accurate, a single model optimization may lead to good steady-state values for the set-points c. Hierarchical optimization methods may be a reasonable choice for certain classes of interconnected system models. These methods are out of the scope of this book, since the theory is well developed and good references are available, see e.g., (Findeisen *et al.*, 1980).

We are interested in situations with uncertainty that should not be neglected, as in previous chapters of the book. In the next sections of this chapter two kinds of optimizing control problem solutions for interconnected systems with uncertainty will be presented. The first one is the Interaction Balance Method with Feedback (IBMF), a suboptimal but relatively simple approach. The second one is the extension of the ISOPE technique to the considered case of an interconnected system, which is more involved (requiring system output mapping derivatives estimation), but yielding true optimal set-point values.

7.2 Iterative Correction by Interaction Balance Method with Feedback

7.2.1 *Presentation of the method*

The performance function for the system optimization may have the general form

$$Q(c, u, y, y^o) = \sum_{i=1}^{N} Q_i(c_i, u_i, y_i, y_i^o) \tag{7.21}$$

involving both interconnection input and output variables. For the development of the method presented in this chapter it will be assumed that the performance function is dependent on c and u only. If it is not the case originally, the output variables can be eliminated from the description as follows (note that due to interconnections $u = Hy$, y and corresponding u represent the same physical variables)

$$q_i(c_i, u_i, \alpha_i) = Q_i(c_i, u_i, F_i(c_i, u_i, \alpha_i), F_i^o(c_i, u_i, \alpha_i)), \quad i = 1, \ldots, N. \tag{7.22}$$

The system model optimization problem (SMOP) can be then stated as follows

$$\begin{array}{ll} \text{minimize} \displaystyle\sum_{i=1}^{N} q_i(c_i, u_i, \alpha_i) & \\ \text{subject to}: \ u_i = \sum_{i=1}^{N} H_{ij} F_j(c_j, u_j, \alpha_j) & \\ \qquad\qquad (c_i, u_i) \in CU_i, \qquad i = 1, \ldots, N. & \end{array} \tag{7.23}$$

Due to the special structure, the problem SMOP can be solved using a hierarchical optimization method, in a decomposition-coordination structure. The structure of SMOP is particularly suitable for an application of the *price method*, see e.g., (Findeisen *et al.*, 1980), called also the *dual method* (Lasdon, 1970) or the *goal coordination method* (Wismer, 1971). The price method relies on application of the Lagrange function, where the Lagrange multipliers ("prices") play the role of coordination variables. The Lagrange function is constructed only with output and interconnection constraints, the local constraints $(c_i, u_i) \in CU_i$ are treated in another, direct way. Therefore, for the problem SMOP the Lagrange function takes the following form

$$\begin{aligned} L(c, u, p, \alpha) &= \sum_{i=1}^{N} q_i(c_i, u_i, \alpha_i) + \sum_{i=1}^{N} p_i^T [u_i - \sum_{j=1}^{N} H_{ij} F_j(c_j, u_j, \alpha_j)] = \\ &= \sum_{i=1}^{N} [q_i(c_i, u_i, \alpha_i) + p_i^T u_i - \sum_{j=1}^{N} p_j^T H_{ji} F_i(c_i, u_i, \alpha_i)] = \\ &= \sum_{i=1}^{N} L_i(c_i, u_i, p, \alpha_i). \end{aligned} \tag{7.24}$$

In the price method, *local optimization problems* are formulated in the form

$$\begin{array}{ll} \text{minimize}_{(c_i, u_i)} L_i(c_i, u_i, p, \alpha_i) & \\ \text{subject to}: \ (c_i, u_i) \in CU_i \ , \qquad i = 1, \ldots, N & \end{array} \tag{7.25}$$

whereas the goal of the coordinator is to find *coordinating prices* $p = \hat{p}$ such that local problems solutions fulfill the *global constraint*, i.e.,

$$\hat{u}_i(\hat{p}) - \sum_{j=1}^{N} H_{ij} F_j(\hat{c}_j(\hat{p}), \hat{u}_j(\hat{p}), \alpha_j) = 0, \qquad i = 1, \ldots, N \tag{7.26}$$

where $(\hat{c}_j(\hat{p}), \hat{u}_j(\hat{p}))$ denote solutions of local optimization problems (7.25) for prices $p = \hat{p}$, $i = 1, ..., N$.

It is not our aim to discuss applicability conditions of the price method and strategies for finding optimal prices $\hat{p}$, it is classics in optimization methods and the reader is referred to good literature, see e.g., (Findeisen *et al.*, 1980) and references therein. We presented the formulation of the price method because it is a basis for presentation of the *interaction balance method with feedback* (IBMF). The IBMF, originally introduced in (Findeisen *et al.,* 1978), see also (Findeisen *et al.,* 1980; Tatjewski, 1988), is an optimizing control method with global feedback, i.e., from the controlled system to the coordinator. It assumes significant uncertainty in the interconnected system modeling assuming differences between the (unknown) true subsystem input-output mappings $F_{*i}(c_i, u_i)$ and their (known) models $F_i(c_i, u_i, \alpha_i)$, $i = 1, ..., N$. The IBMF works with local optimization problems (7.25), i.e., the same as in the price method. However, the *coordination condition* is different than (7.26), it is of the form

$$\hat{u}_i(\tilde{p}) - \sum_{j=1}^{N} H_{ij} F_{*j}(\hat{c}_j(\tilde{p}), \hat{u}_j(\tilde{p})) = 0, \qquad i = 1, ..., N \tag{7.27}$$

where $\tilde{p}$ denotes coordinating values of the price vector. Using the explicit interconnection output mapping (7.15) the condition (7.27) can be easily written in the following equivalent global form

$$\hat{u}(\tilde{p}) - HK_*(\hat{c}(\tilde{p})) = 0. \tag{7.28}$$

The reasoning behind the condition (7.28) is to require that *coordinating values* $\tilde{p}$ of the prices p should be such that interactions $\hat{u}(\tilde{p}) = (\hat{u}_1(\tilde{p}), ..., \hat{u}_N(\tilde{p}))$ calculated in local optimization problems are equal to real interactions in the controlled system $u_*(\tilde{p})$ resulting from application of the set-points $\hat{c}(\tilde{p})$ to the system direct controllers, where from definition $u_*(p) = HK_*(\hat{c}(p))$. Applicability conditions and coordination strategies of the presented IBMF were discussed in several papers, the most comprehensive presentation can be found in the book (Findeisen *et al.,* 1980). The most essential condition is the requirement that the model optimization problem SMOP can be solved using the price method. This implies, in fact, that a saddle point of the Lagrange function (7.24) should exist and that solutions of local optimization problems (7.25) should be unique. These requirements are rather restrictive, limiting applicability of the method.

However, the applicability conditions could be significantly relaxed when applying, instead of the usual Lagrange function (7.24), the following *augmented Lagrange function*

$$L_r(c,u,p,\alpha) = \sum_{i=1}^{N} q_i(c_i,u_i,\alpha_i) + \sum_{i=1}^{N} p_i^T[u_i - \sum_{j=1}^{N} H_{ij}F_j(c_j,u_j,\alpha_j)] + \\ +0.5r\|u_i - \sum_{j=1}^{N} H_{ij}F_j(c_j,u_j,\alpha_j)\|^2 \quad (7.29)$$

where $r > 0$ is a penalty coefficient. Unfortunately, the augmented Lagrange function (7.29) is not directly separable. (Stephanopoulos and Westerberg, 1975) proposed a method to overcome this obstacle. Writing the augmented Lagrange function in the form

$$L_r(c,u,p,\alpha) = \sum_{i=1}^{N}[q_i(c_i,u_i,\alpha_i) + p_i^T u_i - \sum_{j=1}^{N} p_j^T H_{ji}F_i(c_i,u_i,\alpha_i) + \\ +0.5r(\|u_i\|^2 + \|F_i(c_i,u_i,\alpha_i)\|^2) - ru_i^T H_i F(c,u,\alpha)] \quad (7.30)$$

where $H_i = [H_{i1} \; \cdots \; H_{iN}]$ and observing that only the last, mixed terms are the source of nonseparability the following separable approximations of these terms can be proposed

$$u_i^T H_i F(c,u,\alpha) \simeq (u_i^k)^T H_i F(c,u,\alpha) + \\ + u_i^T H_i F(c^k,u^k,\alpha) - (u_i^k)^T H_i F(c^k,u^k,\alpha) \quad (7.31)$$

where (c^k,u^k) is an approximation point. Using this approximation the augmented Lagrange function takes the following approximate, but separable form

$$L_r^k(c,u,p,\alpha) = \sum_{i=1}^{N}[q_i(c_i,u_i,\alpha_i) + p_i^T u_i - \sum_{j=1}^{N} p_j^T H_{ji}F_i(c_i,u_i,\alpha_i) + \\ -r\sum_{j=1}^{N}(u_j^k)^T H_{ji}F_i(c_i,u_i,\alpha_i) - ru_i^T H_i F(c^k,u^k,\alpha) + \\ +0.5r(\|u_i\|^2 + \|F_i(c_i,u_i,\alpha_i)\|^2)] \\ = \sum_{i=1}^{N} L_{ri}^k(c_i,u_i,p,\alpha,c^k,u^k). \quad (7.32)$$

Certainly, application of approximation points means necessity to iterate not only the prices (Lagrange multipliers), but also the approximation points. First algorithm of the presented *augmented interaction balance method with feedback* (AIBMF) was originally proposed by (Tatjewski, 1985), versions with more elaborate coordination strategy were derived in (Tatjewski, 1988).

The *structure of the AIBMF algorithm* is as follows:

Start. Given initial point (c^0, u^0), select initial value of multipiers p^0, penalty coefficient $r > 0$, outer and inner loop tolerances $\varepsilon > 0$, $\varepsilon_a > 0$ and relaxation coefficient $\gamma > 0$. Set $n := 0$, $k := 0$.

Step 1. For $p = p^n$ perform, with accuracy ε_a, the following (inner) *approximation loop*:

Step 1a. Set $(c^0, u^0) := (c^k, u^k)$, set $k := 0$.

Step 1b. Solve *local optimization problems*

$$\begin{aligned}&\text{minimize}_{(c_i,u_i)} L^k_{ri}(c_i, u_i, p, \alpha, c^k, u^k)\\&\text{subject to: } (c_i, u_i) \in CU_i, \qquad i = 1, ..., N\end{aligned} \tag{7.33}$$

denoting the solutions by $(\hat{c}_i^k, \hat{u}_i^k)$, $i = 1, ..., N$.

Step 1c. If

$$\|(c^k, u^k) - (\hat{c}^k, \hat{u}^k)\| \leq \varepsilon_a \tag{7.34}$$

then set $(\hat{c}(p^n), \hat{u}(p^n)) := (\hat{c}^k, \hat{u}^k)$ and go to Step 2, else improve the approximation point using the relaxation scheme

$$(c^{k+1}, u^{k+1}) := (c^k, u^k) + \gamma[(\hat{c}^k, \hat{u}^k) - (c^k, u^k)] \tag{7.35}$$

set $k := k + 1$ and go to Step 1b.

Step 2. Apply the set-points $\hat{c}(p^n)$ to the controlled system, measure (after the transients have died) the corresponding interactions $u_*(p^n) = HK_*(\hat{c}(p^n))$ and send to the coordinator.

Step 3. Coordination: if

$$d^n_* = \|\hat{u}(p^n) - u_*(p^n)\| \leq \varepsilon \tag{7.36}$$

then terminate (coordination condition satisfied), else modify the prices

$$p^{n+1} := M_*(p^n, \hat{c}(p^n), \hat{u}(p^n), u_*(p^n)) \tag{7.37}$$

where M_* denotes the operator of a chosen coordination strategy, set $n := n + 1$ and go to Step 1.

The presented algorithm structure of the AIBMF differs from the original IBMF structure in the existence of the *approximation loop* in the Step 1 (Step 1a - Step 1c), in place of a single solution of the local optimization problems (7.25). The goal of the approximation loop is to find, in a decentralized way, a solution $(\hat{c}(p^n), \hat{u}(p^n))$ of the not separable optimization problem

$$\begin{aligned} &\text{minimize}_{(c,u)} L_r(c, u, p^n, \alpha) \\ &\text{subject to: } \ (c_i, u_i) \in CU_i \ , \quad i = 1, ..., N. \end{aligned} \tag{7.38}$$

The idea of the approximation loop stems originally from the augmented price method of nonconvex optimization (Tatjewski, 1986, 1988, 1989), the updating formula for approximation points (7.35) was proposed and investigated there, together with the (rather natural) conditions assuring existence and uniqueness of solutions of the local optimization problems without convexity assumptions (as it is the case in standard price method and in the IBMF), provided the value of the penalty coefficient r is sufficient. Note that the approximation loop is performed *on the system model only,* it can be therefore calculated quickly and with great accuracy, not affecting the most time consuming set-points change.

The *coordination condition* of the AIBMF is certainly the same as the condition (7.28) in the IBMF: to find a *coordinating price* $\tilde{p}_r$ such that

$$\hat{u}(\tilde{p}_r) - HK_*(\hat{c}(\tilde{p}_r)) = 0. \tag{7.39}$$

Reasoning analogously as in the proof of theorem 3.5 in (Findeisen *et al.*, 1980) existence of the coordinating price $\tilde{p}_r$ can be proved under similar, rather natural conditions, see (Tatjewski, 1988). It is interesting that for model-reality difference, $F(c, u, \alpha) \neq F_*(c, u)$, the coordinating value $\tilde{p}_r$ does depend on the penalty coefficient value r and on the model-reality difference, whereas the corresponding optimal set-point value $\hat{c}(\tilde{p}_r)$ does not depend on r. To see that let us define

$$h(c, u, \alpha) = u - HF(c, u, \alpha) \tag{7.40}$$

and denote by $(\tilde{c}_r, \tilde{u}_r)$ a solution $(\hat{c}(\tilde{p}_r), \hat{u}(\tilde{p}_r))$ of the problem

$$\begin{aligned} &\text{minimize}_{(c,u)} L_r(c, u, \tilde{p}_r, \alpha) \\ &\text{subject to}: \ (c_i, u_i) \in CU_i \ , \quad i = 1, ..., N \end{aligned} \tag{7.41}$$

(existing without convexity assumptions provided r is sufficiently great). Necessary optimality conditions for (7.41) together with the coordinating condition are

$$q'(\tilde{c}_r, \tilde{u}_r, \alpha)^T + h'(\tilde{c}_r, \tilde{u}_r, \alpha)^T \tilde{p}_r + \\ + rh'(\tilde{c}_r, \tilde{u}_r, \alpha)^T h(\tilde{c}_r, \tilde{u}_r, \alpha) = 0 \tag{7.42a}$$

$$\tilde{u}_r - HK_*(\tilde{c}_r) = 0 \tag{7.42b}$$

where the local constraints $(c_i, u_i) \in CU_i \ , \ i = 1, ..., N$ have been omitted, to simplify the formulae and without loss of generality. Next, denote by $\tilde{p} = \tilde{p}_0$ the value of prices satisfying the conditions (7.42) for $r = 0$ and by $(\tilde{c}, \tilde{u})$ the corresponding optimal point:

$$q'(\tilde{c}, \tilde{u}, \alpha)^T + h'(\tilde{c}, \tilde{u}, \alpha)^T \tilde{p} = 0 \tag{7.43a}$$

$$\tilde{u} - HK_*(\tilde{c}) = 0. \tag{7.43b}$$

Equations (7.43) are necessary optimality conditions of local optimization problems and the coordinating condition for the standard IBMF. Let us write now (7.42) in a slightly reformulated form

$$q'(\tilde{c}_r, \tilde{u}_r, \alpha)^T + h'(\tilde{c}_r, \tilde{u}_r, \alpha)^T [\tilde{p}_r + rh(\tilde{c}_r, \tilde{u}_r, \alpha)] = 0 \tag{7.44a}$$

$$\tilde{u}_r - HK_*(\tilde{c}_r) = 0. \tag{7.44b}$$

Comparing this set of equations with (7.43) one easily concludes that

$$\tilde{p}_r = \tilde{p} - rh(\tilde{c}, \tilde{u}, \alpha) \tag{7.45}$$

and, moreover, $(\tilde{c}_r, \tilde{u}_r) = (\tilde{c}, \tilde{u})$, does not depend on the value of r.

Taking into account that

$$\tilde{u} = HK_*(\tilde{c}) = HF_*(\tilde{c}, HK_*(\tilde{c})) = HF_*(\tilde{c}, \tilde{u}) \tag{7.46}$$

we get the final AIBMF *coordinating price formula*

$$\tilde{p}_r = \tilde{p} - rH[F_*(\tilde{c}, \tilde{u}) - F(\tilde{c}, \tilde{u}, \alpha)]. \tag{7.47}$$

Note that if $F_*(\tilde{c}, \tilde{u}) = F(\tilde{c}, \tilde{u}, \alpha)$, i.e., the model is perfect at the equilibrium point, then the coordinating price $\tilde{p}_r$ does not depend on r and is equal to $\tilde{p}$, the coordinating value for the IBMF — if it exists. The last remark is essential — the IBMF is usually not applicable for nonconvex problems, whereas the AIBMF is applicable for a wide class of such problems due to convexifying properties of the penalty term in the augmented Lagrange function.

7.2.2 *Coordination strategies*

Efficient coordination strategy M_* (see Step 3 of the algorithm) capable to approach coordinating values of the prices in small number of iterations is crucial. Iterating for coordinating prices can be seen as solving iteratively equations of the coordination condition (7.39)

$$\hat{u}(\tilde{p}_r) - HK_*(\hat{c}(\tilde{p}_r)) = 0. \tag{7.48}$$

Denoting

$$R_*(p) = \hat{u}(p) - HK_*(\hat{c}(p)) \tag{7.49}$$

we shall look for an efficient iterative algorithm solving the equation $R_*(p) = 0$. It is known that Newton or quasi-Newton algorithms can be very effective in solving sets of nonlinear equations. Therefore, our goal will be to construct the following quasi-Newton algorithm

$$p^{n+1} = p^n - [R^{'}(p^n)]^{-1} R_*(p^n) \tag{7.50}$$

where $R(p)$ is a differentiable approximation of the operator $R_*(p)$. Recall that the latter is unknown as a true system operator, see (7.49).

Aiming at derivation of a reasonable formula for $R(p)$ let us try first to evaluate the formula for $R^{'}_*(p)$ assuming, temporarily, that $R_*(p)$ is known and differentiable. Denoting $R_*(p) = \bar{R}_*(\hat{x}(p))$ where $\hat{x}(p) = (\hat{c}(p), \hat{u}(p))$ we have

$$R^{'}_*(p) = \bar{R}^{'}_*(\hat{x}(p)) \cdot \hat{x}^{'}(p) \tag{7.51}$$

Because $\bar{R}_*(x) = u - HK_*(c)$, then applying the implicit function theorem yields

$$\bar{R}^{'}_*(\hat{x}(p)) = -[-(I - H(F_*)^{'}_u(\hat{x}(p)))^{-1} H(F_*)^{'}_c(\hat{x}(p)) \quad I]. \tag{7.52}$$

On the other hand, necessary optimality conditions for the problem (7.38) imply that

$$\hat{x}^{'}(p) = -[I\ 0]\begin{bmatrix} \nabla^2_{xx}\mathcal{L}_r(\hat{x}(p),p,\alpha,\hat{\mu}(p)) & G_A(\hat{x}(p))^T \\ G_A(\hat{x}(p)) & 0 \end{bmatrix}^{-1}\begin{bmatrix} h^{'}(\hat{x}(p),\alpha)^T \\ 0 \end{bmatrix} \tag{7.53}$$

where

$$\mathcal{L}_r(x,p,\alpha,\mu) = q(x,\alpha) + p^T h(x,\alpha) + 0.5r\|h(x,\alpha)\|^2 + \mu^T g(x) \tag{7.54}$$

$h(x,\alpha) = u - HF(c,u,\alpha)$ and $G_A(x) = g^{'}_A(x)$ is the matrix consisting of gradients of active local constraints (see (7.6), 7.12)). Combining (7.52) with (7.53) yields finally

$$R^{'}_*(p) = [(I - H(F_*)^{'}_u)^{-1}H(F_*)^{'}_c\ \ I][I\ 0]\begin{bmatrix} \nabla^2_{xx}\mathcal{L}_r & G_A^T \\ G_A & 0 \end{bmatrix}^{-1}\begin{bmatrix} (h^{'})^T \\ 0 \end{bmatrix} \tag{7.55}$$

where arguments on the right hand side have been omitted to shorten the description.

The most accurate available approximation $R^{'}(p)_a$ of the operator $R^{'}_*(p)$ will be obtained by replacing in (7.55) unknown mappings $(F_*)^{'}_u(\hat{x}(p))$ and $(F_*)^{'}_c(\hat{x}(p))$ by the models $F^{'}_u(\hat{x}(p),\alpha)$ and $F^{'}_c(\hat{x}(p),\alpha)$. Further, multiplying (7.55) by $I = [I - HF^{'}_u]^{-1}[I - HF^{'}_u]$ and using the block-matrix inverse formula the following form of the available approximation is obtained

$$R^{'}(p)_a = -[I - HF^{'}_u]^{-1}h^{'}[A_r^{-1} - A_r^{-1}G_A^T(G_A A_r^{-1}G_A^T)^{-1}G_A A_r^{-1}](h^{'})^T \tag{7.56}$$

where $A_r = \nabla^2_{xx}\mathcal{L}_r$ to shorten the presentation.

Let us denote

$$d^{''}_r(p) = h^{'}[A_r^{-1} - A_r^{-1}G_A^T(G_A A_r^{-1}G_A^T)^{-1}G_A A_r^{-1}](h^{'})^T. \tag{7.57}$$

It can be shown (Tatjewski, 1988) that

$$[d^{''}_r(p)]^{-1} = [D_K(p)]^{-1} - rI \tag{7.58}$$

where

$$D_K(p) = h^{'}[A_K^{-1} - A_K^{-1}G_A^T(G_A A_K^{-1}G_A^T)^{-1}G_A A_K^{-1}](h^{'})^T \tag{7.59}$$

$$A_K(\hat{x}(p),p,\alpha,\hat{\mu}(p)) = \nabla^2_{xx}\mathcal{L}_0(\hat{x}(p),p,\alpha,\hat{\mu}(p)) + rh^{''}(\hat{x}(p))h(\hat{x}(p)) \tag{7.60}$$

and $\mathcal{L}_0$ denotes $\mathcal{L}_r$ for $r = 0$. Therefore, the inverse of (7.56) can be presented in the form

$$[R^{'}(p)_a]^{-1} = [D_K(p)^{-1} - rI][I - HF_u^{'}(\hat{x}(p), \alpha)\ ]. \tag{7.61}$$

For the coordinating value of the pice vector $\tilde{p}_r$ the following holds

$$\begin{aligned} A_K(\hat{x}(\tilde{p}_r), \tilde{p}_r, \alpha, \hat{\mu}(\tilde{p}_r)) &= A_0(\hat{x}(\tilde{p}_r), \tilde{p}, \alpha, \hat{\mu}(\tilde{p}_r)) \\ &= \nabla^2_{xx}\mathcal{L}_0(\hat{x}(\tilde{p}_r), \tilde{p}, \alpha, \hat{\mu}(\tilde{p}_r)) \end{aligned} \tag{7.62}$$

where $\tilde{p}_r = \tilde{p} - rh(\hat{x}(\tilde{p}_r))$, see (7.47). Because $\tilde{p}$ does not depend on r and, as it has been shown in the previous subsection, also $\hat{x}(\tilde{p}_r)$ does not (and consequently $\hat{\mu}(\tilde{p}_r)$, which must be taken into account in the presence of local constraints, does not too) then the matrix $D_K(\tilde{p}_r)$ does not depend on r. This has significant practical implications. It implies that for sufficiently large value of r the rather complex formula (7.61) can be *well approximated by a very simple formula*

$$[R^{'}(p)_a]^{-1} \cong [R^{'}(p)]^{-1} = -r[I - HF_u^{'}(\hat{x}(p), \alpha)\ ] \tag{7.63}$$

locally in a neighbourhood of $\tilde{p}_r$.

Applying $[R^{'}(p)]^{-1}$ as an implementable form of the approximation of the operator $[R^{'}_*(p)]^{-1}$ results in the following final form of *the AIBMF coordination strategy*

$$p^{n+1} = p^n + r[I - HF_u^{'}(\hat{c}(p^n), \hat{u}(p^n), \alpha)][\hat{u}(p^n) - u_*(p^n)] \tag{7.64}$$

which is a very good approximate Newton formula for solving the coordination condition, the larger r the better the approximation. It should be stressed that such simple and accurate approximation of the operator (7.61) as given by the formula (7.63) is possible only when using the augmented Lagrange function, i.e. only in AIBMF. It is not possible when using IBMF which is based on the standard Lagrange function. That is why also for convex problems, the case when the original IBMF is usually applicable, it is advised to use it in the AIBMF version. The reason is that the coordination strategy (7.64) is more effective than any known IBMF coordination strategy, in the sense that it needs a smaller number of price iterations, and hence set-points iterations applied to the controlled system to reach the coordinating condition. Certainly, the number of local problems optimizations may be and usually is larger in AIBMF due to the existence of the approximation loop, but *the approximation loop is performed on the model only,* thus not affecting practical effectiveness of the method measured in the number of required set-point iterations.

Nevertheless, if one applies the IBMF then the following strategy can be advised as most effective (Tatjewski, 1988)

$$\begin{aligned} p^{n+1} &= p^n - k_p[h'_{(c,u)}(\hat{c}(p^n), \hat{u}(p^n), \alpha) B h'_{(c,u)}(\hat{c}(p^n), \hat{u}(p^n), \alpha)^T] \times \\ &\quad \times [I - HF'_u(\hat{c}(p^n), \hat{u}(p^n), \alpha)\ ][\hat{u}(p^n) - u_*(p^n)] \quad (7.65) \end{aligned}$$

when B is a certain matrix to be chosen by the user (the simplest choice is $B = I$) and k_p is a step coefficient. This strategy corresponds also to certain approximation of the Newton iteration formula, where $k_p B$ replaces the term $[A_0^{-1} - A_0^{-1} G_A^T (G_A A_0^{-1} G_A^T)^{-1} G_A A_0^{-1}]$, difficult for practical calculations by its complex form and possible structural changes corresponding to changes in local constraints activity.

It should be remembered that IBMF (and its AIBMF version) is a *suboptimal* algorithm, iterating set-points to satisfy the coordination condition (7.28). This does not imply the true system optimum, although simulation studies indicate that application of the method leads to significant improvement in the system performance (Findeisen *et al.*, 1980; Tatjewski, 1988). The ISOPE type algorithms yielding the true system optimum will be presented in the next section. However, these algorithms are much more complex and require more information from the system.

An extension of the IBMF like algorithms to the steady-state systems with disturbance inputs varying fast enough so that the IBMF algorithm is not able to reach convergence can be found in (Findeisen *et al.*, 1980; Brdyś, 1980). A decentralized counterpart of the IBMF with local feedbacks was investigated by (Brdyś and Ulanicki, 1978; Brdyś and Malinowski, 1979) and (Brdyś *et al.*, 1980) for slowly-varying disturbances and by (Brdyś and Michalak, 1978; Brdyś, 1980; Brdyś *et al.*, 1982) for fast disturbances.

7.2.3 *Simulation studies*

Example 7.1

A simple system consisting of two interconnected subsystems structured as depicted in Fig. 7.5 will first be considered (Tatjewski, 1988). True

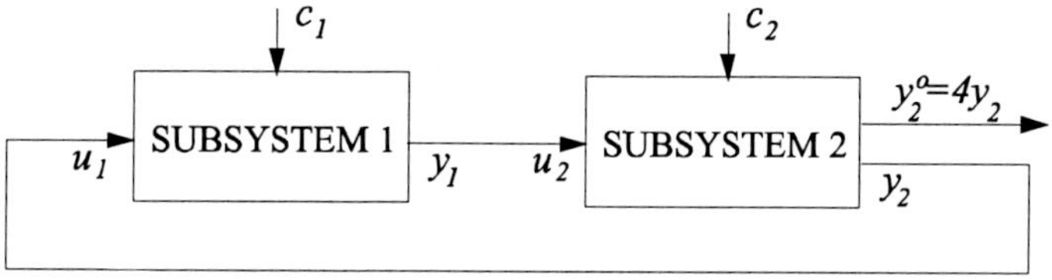

Fig. 7.5 Structure of the system in the example problem 7.1.

subsystem output mappings are as follows

$$\begin{aligned} y_1 &= F_{*1}(c_1, u_1) = 2.1c_1 + u_1 + \beta c_1 u_1 && (7.66a) \\ y_2 &= F_{*2}(c_2, u_2) = 0.6c_2 + 0.55u_2 && (7.66b) \\ y_2^o &= F_{*2}^o(c_2, u_2) = 4F_{*2}(c_2, u_2) && (7.66c) \end{aligned}$$

where β is a parameter with nominal value $\beta_n = 0.5$. Models of the output mappings are as follows

$$\begin{aligned} y_1 &= F_1(c_1, u_1) = 2c_1 + u_1 && (7.67a) \\ y_2 &= F_2(c_2, u_2) = 0.5c_2 + 0.5u_2 && (7.67b) \\ y_2^o &= F_2^o(c_2, u_2) = 4F_2(c_2, u_2) && (7.67c) \end{aligned}$$

whereas local performance functions are

$$\begin{aligned} Q_1(c_1, u_1, y_1) &= 32(c_1)^2 - 16c_1 + (y_1 - 1)^2 && (7.68a) \\ Q_2(c_2, u_2, y_2) &= 10(c_2)^2 + 4c_2u_2 - 0.5(y_2^o)^2 && (7.68b) \end{aligned}$$

where all variables are one-dimensional, $c_1 \in \mathbb{R}^1$, $u_1 \in \mathbb{R}^1$,*etc.* Let us note that originally there was one output from the second subsystem, with 20% of it fed back to the first subsystem in a recycle. Therefore, to formulate the structure in the general form used in this chapter (with structure matrix composed of ones and zeros only) the output stream was split into two resulting in the presented required equivalent structure.

There is also an inequality constraint in the first subsystem

$$CU_1 = \{(c_1, u_1) : \ 2c_1 + u_1 - 2.25 \leq 0\}, \quad CU_2 = \mathbb{R}^2. \qquad (7.69)$$

The presented problem is nonconvex, with nonconvex model optimization problem with a duality gap. Solving it, i.e., minimizing the performance function $Q(c,u,y) = Q_1(c_1,u_1,y_1)+Q_2(c_2,u_2,y_2)$ with respect to the equality model output constraints (7.67a),(7.67b) and the inequality constraint results in the model-optimal point $\hat{c}_1 = 0.5$, $\hat{c}_2 = 0.25$, $\hat{u}_1 = 1.25$, $\hat{u}_2 = 2.25$.

Applying the model-optimal set-points $\hat{c}_1$ and $\hat{c}_2$ to the true controlled system (with $\beta = \beta_n = 0.5$) results in different than model-optimal interaction values $u_*(\hat{c}) = HK_*(\hat{c})$ and in violation of the inequality constraint, $2\hat{c}_1 + u_{*1}(\hat{c}) - 2.25 = 1.078$, due to model-reality difference. Even from this reason only application of the optimizing set-point correction mechanism is necessary. Application of the AIBMF algorithm results in the set-point values $\hat{c}_1(\tilde{p}_r) = 0.3793$, $\hat{c}_2(\tilde{p}_r) = 0.1293$ and corresponding interaction values $u_{*1}(\hat{c}(\tilde{p}_r)) = 1.4915$, $u_{*2}(\hat{c}(\tilde{p}_r)) = 2.5708$, satisfying the inequality constraint. The corresponding true performance function value is $Q(\hat{c}(\tilde{p}_r), HK_*(\hat{c}(\tilde{p}_r), K_*(\hat{c}(\tilde{p}_r)) = -15.298$ and occurs to be close to the true optimal value $Q(\hat{c}_*, HK_*(\hat{c}_*), K_*(\hat{c}_*)) = -15.447$. Data describing a single run of the AIBMF algorithm with $r = 32$, $\epsilon = 0.0001$, $\epsilon_a = 0.0001$ and the coordination strategy

$$p^{n+1} = p^n + k_p r[I - HF_u^{'}(\hat{c}(p^n), \hat{u}(p^n), \alpha)][\hat{u}(p^n) - u_*(p^n)] \tag{7.70}$$

i.e., the strategy (7.64) with added possibility to adjust the steplength by adjusting k_p are presented in Table 7.1, for $k_p = 0.9$.

n	$k(n)$	$\|d_*^n\|$	Q_*^n	constraint
0	1	2.02143	−30.01004	1.07800
1	18	0.45842	−13.72031	−0.13829
2	14	0.08383	−15.69545	0.03593
3	10	0.01439	−15.24209	−0.00506
4	8	0.00265	−15.31006	0.00112
5	4	0.00045	−15.29538	−0.00020
6	2	0.00005	−15.29816	0.00004

Table 7.1 Data of a single run of the AIBMF algorithm for the example problem.

The initial point for the simulation presented in the table was the model-optimal point given above. In the first column of the table current number n of the main algorithm iteration is given, in the second number of approximation loop (inner) iterations $k(n)$ needed to satisfy the stop criterion (7.34) in the n-th iteration and in the following columns the corresponding

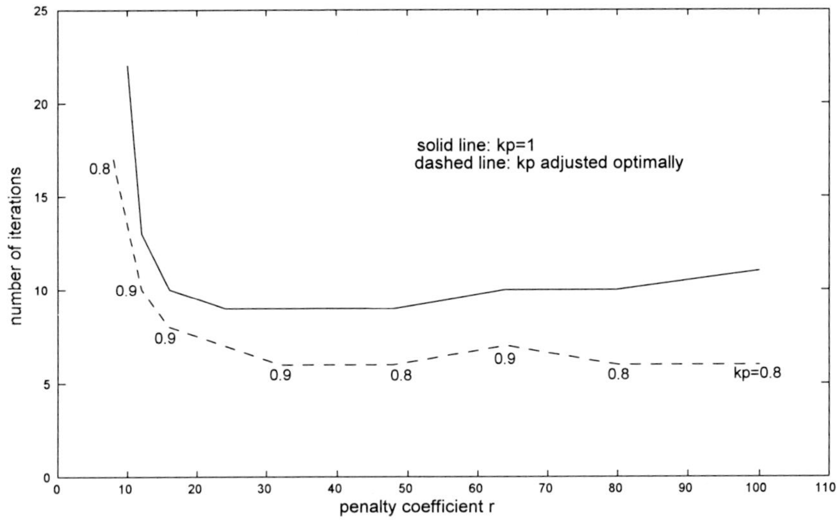

Fig. 7.6 Dependence of the number of iterations on penalty coefficient r, example 7.1 (with $\beta = 0.5$).

values of the coordination condition $||d_*^n||$, of the true performance function $Q_*^n = Q(\hat{c}(\tilde{p}^n), HK_*(\hat{c}(\tilde{p}^n), K_*(\hat{c}(\tilde{p}^n))$ and the value of the left-hand side of the inequality constraint. It is interesting to note that constraint violation is rapidly reduced (see last column).

To investigate the influence of the penalty coefficient r on the algorithm convergence several simulation runs were made. The results are summarized in Figure 7.6 for two cases: the coordination strategy (7.64) with unit step size (solid line) and the coordination strategy (7.70) with optimally chosen step-size $k_p = \hat{k}_p$ (adjusted with an increment 0.1, dashed line — optimal values of k_p are given by the points defining the line). The results fully confirm the theoretical analysis of the AIBMF presented in the previous subsection: if the penalty coefficient is over certain threshold (about 15 in the considered non-convex example) then the algorithm becomes effective and practically not sensitive to the value of r — the number of iterations needed to achieve the desired accuracy ($\epsilon = 0.0001$) is almost constant. The results also show that adjusting k_p allows to improve the convergence, although the improvement is not substantial.

In Fig. 7.7 simulation results made for different values of the nonlinearity parameter β are presented, also for the coordination strategy (7.64) with unit step-size (solid line) and the coordination strategy (7.70) with optimally chosen step-size $k_p = \hat{k}_p$ (dashed line). The aim of the analysis is to investigate the influence of the model-reality difference on the algorithm convergence — the larger β the larger the structural and the parametric difference between F and F_*. The obtained results indicate that the convergence becomes slower with the increase in the difference and that additional step-size adjustment is essential for larger differences. Even slight damping of the step-size results then with much lower number of iterations needed to achieve the desired accuracy.

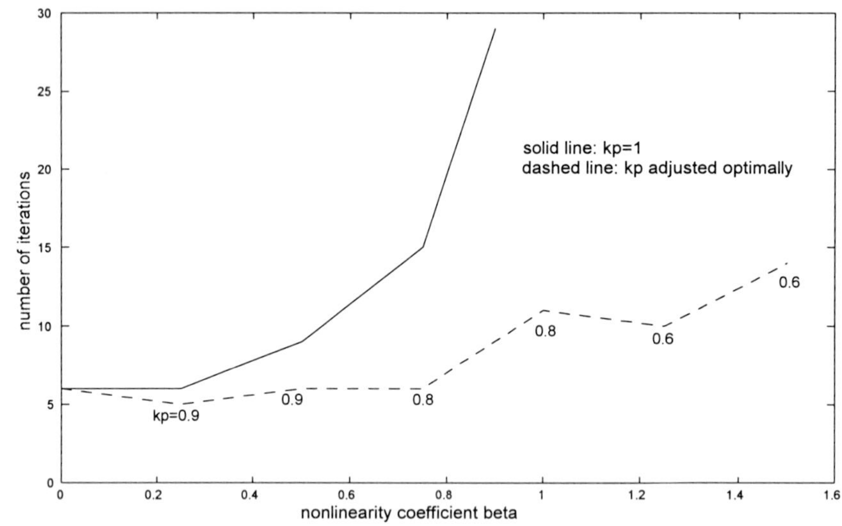

Fig. 7.7 Dependence of the number of iterations on nonlinearity coefficient β, example 7.1 (with $r = 32$).

□

Example 7.2

A three-subsystem example investigated in (Findeisen *et al.*, 1980; Shao and Roberts, 1983) will be considered, with structure presented in Fig. 7.8.

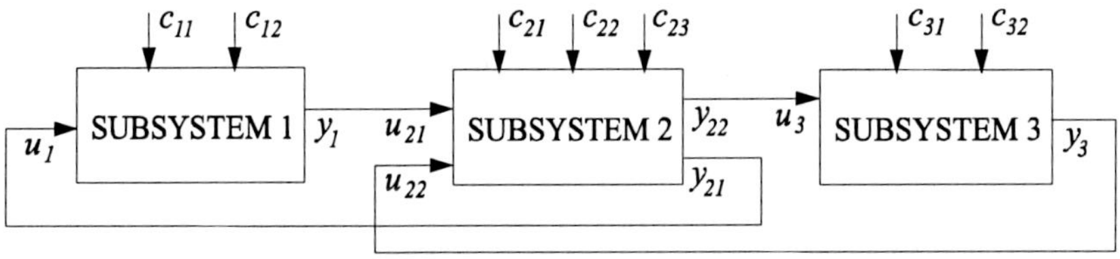

Fig. 7.8 Structure of the system in the example problem 7.2.

The true subsystem equations are as follows

$$y_1 = F_{*1}(c_1, u_1) = 1.3c_{11} - c_{12} + 2u_1 + 0.5c_{11}u_1 \quad (7.71a)$$

$$y_{21} = F_{*21}(c_2, u_2) = c_{21} - c_{22} + 1.2u_{21} - 3u_{22} + 0.1(c_{22})^2 \quad (7.71b)$$

$$y_{22} = F_{*22}(c_2, u_2) = 2c_{22} - 1.25c_{23} - u_{21} + u_{22} + \\ +0.25c_{22}c_{23} + 0.1 \quad (7.71c)$$

$$y_3 = F_{*3}(c_3, u_3) = 0.8c_{31} + 2.5c_{32} - 4.2u_3 \quad (7.71d)$$

whereas equations of their models are

$$F_1(c_1, u_1) = c_{11} - c_{12} + 2u_1 \quad (7.72a)$$

$$F_{21}(c_2, u_2) = c_{21} - c_{22} + u_{21} - 3u_{22} \quad (7.72b)$$

$$F_{22}(c_2, u_2) = 2c_{22} - c_{23} - u_{21} + u_{22} \quad (7.72c)$$

$$F_3(c_3, u_3) = c_{31} + 2.5c_{32} - 4u_3. \quad (7.72d)$$

Performance functions and constraint sets are as follows

$$Q_1(c_1, u_1) = 5(c_{11} + c_{12} - 1)^2 \quad (7.73a)$$

$$Q_2(c_2, u_2) = 2(c_{12} - 2)^2 + (c_{22})^2 + 3(c_{23})^2 + \\ +4(u_{21})^2 + (u_{22})^2 \quad (7.73b)$$

$$Q_3(c_3, u_3) = (c_{31} + 1)^2 + 2.5(c_{32})^2 + (u_3 - 1)^2 \quad (7.73c)$$

$$Q(c, u) = \sum_{i=1}^{3} Q_i(c_i, u_i) \quad (7.73d)$$

$$CU_1 = \{(c_1, u_1) : \quad (c_{11})^2 + (c_{12})^2 \leq 1, \quad 0 \leq u_{11} \leq 0.5\} \tag{7.74}$$

$$CU_2 = \{(c_2, u_2) : \; 0.5c_{21} + c_{22} + 2c_{23} \leq 1, \quad 4(c_{21})^2 + 2c_{21}u_{21} + \\ +0.4u_{21} + c_{21}c_{23} + 0.5(c_{23})^2 + (u_{21})^2 \leq 4\} \tag{7.75}$$

$$CU_3 = \{(c_3, u_3) : \quad c_{31} + u_3 + 0.5 \geq 0, \quad 0 \leq c_{31} \leq 1\}. \tag{7.76}$$

The subsystem models are linear and performance functions convex, therefore the price method can be successfully applied to the optimization of the model of the presented example problem and the IBMF method to the optimizing control of the interconnected system, despite complexity of the constraints, see (Findeisen *et al.*, 1980). Therefore, a comparison of the effectiveness of the IBMF and AIBMF algorithms is also possible.

First, convergence of the AIBMF algorithm with coordination strategy (7.64) were investigated for various values of the penalty coefficient r, the results are presented in Fig. 7.9.

The required tolerances were $\epsilon = 0.0001$, $\epsilon_a = 0.0001$. The algorithm converged to the value of the performance function of the controlled system $Q(\hat{c}(\tilde{p}_r), HK_*(\hat{c}(\tilde{p}_r), K_*(\hat{c}(\tilde{p}_r)) = 6.3356$ that occurs to be close to the true

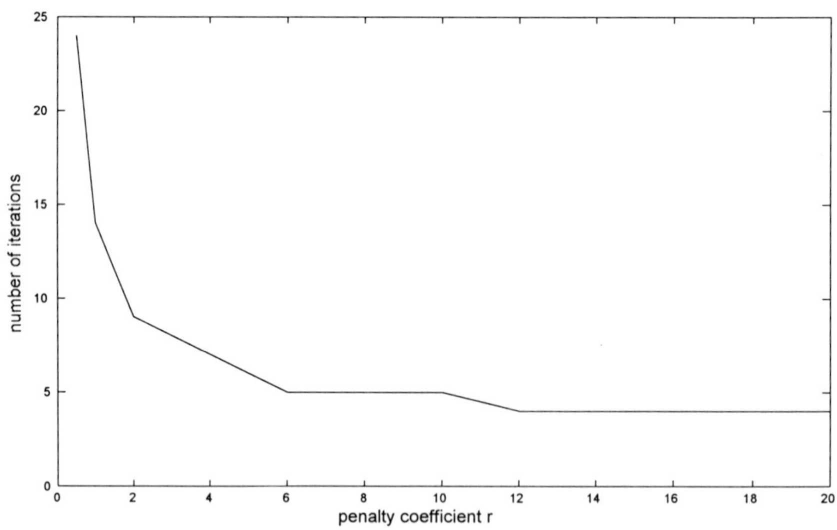

Fig. 7.9 Dependence of the number of iterations on penalty coefficient r, example 7.2 (with $k_p = 1$).

optimal value $Q(\hat{c}_*, HK_*(\hat{c}_*), K_*(\hat{c}_*)) = 6.3266$. The results presented in Fig. 7.9 fully confirm the theoretical analysis: with the increase of r the algorithm approaches high efficiency of the Newton type iteration.

A comparison of the AIBMF and IBMF algorithms is presented in Fig. 7.10. The AIBMF algorithm used the coordination strategy (7.64), i.e., with the unit step-size, and the penalty coefficient value $r = 12$. The IBMF algorithm was applied using the best advisable coordination strategy (7.65) with $B = I$ and optimally chosen step-size parameter $k_p = 3.75$ (i.e., with $k_pB = 3.75I$). The comparison shows behavior of the norm of the coordination constraint $d_*^n = \|\hat{u}(p^n) - u_*(p^n)\|$ during the iteration process commencing after the threshold value $\epsilon = 0.0001$. The comparison shows clearly very high effectiveness of the AIBMF method, much better than in the case of the IBMF. A broader comparison, also with another approaches known in the literature, can be found in (Tatjewski, 1988). It is omitted here, because none of the method investigated there was found to be competitive with the AIBMF approach, both in the convergence rate and in the applicability conditions.

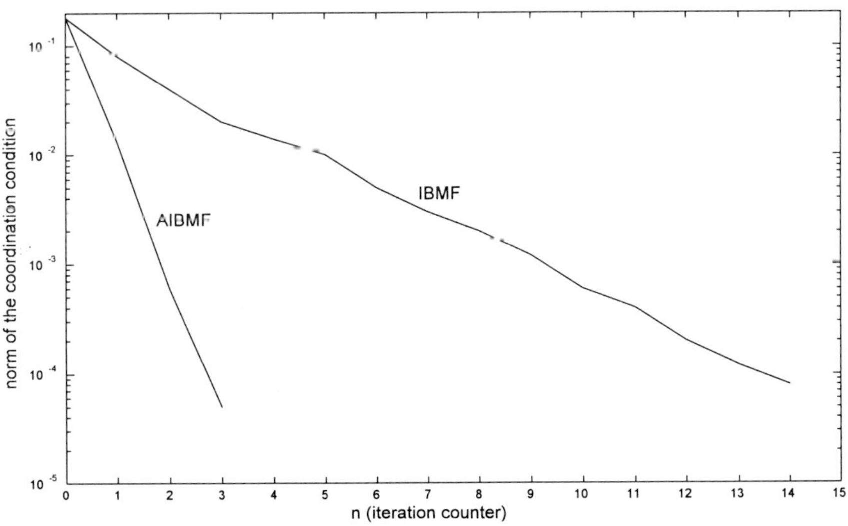

Fig. 7.10 Comparison of the convergence rate, AIBMF versus IBMF.

7.3 ISOPE Double-Loop Iterative Algorithms with Input-Output Measurements

7.3.1 *ISOPE double-loop iterative algorithms*

Although the IBMF offers a very attractive way of handling model-reality differences the solution obtained is suboptimal in general. In this section the ISOPE mechanism is employed in order to achieve optimal steady state controls. Similarly as in Chapter 4, it is assumed that models (7.3a) and (7.3b) of the unknown subsystem input-output mappings are point parametric on the sets CU_i, $i = 1, .., N$. This means that for any $(\bar{c}_i, \bar{u}_i) \in CU_i$ there is, possibly not unique, parameter value $\bar{\alpha}_i$ so that

$$F_{*i}(\bar{c}_i, \bar{u}_i) = F_i(\bar{c}_i, \bar{u}_i, \bar{\alpha}_i) \tag{7.77a}$$

$$F^o_{*i}(\bar{c}_i, \bar{u}_i) = F^o_i(\bar{c}_i, \bar{u}_i, \bar{\alpha}_i) \quad i = 1, ...N. \tag{7.77b}$$

Clearly, the parameter value depend on the inputs. With the point parametric models a search in the system output space that uses unknown system input-output mappings can be carried out by varying control and interaction inputs and parameters in the models. If the model structure is perfect then there is a unique parameter value so that with this value (7.77a) and (7.77b) hold for any $(c_i, u_i) \in CU_i$. The subsystem models are then *parametric* and if they are *identifiable* the unknown parameter value can be calculated form input-output data. The optimal controls can then be produced entirely based on the identified models by employing one of the hierarchical optimization algorithms, e.g., the price method (Findeisen *et al.*, 1980).

It is easy to verify that the mappings $K(\cdot, \cdot)$ and $K^0(\cdot, \cdot)$ (see (7.17) and (7.20)) are point parametric models of the system explicit interconnection output mapping and $K_*(\cdot)$ (see (7.15)) and the system explicit external output mapping $K^0_*(\cdot)$ (see (7.18)), respectively

Assuming that the objective is to minimize the performance function in the general form

$$Q(c, u, y, y^o) = \sum_{i=1}^{N} Q_i(c_i, u_i, y_i, y^o_i) \tag{7.78}$$

the System Optimizing Control Problem (SOCP) can be formulated as follows

$$
\begin{aligned}
&\text{minimize} \sum_{i=1}^{N} Q_i(c_i, u_i, y_i, y_i^o) \\
&\text{subject to}: \; y_i = F_{*i}(c_i, u_i) \qquad\qquad (7.79)\\
&\qquad\qquad y_i^o = F_{*i}^o(c_i, u_i) \\
&\qquad\qquad u_i = \textstyle\sum_{i=1}^{N} H_{ij} y_j \\
&\qquad\qquad (c_i, u_i) \in CU_i, \qquad i = 1, \ldots, N
\end{aligned}
$$

(analogous to the OCP defined for a single plant in Section 2.1.4).

Using the above point parametric models the System Optimizing Control Problem can be written in an equivalent form SOCP1 as:

$$\text{minimize}_{c,u,\alpha} q(c, u, \alpha) \qquad (7.80a)$$

$$
\begin{aligned}
\text{subject to:} \quad F^o(c, u, \alpha) &= K_*^o(c) && (7.80b)\\
F(c, u, \alpha) &= K_*(c) && (7.80c)\\
u &= HF(c, u, \alpha) && (7.80d)\\
(c, u) &\in CU && (7.80e)\\
\alpha &\in A && (7.80f)
\end{aligned}
$$

where

$$q(c, u, \alpha) = Q(c, u, F(c, u, \alpha), F^o(c, u, \alpha)). \qquad (7.80g)$$

A rigorous proof of the equivalence was given by (Brdyś, 1983). Here we shall present only intuitive arguments. Consider a control input that is feasible, hence satisfying the constraints (7.80b),..., (7.80f). This control input forces certain interconnection and external outputs and the interaction input in the interconnected systems according to the mappings $K_*(\cdot), K_*^o(\cdot), HK_*(\cdot)$. The equations (7.80b) and (7.80c) ensure that the outputs are produced by the model. The equations (7.80c) and (7.80d) ensure that the interaction input is also forced by the model. Hence, the optimized performance value corresponding to the considered control input is the same as in reality.

We shall assume throughout this section that:

Assumption 1: All the input-output mappings and their models are continuously *Fréchet* differentiable on their domains.

The SOCP1 formulation is a base for developing the double-loop iterative optimizing control strategies. Let us note that the formulation is a mix of knowledge provided by the models and gathered from the measurements $(K_*(\cdot), K_*^o(\cdot))$. A thorough analysis of possible hierarchical control strategies of the ISOPE type was given by (Brdyś and Roberts, 1986). It was found there that the strategies with specially nested iterative loops that heavily exploit the system models managed to achieve optimal controls with a smaller number of iterations carried out on a real system than the other strategies did. As computing power is not a problem these days these strategies called *model based double-loop* are of primary importance from a practical point of view. However, applicability conditions of the model based double-loop control algorithms are rather restricted by their convergence properties. Weaker conditions for the convergence can be derived for so called *system based double-loop* strategies. This is achieved at a cost of increased number of iterations carried out on a real system. Both types of double-loop strategies are presented in this section.

Depending on how the measurements are utilized by the double-loop strategies we shall distinguish between the strategies using only output measurements and the strategies using both the output and the interaction input measurements. Availability of the output measurements is necessary for implementation of ISOPE type of algorithm. However, using the interaction input measurements is optional. On one hand these measurements bring new information to the controller. On the other hand the ISOPE algorithms need to produce derivatives of the interaction inputs in order to utilize this information. A quality of the derivatives depends on the measurement noise structure and its level. Therefore, it may happen that it is better to abandon the measurements and predict the interaction inputs by using models.

For each of the model based double-loop strategies two possibilities of the inner loop synthesis will be exploited bearing in mind convergence properties of the inner loop algorithm. Tightening the inner loop reduces number of iteration carried out on a real system. However, this makes more difficult to ensure convergence of an algorithm.

7.3.2 *Double-loop iterative algorithms with input-output measurements*

Following development of basic ISOPE algorithm in Chapter 4 a new variable $v \in \mathbb{R}^{n_c}$ is introduced in order to separate the parameter estimation and modified optimization tasks. Also the interaction input measurements are introduced into the equations (7.80b) and (7.80c) Hence, both the output measurements and the interaction input measurements will be used to estimate the model parameters. The SOCP1 is then transformed into the equivalent form SOCP2 as:

$$\text{minimize}_{c,v,u,\alpha} q(v, u, \alpha) \tag{7.81a}$$

$$\begin{aligned} \text{subject to: } F^o(c, HK_*(c), \alpha) &= K_*^o(c) && (7.81b)\\ F(c, HK_*(c), \alpha) &= K_*(c) && (7.81c)\\ u &= HF(v, u, \alpha) && (7.81d)\\ g(v, u) &\leq 0 && (7.81e)\\ \alpha &\in A && (7.81f)\\ v &= c. && (7.81g) \end{aligned}$$

The above problem is to be further modified maintaining its equivalent form. Firstly, the equations (7.81b) and (7.81c) are combined as:

$$\begin{aligned} \overline{F}(c, HK_*(c), \alpha)^T &= [F^o(c, HK_*(c), \alpha)^T, F(c, HK_*(c), \alpha)^T]^T && (7.82a)\\ \overline{K}_*(c)^T &= [K_*^o(c)^T, K_*(c)^T]^T && (7.82b)\\ \overline{F}(c, HK_*(c), \alpha) &= \overline{K}_*(c). && (7.82c) \end{aligned}$$

Secondly, the problem is convexified similarly as in Chapter 4. However, there is now a new variable that is the interaction input and this requires an additional treatment. The convexified performance function reads (Brdyś and Roberts, 1986; Brdyś *et al.*, 1990a; Tatjewski *et al.*, 1990):

$$q(v, u, \alpha) + \rho_c\|v - c\|^2 + \rho_u\|u - HK_*(c)\|^2. \tag{7.83}$$

The coefficients ρ_c, ρ_u need to be selected sufficiently large so that required convex structure of the optimization problem can be achieved. Note that the interaction input measurements have now been introduced into the performance function. The equivalent form SOCP3 of SOCP2 can now be stated as follows:

$$\text{minimize}_{c,v,u,\alpha} \left\{q(v, u, \alpha) + \rho_c\|v - c\|^2 + \rho_u\|u - HK_*(c)\|^2\right\} \tag{7.84a}$$

$$\begin{aligned}\text{subject to}: \overline{F}(c, HK_*(c), \alpha) &= \overline{K}_*(c) && (7.84b)\\ u &= HF(v, u, \alpha) && (7.84c)\\ g(v, u) &\leq 0 && (7.84d)\\ \alpha &\in A && (7.84e)\\ v &= c. && (7.84f)\end{aligned}$$

The Lagrange function for the SOCP3 reads:

$$\begin{aligned}L(c, \nu, u, \alpha, \lambda, p, \xi, \mu) &= q(\nu, u, \alpha) + \rho_c \|\nu - c\|^2 + \rho_u \|u - HK_*(c)\|^2 + \\ &+ \lambda^T (c - \nu) + p^T (u - HF(\nu, u, \alpha) + \mu^T g(\nu, u) + \\ &+ \xi^T (\overline{F}(c, HK_*(c), \alpha) - \overline{K}_*(c)) \qquad (7.85)\end{aligned}$$

where λ, p, ξ, μ are the Lagrange multipliers associated with the constraints (7.84f), (7.84c), (7.84b) and (7.84d), respectively.

Let us consider the Kuhn-Tucker necessary optimality conditions for SOCP3:

$$L'_{[\nu,u]}(c, \nu, u, \alpha, \lambda, p, \xi, \mu)^T = q'_{[\nu,u]}(\nu, u, \alpha)^T + 2\rho_c \begin{bmatrix} \nu - c \\ 0 \end{bmatrix} +$$

$$+2\rho_u \begin{bmatrix} 0 \\ u - HK_*(c) \end{bmatrix} - \begin{bmatrix} \lambda \\ 0 \end{bmatrix} + \{(u - HF(\nu, u, \alpha))'_{[\nu,u]}\}^T p + g'(\nu, u)^T \mu = 0$$

$$g(\nu, u) \leq 0, \ \mu \geqslant 0, \ \mu^T g(\nu, u) = 0 \qquad (7.86a)$$

$$L'_c(c, \nu, u, \alpha, \lambda, p, \xi, \mu)^T = 2\rho_c (c - \nu) - 2\rho_u K'_*(c)^T H^T (u - HK_*(c)) +$$

$$+\lambda + ([\overline{F}(c, HK_*(c), \alpha)]_c^{'T} - \overline{K}'_*(c)^T)\xi = 0 \qquad (7.86b)$$

$$L'_\alpha(c, \nu, u, \alpha, \lambda, p, \xi, \mu)^T = q'_\alpha(\nu, u, \alpha)^T - F'_\alpha(\nu, u, \alpha)^T H^T p +$$

$$+\overline{F}'_\alpha(c, HK_*(c), \alpha)^T \xi = 0 \qquad (7.86c)$$

$$\begin{aligned}L'_\lambda(c, \nu, u, \alpha, \lambda, p, \xi, \mu)^T &= \nu - c = 0 && (7.86d)\\ L'_p(c, \nu, u, \alpha, \lambda, p, \xi, \mu)^T &= u - HF(\nu, u, \alpha) = 0 && (7.86e)\\ L'_\xi(c, \nu, u, \alpha, \lambda, p, \xi, \mu)^T &= \overline{F}(c, HK_*(c), \alpha) - \overline{K}_*(c) && (7.86f)\end{aligned}$$

where $[\overline{F}(c, HK_*(c), \alpha)]_c^{'T}$ (see (7.86b) denotes transpose of the derivative with respect to c of the mapping $\varphi_* : \mathbb{R}^{n_c} \times \mathbb{R}^s \mapsto \mathbb{R}^{m+m_0}$

$$\varphi_*(c, \alpha) = \overline{F}(c, HK_*(c), \alpha). \tag{7.87}$$

The double-loop ISOPE algorithm with input-output measurements will be derived as a double-loop structured iterative method for solving the stated above necessary optimality conditions corresponding to the SOCP3. Hence, we shall now proceed analogously as in Chapter 4. From (7.86c) the multiplier ξ can be calculated as:

$$\begin{aligned}\xi(c, \nu, u, \alpha, p) = {} & [\overline{F}'_\alpha(c, HK_*(c), \alpha)\overline{F}'_\alpha(c, HK_*(c), \alpha)^T]^{-1}\times \\ & \times\overline{F}'_\alpha(c, HK_*(c), \alpha)[F'_\alpha(\nu, u, \alpha)^T H^T p - q'_\alpha(\nu, u, \alpha)^T]. \end{aligned} \tag{7.88}$$

The inverse in (7.88) exists for reasonably parameterized models.

Substituting the expression for ξ into the equation (7.86b) yields an expression for the multiplier λ in terms of c, ν, u, α and p as:

$$\begin{aligned}\lambda(c, \nu, u, \alpha, p) &= 2\rho_c(c - \nu) + 2\rho_u K'_*(c)^T H^T(u - HK_*(c)) + \\ &\quad +(\overline{K}'_*(c)^T - \varphi'_*(c)^T)\xi \end{aligned} \tag{7.89}$$

where ξ is given by (7.88).

It follows from (7.86d) that at a solution of the optimality conditions $c = \nu$. Moreover, it can be easily checked by using (7.81c) that the interaction input u that is equal to $HK_*(c)$ satisfies the equation (7.81d) with $c = \nu$. Hence, at the solution $u = HK_*(c)$ and the expression (7.89) for λ simplifies to:

$$\begin{aligned}\lambda(c, \alpha, p) \triangleq {} & \lambda(c, c, HK_*(c), \alpha, p) = \\ & (\overline{K}'_*(c)^T - \varphi'_*(c)^T)\xi(c, HK_*(c), \alpha, p). \end{aligned} \tag{7.90}$$

Let us now examine the formula (7.88) for ξ at the solution of the optimality conditions. Let us define a combined output vector and the corresponding performance function as:

$$\overline{y} = [y^{oT}, y^T]^T \tag{7.91}$$

$$\overline{Q}(c, u, \overline{y}) = Q(c, u, y, y^o). \tag{7.92}$$

Then the following holds:

$$q'_\alpha(c, u, \alpha)^T = \overline{F}'_\alpha(c, u, \alpha)^T \overline{Q}_{\overline{y}}(c, u, \overline{F}(c, u, \alpha))^T. \tag{7.93}$$

Substituting (7.93) with $u = HK_*(c)$ into (7.88) yields:

$$\xi(c,\alpha,p) \triangleq \xi(c, HK_*(c),\alpha,p) = -\overline{Q}_{\overline{y}}(c, HK_*(c), \overline{F}(c, HK_*(c),\alpha))^T + \\ +[\overline{F}'_\alpha(c, HK_*(c),\alpha)\overline{F}'_\alpha(c, HK_*(c),\alpha)^T]^{-1}\times \\ \times\overline{F}'_\alpha(c, HK_*(c),\alpha)F'_\alpha(c, HK_*(c),\alpha)^T H^T p. \tag{7.94}$$

Let us notice that in case of not interconnected system the terms $HK_*(c), p, H$ disappear and also $\overline{F}(\cdot) = \overline{F}^o(\cdot) = F(\cdot)$. The formula (7.94) becomes then the same as the formula (4.7) in Chapter 4. Also, substituting the multiplier ξ into (7.90) gives an expression for λ that is exactly the same as derived in Chapter 4 (see (4.10)).

The second term in (7.94) looks complicated. However, this is only superficial. Indeed, suppose that all the outputs are of the interconnection type. Then $y = y^o$. The second term in (7.94) becomes then equal to $H^T p$, the expression (7.94) reads

$$\xi(c,\alpha,p) = -Q'_y(c, HK_*(c), F(c, HK_*(c),\alpha))^T + H^T p \tag{7.95}$$

yielding the expression for λ (see (7.90) as

$$\lambda(c,\alpha,p) = (\varphi'_*(c)^T - K'_*(c)^T)[Q'_y(c, HK_*(c), F(c, HK_*(c),\alpha))^T - H^T p]. \tag{7.96}$$

A structure of the formula (7.96) is very clear. The modifier λ needs to cater not only for model-reality differences but also for meeting the interconnections.

So far we have managed to solve two out of six necessary optimality conditions that is (7.86b), (7.86c) and and in the solution the condition (7.86d) has been imbedded. The condition (7.86f) allows estimation of the model parameter vector α given the output and interaction input measurements corresponding to control input c.

Hence, the *parameter estimation problem* PEP is defined as of finding for given control input c such parameter value $\hat{\alpha}(c)$ that

$$F^o(c, HK_*(c), \hat{\alpha}(c)) = K^o_*(c) \tag{7.97a}$$

and

$$F(c, HK_*(c), \hat{\alpha}(c)) = K_*(c) \tag{7.97b}$$

where $\alpha \in A$.

The solution, possibly not unique, defines a point-to-set mapping $\hat{\alpha}(\cdot)$. Such formulation of the PEP is purely deterministic as a formulation of the optimizing control problem is also deterministic. In practise, the measurement noise will be handled by such well known techniques as for example Weighted Least Squares or Set-Bounded Parameter Estimation (Walter and Pronzato, 1997) or Set-Bounded Parameter Estimation with Moving Measurement Window (Brdyś, 1999). Alternatively, simple filtering of the measurements may be sufficient. A non-deterministic formulation of ISOPE still remains a challenge.

Finally, the remaining conditions (7.86a), (7.86e) and still (7.86d) are met as follows. First a *modified model-based optimization problem* MMOP is defined for prescribed vectors c and p as:

$$\begin{aligned} &\text{minimize}_{\nu,u}\{q(\nu,u,\alpha)+\rho_c\|\nu-c\|^2+\rho_u\|u-HK_*(c)\|^2+ \\ &\qquad\qquad -\lambda^T\nu+p^T(u-HF(\nu,u,\alpha))\} \\ &\text{subject to: } g(\nu,u)\leq 0 \end{aligned} \tag{7.98}$$

where $\alpha=\hat{\alpha}(c)$ and λ is defined by (7.90) and (7.94) under $\alpha=\hat{\alpha}(c)$.

Hence, the MMOP is defined for prescribed values of the control inputs c and prices p. These are the MMOP parameters. The control inputs are applied to the system, the corresponding interaction inputs are measured and with this information PEP is solved to produce estimated parameter values $\hat{\alpha}(c)$.Finally, the modifier is calculated by using the values of c, p nad $\hat{\alpha}(c)$ in (7.90) and (7.94).

Let us denote the MMOP solution by

$$\hat{\nu}(c,p) \text{ and } \hat{u}(c,p). \tag{7.99}$$

Second, meeting the conditions (7.86a) and (7.86e) and completing the meeting of the condition (7.86d) is achieved by solving the following set of equations:

$$\hat{\nu}(c,p) = c \tag{7.100}$$

$$\hat{u}(c,p) = HF(\hat{\nu}(c,p),\hat{u}(c,p),\hat{\alpha}(c)). \tag{7.101}$$

Indeed, writing down Kuhn-Tucker necessary conditions for optimality for MMOP yields the condition (7.86a) with

$$c,\nu=\hat{\nu}(c,p),\quad u=\hat{u}(c,p),\quad \alpha=\hat{\alpha}(c) \tag{7.102}$$

and λ as taken in the MMOP. Moreover, the equations (7.100) and (7.101) represent nothing else but the conditions (7.86d) and (7.86e), respectively with c, u, α as given by (7.102).

The equations (7.100) and (7.101) form a base for deriving ISOPE double-loop iterative strategies utilizing input-output measurements. The optimality properties of the solutions produced immediately emerge from the strategy derivation. Namely, the solutions satisfy Kuhn-Tucker necessary conditions for optimality of SOCP. There are two possible ways of double-loop structuring an algorithm for solving these two equations that lead to a *system-based double-loop algorithm* and a *model-based double-loop algorithm.*

7.3.3 *System-based double-loop algorithm*

The method was first proposed in (Brdys *et al.*, 1990b). The system-based double-loop algorithm is made up of two nested iterative loops, each having different frequency of intervention: an *inner loop* and an *outer loop.* The outer loop task is to solve equation (7.101) by iterating price variable p while the inner loop task is to solve equation (7.100) by iterating control input c under p prescribed by the outer loop. Let us denote the inner loop solution as $\hat{c}(p)$. The corresponding interaction input and parameter are

$$\hat{u}(p) \triangleq \hat{u}(\hat{c}(p), p) \tag{7.103a}$$

and

$$\hat{\alpha}(p) \triangleq \hat{\alpha}(\hat{c}(p)) \tag{7.103b}$$

respectively.

As in the inner loop solution $\hat{c}(p) = \hat{\nu}(\hat{c}(p), p)$ (see (7.100) then the outer-loop task can be stated as of finding such value of p that

$$\hat{u}(p) = HF(\hat{c}(p), \hat{u}(p), \hat{\alpha}(p)). \tag{7.104}$$

The strategy is called system-based because the inner loop utilizes both the system model and the measurements (see (7.98)) while the outer loop does not directly utilize the measurements.

The *system-based double-loop algorithm* can now be formulated:

Start. Given initial points $c^{0,0}, p^0$, inner and outer loop algorithmic mappings $\{\Psi_{c_i}\}_{i\in\overline{1:N}}$, Ψ_p, respectively, convexifying coefficients ρ_c

and ρ_u and the loop solution accuracies $\varepsilon_c = [\varepsilon_{c_1}, ...\varepsilon_{c_N}], \varepsilon_{c_i} > 0, \varepsilon_p > 0$. Set $n := 0$ and $l = 0$; n and l denote the outer and inner loop iteration numbers, respectively.

Step 1. Apply $c^{l,n}$ to the controlled plant and measure the outputs $y^{l,n} = K_*(c^{i,n})$, $y^{ol,n} = K_*^o(c^{l,n})$ and interaction inputs $u^{l,n} = HK_*(c^{l,n})$. Calculate the plant output mapping derivatives $K_*^{'}(c^{l,n}), K_*^{o'}(c^{l,n})$ by using additional perturbation of $c^{l,n}$ or by employing previous output measurements as it is done by ISOPED.

Step 2. Solve the *parameter estimation problem* PEP defined by (7.97a) and (7.97b). The PEP decomposes into N independent *local parameter estimation problems* PEP$_i$, $i \in \overline{1,N}$ yielding the parameter estimates $\alpha_i^{l,n} = \hat{\alpha}_i(c^{l,n})$ satisfying

$$\begin{aligned} F_i^o(c_i^{l,n}, H_i K_*(c^{l,n}), \alpha_i^{l,n}) &= K_{*i}^o(c^{l,n}) \\ F_i(c_i^{l,n}, H_i K_*(c^{l,n}), \alpha_i^{l,n}) &= K_{*i}(c^{l,n}). \end{aligned} \tag{7.105}$$

Step 3. For each subsystem calculate modifier $\lambda_i^{l,n} = \lambda_i(c^{l,n}, \alpha^{l,n}, p^n)$, $i \in \overline{1:N}$ by applying formulae (7.90) and (7.94). This can not be done independently for each subsystem due to the interconnection. Hence, an *information exchange* between *local subsystem controllers* is needed in order to perform this step.

Step 4. For $c = c^{l,n}$, $p = p^n$ solve the *modified model-based optimization problem* MMOP defined by (7.98) with $\lambda = \lambda^{l,n}$ and $\alpha = \alpha^{l,n}$. The MMOP decomposes into N independent *local modified model-based optimization problems* MMOP$_i$, $i \in \overline{1,N}$ (see (7.98) and (7.4)

$$\begin{aligned} &\text{minimize}_{\nu_i,u_i}\{q_i(\nu_i, u_i, \alpha_i^{l,n}) + \rho_c\|\nu_i - c_i^{l,n}\|^2 + \\ &\quad +\rho_u\|u_i - H_i K_*(c^{l,n})\|^2 - (\lambda_i^{l,n})^T \nu_i + (p_i^n)^T u_i + \\ &\quad - \sum_{j=1}^{N} (p_j^n)^T H_{ji} F_i(\nu_i, u_i, \alpha_i^{l,n})\} \\ &\text{subject to: } g_i(\nu_i, u_i) \leq 0. \end{aligned} \tag{7.106}$$

Solving the MMOP$_i$ $i \in \overline{1:N}$ produces $\nu_i^{l,n} = \hat{\nu}_i(c_i^{l,n}, p^n)$ and $u_i^{l,n} = \hat{u}_i(c_i^{l,n}, p^n)$ (see (7.99). If for $i \in \overline{1:N}$

$$\| c_i^{l,n} - \nu_i^{l,n} \| \leq \varepsilon_{c_i} \tag{7.107}$$

then terminate the inner loop iterations.

The inner loop problem solution $c^n = [(c_1^n)^T, .., (c_N^n)^T] = \hat{c}(p^n)$ has now been found ((7.100) is satisfied). Continue from step 6.

Step 5. Set

$$c_i^{l+1,n} := \Psi_{c_i}(c_i^{l,n}, \nu_i^{l,n}) \tag{7.108}$$

set $l := l + 1$ and continue from Step 1.
The control input iteration according to (7.108) are completely decentralized. Hence they are performed at the subsystem level. However, as the modifier λ depends on all the components of vector c(see (7.90) then so does the mapping $\hat{\nu}_i(\cdot)$ (see (7.99) and (7.106)) Hence, applying a completely decentralized algorithmic mapping at the inner loop in order to solve (7.119) or in other words to find a fixed point of the mapping $\hat{\nu} : \mathbb{R}^{n_c} \longmapsto \mathbb{R}^{n_c}$

$$\hat{\nu}(c) = c \tag{7.109}$$

limits convergence of the iterations.

Step 6. The interaction inputs and model parameters determined at the inner loop problem solutions are (see (7.103a) and (7.103b)): $u^n = \hat{u}(c^n, p^n)$ and $\alpha^n = \hat{\alpha}(c^n)$. If

$$\| u^n - HF(c^n, u^n, \alpha^n) \| \leq \varepsilon_p \tag{7.110}$$

then terminate the outer loop iterations.
The outer loop problem solution has now been found ((7.104) is satisfied).

Step 7. Set

$$p^{n+1} := \Psi_p(p^n, u^n - HF(c^n, u^n, \alpha^n)) \tag{7.111}$$

set $n := n + 1$ and continue from step 1.

An information structure of the system based double loop algorithm is two-level hierarchical with information exchange between local subsystem controllers and it is illustrated in Fig. 7.11.

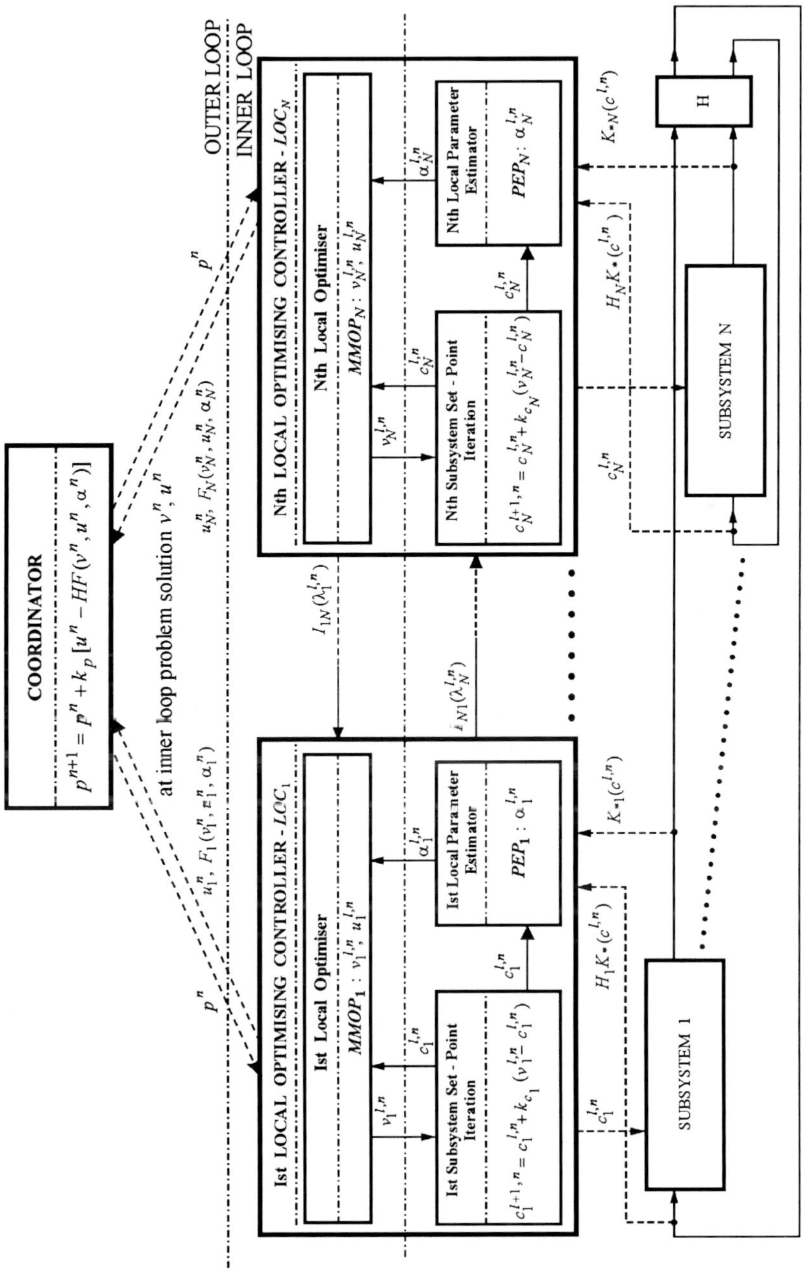

Fig. 7.11 Information structure of AISOPE system-based double-loop iterative strategy with input-output measurements.

The outer loop task is solved by the *coordinator*. The inner loop task is solved in parallel by the local subsystem optimizing controllers - *local optimizing controllers* (LOC).

The local controllers need to exchange information in order to calculate the modifiers. The information required by the ith local controller from the jth control subsystem in order to calculate at nth outer loop iteration and lth inner loop iteration the modifier $\lambda_i^{l,n}$ is denoted in Figure 7.11 by $I_{ij}(\lambda_i^{l,n})$.

Each local optimizing controller LOC_i, $i \in \overline{1:N}$ performs during each iteration four major tasks under price variable value prescribed by the coordinator in order to produce new control input. The tasks are carried out in sequence as:

- estimation of a subsystem model parameters by solving PEP_i
- information exchange with other optimizing controllers to provide these controllers with $I_{ij}(\lambda_i^{l,n})$, $j \in \overline{1:N}$, $j \neq i$
- solving the MMOP_i
- producing new control input that is applied to the ith subsystem of the interconnected system in order to evaluate quality of this new control input from a point of view of the optimized subsystem performance function. In addition, information at the subsystem level is gathered that allow the coordinator to assess optimality of its coordination instrument that is the price vector p.

The control algorithm as a whole can be viewed as a combination of open-loop interaction balance method — IBM (see this Chapter and (Findeisen *et al*, 1980)) with the ISOPE method. The latter method is employed at the subsystem level of the IBM in order to handle model-reality differences while the former one is applied in order to handle the subsystem interactions by satisfying the interaction balance condition described in (7.104).

The inner and outer loop algorithmic mappings Ψ_{c_i} and Ψ_p, respectively need to be designed so that the loop convergence is achieved. Clearly, faster convergence better mappings. The inner loop problem is a classic problem of finding a fix point of a mapping. Decentralization of the inner loop algorithmic mapping allows to iterate the control inputs independently for each controlled subsystem. Limits on the convergence properties is the price to be paid. As it was discussed in the previous chapters a relaxation algorithm with a scalar relaxation gain is an attractive possibility. Hence, the corresponding inner loop algorithmic mappings are:

$$\Psi_{c_i} : CU_i \mapsto \mathbb{R}^{n_{c_i}}$$
$$\Psi_{c_i}(c_i, v_i) = c_i + k_{c_i}(\nu_i - c_i), \quad i \in \overline{1:N} \tag{7.112}$$

where $k_{c_i} > 0$ is the ith inner loop relaxation mapping scalar gain. The corresponding inner loop iterative algorithms read:

$$c_i^{l+1,n} = c_i^{l,n} + k_{c_i}(\nu_i^{l,n} - c_i^{l,n}), \quad i \in \overline{1:N}. \tag{7.113}$$

As already stated in section (4.1) a feasibility of control inputs generated by ISOPE algorithm with the relaxation mapping is extremely important from practical point of view. Unfortunately this can be guaranteed only by the final solution as only then the interaction inputs that are predicted by LOCs are equal to the interaction inputs in the interconnected system. This is well known drawback of open-loop hierarchical control technology that applies price as the coordination instrument. However, the considered double-loop system based hierarchical algorithm is closed-loop with a feedback from the interaction inputs directly introduced in both the PEP and MMOP. The feedback in the MMOP is represented by the term $\rho_u \|u_i - H_i K_*(c^{l,n})\|^2$ (see (7.106)). Although a convexification of the MMOP was the main reason of introducing this term (see (7.83) it can be now used to significantly improve a feasibility of the control inputs generated by the LOCs. Indeed, assume that $c^{l,n}$ is feasible. Thus, $g_i(c_i^{l,n}, H_i K_*(c^{l,n})) \leq 0$ for $i \in \overline{1:N}$. With ρ_u large the interaction input $u_i^{l,n}$ resulting from solution of MMOP$_i$ will not differ much from $H_i K_*(c^{l,n})$ and this can be controlled by proper selection of ρ_u. Hence, $g_i(c_i^{l,n}, u_i^{l,n}) \leq 0$. The new control inputs $c_i^{l+1,n}$ can be maintained close enough to $v_i^{l,n}$ by suitable choice of the relaxation gains k_{c_i}. Consequently, $g_i(c_i^{l+1,n}, H_i K_*(c_i^{l+1,n})) \leq 0$, for $i \in \overline{1:N}$, what means that $c^{l+1,n}$ is feasible. In summary, the algorithm will generate feasible control inputs during its operation if the initial control input $c^{0,0}$ is feasible and ρ_u is sufficiently large.

The the outer loop algorithmic mapping is proposed being driven by results regarding the classic open-loop IBM. The outer loop problem (7.104) is very similar to the coordination condition of IBM. Indeed, eq. (7.104) implies that $\hat{u}(p) = HK_*(\hat{c}(p))$. Hence, at the outer loop solution the interaction inputs predicted by LOCs are equal to the real system interactions produced by control inputs that are proposed by the LOCs. The parameter updates in (7.104) are not present in IBM and this is the only difference between the IBM coordination condition and the outer loop condition. However, although this is an important technical difference an essence of the both conditions is the same: find the price p value under which the interaction balance is achieved.

Therefore, it is natural to choose a gradient type of outer loop algorithmic mapping as:

$$\begin{aligned} \Psi_p &: \mathbb{R}^{n_u} \mapsto \mathbb{R}^{n_u} \\ \Psi_p(p, u - HF(c, u, \alpha) &= p + k_p[u - HF(c, u, \alpha)]. \end{aligned} \tag{7.114}$$

With this outer loop algorithmic mapping the corresponding outer loop iterative algorithm reads:

$$p^{n+1} = p^n + k_p[u^n - HF(c^n, u^n, \alpha^n)] \tag{7.115}$$

where $k_p > 0$ is the outer loop gradient algorithm step coefficient.

Convergence properties of the double-loop system-based algorithm with the output-interaction input measurements will not be rigorously analyzed for the algorithm with input-output measurements. This will be done for a *system-based double-loop algorithm with output measurements* in a subsequent section of this chapter.

7.3.4 *Model-based double-loop algorithms*

An inner loop in the system-based double-loop algorithm iterates control inputs in order to meet the condition expressed by (7.100) under any price vector that is prescribed by the outer loop. Only at the solution of the outer loop problem the interaction balance is achieved and optimal control inputs found. Hence, all the inner loop iterations imposing changes of the control inputs that are applied to the system may seem to be on any value. Clearly it is not so as in the system-based double-loop algorithm an optimality of p can be only assessed based on the corresponding solution of the inner loop In other words given p, one can judge if this is an optimal price vector by solving the inner loop problem under this p and checking the interaction balance condition (7.110). However, being inspired by the above observation let us try the other way around. What we really need is to assess an optimality of the control inputs and let us do it straight.

Given c and we would like to know if this optimal control input vector. Let us then solve (7.101) with respect to p and obtain the price $\hat{p}(c)$. This implies a number of iterations to be performed on p and each of them involves solving the MMOP defined by (7.98). Clearly, as it follows from a formulation of the MMOP for a double-loop strategy with input-output measurements the interaction balance condition (7.101) is now forced to be satisfied. Hence, the price $\hat{p}(c)$ seems to be of a good quality. Notice, that in solving the MMOP no iterations of c have been required. Hence, the price has been calculated entirely based on the model iterations As the (7.101) is satisfied with $p = \hat{p}(c)$ an optimality of c can be assessed by checking if the second core condition (7.100) is met. This can be easily done as the vector $\hat{\nu}(c, \hat{p}(c))$ (see (7.99)) is known for the MMOP solution. This appealing property of the price vector obtained without doing any iteration on a real system let us think of formalizing the above idea in a form of a *double-loop model-based iterative strategy.* It is hoped that this strategy will achieve an optimal control input under reduced number of iterations to be carried out on a real system.

As the solution of the interaction balance condition (7.101) is not granted for every $c \in \mathbb{R}^{n_c}$ it may be necessary to relax a bit this condition during search for optimal control input vector. This will result in a *model-based double-loop algorithm with relaxed inner loop* as opposed to a *model-based double-loop algorithm with tight inner loop* where the original interaction balance condition can be achieved for a sufficiently wide range of control inputs.

7.3.4.1 *Model-based double-loop algorithm with tight inner loop*

The method was first proposed in (Brdyś *et al.*, 1989) for standard ISOPE algorithm and then generalized in (Brdyś *et al.*, 1990a) to produce the structure and algorithm of the AISOPE presented bellow. The model-based double-loop algorithm consists of two nested iterative loops, each having different frequency of interventions: an *inner loop* and an *outer loop.* The inner loop task is to solve equation (7.101) by iterating price variable p under control input c prescribed by the outer loop while the outer loop task is to solve equation (7.100) by iterating control input c. Let us denote the inner loop problem solution as $\hat{p}(c)$. The corresponding interaction input and auxiliary variable ν are

$$\hat{u}(c) \stackrel{\triangle}{=} \hat{u}(c, \hat{p}(c)) \tag{7.116}$$

and

$$\hat{\nu}(c) \stackrel{\triangle}{=} \hat{\nu}(c, \hat{p}(c)) \tag{7.117}$$

respectively. The inner-loop task is then to find such value of p that

$$\hat{u}(c) = HF(\hat{\nu}(c), \hat{u}(c), \hat{\alpha}(c)) \tag{7.118}$$

while the outer loop task is to find such value of c that

$$\hat{\nu}(c) = c. \tag{7.119}$$

The strategy is called model-based because the inner loop entirely utilizes system model.

The *model-based double-loop algorithm with tight inner loop* can now be formulated:

Start. Given initial points $c^0, p^{0,0}$, inner and outer loop algorithmic mappings $\Psi_p, \{\Psi_{c_i}\}_{i\in\overline{1:N}}$, respectively, convexifying coefficients ρ_c and ρ_u and the loop solution accuracies $\varepsilon_c = [\varepsilon_{c_1}, ...\varepsilon_{c_N}], \varepsilon_{c_i} > 0, \varepsilon_p > 0$. Set $n := 0$ and $l = 0$; n and l denote the outer and inner loop iteration numbers, respectively.

Step 1. Apply c^n to the controlled plant and measure the outputs $y^n = K_*(c^n)$, $y^{on} = K_*^o(c^n)$ and interaction inputs $u^n = HK_*(c^n)$. Calculate the plant output mapping derivatives $K_*^{'}(c^n), K_*^{o'}(c^n)$ by using additional perturbation of c^n or by employing previous output measurements as it is done by ISOPED.

Step 2. Solve the *parameter estimation problem* PEP defined by (7.97a) and (7.97b). The PEP decomposes into N independent *local parameter estimation problems* $\text{PEP}_i, i \in \overline{1, N}$ yielding the parameter estimates $\alpha_i^n = \hat{\alpha}_i(c^n)$ satisfying

$$F_i^o(c_i^n, H_i K_*(c^n), \alpha_i^n) = K_{*i}^o(c^n) \tag{7.120a}$$

$$F_i(c_i^n, H_i K_*(c^n), \alpha_i^n) = K_{*i}(c^n). \tag{7.120b}$$

Step 3. For each subsystem calculate modifier $\lambda_i^{l,n} = \lambda_i(c^n, \alpha^n, p^{l,n})$, $i \in \overline{1:N}$ by applying formulae (7.90) and (7.94). This can not be done independently for each subsystem due to interconnections between subsystems. Hence, an *information exchange* between *local subsystem controllers* is needed in order to perform this step.

Step 4. For $c := c^n$, $p := p^{l,n}$ solve the *modified model-based optimization problem* MMOP defined by (7.98) with $\lambda = \lambda^{l,n}$ and $\alpha = \alpha^n$. The MMOP decomposes into N independent *local modified model-based optimization problems* $MMOP_i$, $i \in \overline{1,N}$ (see (7.98) and (7.4)

$$\begin{aligned} &\text{minimize}_{\nu_i,u_i}\{q_i(\nu_i,u_i,\alpha_i^n) + \rho_c\|\nu_i - c_i^n\|^2 + \\ &+\rho_u\|u_i - H_iK_*(c^n)\|^2 - (\lambda_i^{l,n})^T\nu_i + (p_i^{l,n})^T u_i + \\ &- \sum_{j=1}^{N}(p_j^{l,n})^T H_{ji}F_i(\nu_i,u_i,\alpha_i^n)\} \\ &\text{subject to: } g_i(\nu_i,u_i) \leq 0. \end{aligned} \tag{7.121}$$

Solving the MMOP$_i$ $i \in \overline{1:N}$ produces $\nu_i^{l,n} = \hat{\nu}_i(c_i^n, p^{l,n})$ and $u_i^{l,n} = \hat{u}_i(c_i^n, p^{l,n})$ (see (7.99). If for $i \in \overline{1:N}$

$$\| u_i^{l,n} - H_iF(\nu^{l,n}, u^{l,n}, \alpha^n) \| \leq \varepsilon_p \tag{7.122}$$

then terminate the inner loop iterations.
The inner loop problem solution $p^n = [(p_1^n)^T, .., (p_N^n)^T]^T = p(c^n)$ has now been found ((7.118) is satisfied). Continue from Step 6.
The condition (7.122) can not be verified at the subsystem level without information exchange between local controllers. Typically, an *inner loop coordinator* is introduced to carry out this operation.

Step 5. Set

$$p^{l+1,n} := \Psi_p(p^{l,n}, u^{l,n} - HF(\nu^{l,n}, u^{l,n}, \alpha^n)) \tag{7.123}$$

set $l := l + 1$ and continue from Step 4.
This inner loop price iteration can not be carried out at the subsystem level that is in a decentralized manner and it is performed by the inner loop coordinator.

Step 6. The interaction inputs and auxiliary variable determined at the inner loop problem solutions are (see (7.116) and (7.117)): $u^n = \hat{u}(c^n, p^n)$ and $\nu^n = \hat{\nu}(c^n, p^n)$. If

$$\| c_i^n - \nu_i^n \| \leq \varepsilon_{c_i} \tag{7.124}$$

then terminate the outer loop iterations.
The outer loop problem solution has now been found ((7.119) is satisfied).

Step 7. Set

$$c_i^{n+1} := \Psi_{c_i}(c_i^n, \nu_i^n), \quad i \in \overline{1:N} \tag{7.125}$$

set $n := n + 1$ and continue from step 1.

The control input iteration according to (7.125) are completely decentralized. Hence they are performed at the subsystem level and this is a key benefit of decentralization. However, as the modifier λ_i depends on all the components of vector c (see (7.90) then so does the mapping $\hat{\nu}_i(\cdot)$ (see (7.117) Hence, applying a completely decentralized algorithmic mapping at the outer loop in order to solve (7.119) or in other words to find a fixed point of the mapping $\hat{\nu} : \mathbb{R}^{n_c} \longmapsto \mathbb{R}^{n_c}$

$$\hat{\nu}(c, p) = c \tag{7.126}$$

where p is fixed, limits convergence of the iterations (7.125).

An information structure of the *double-loop model-based with tight inner loop algorithm* is two-level hierarchical with information exchange between local subsystem controllers and it is illustrated in Figure 7.12.

Comparison of Fig. 7.12 with the Fig. 7.11 illustrating the double-loop system-based algorithm shows that exactly the same units are present and exactly the same data are processed in exactly the same manner. The only difference is in the unit intervention frequency that implies different location of the inner and outer loops in the figures. An operation of the *double-loop model-based with tight inner loop structure* can be summarized as:

- The local controllers communicate with the *coordinator* more frequently than with the controlled interconnected system and this is carried out by inner loop iterations.
- Once a balance between the predicted and real interactions has been achieved (see (7.118) an optimality of the corresponding modified model-based optimized control inputs is assessed by using the outer loop fixed point condition (7.119).
- In order to assess the control inputs optimality, the predicted optimized control inputs ν^n are compared with the currently available ones c^n.

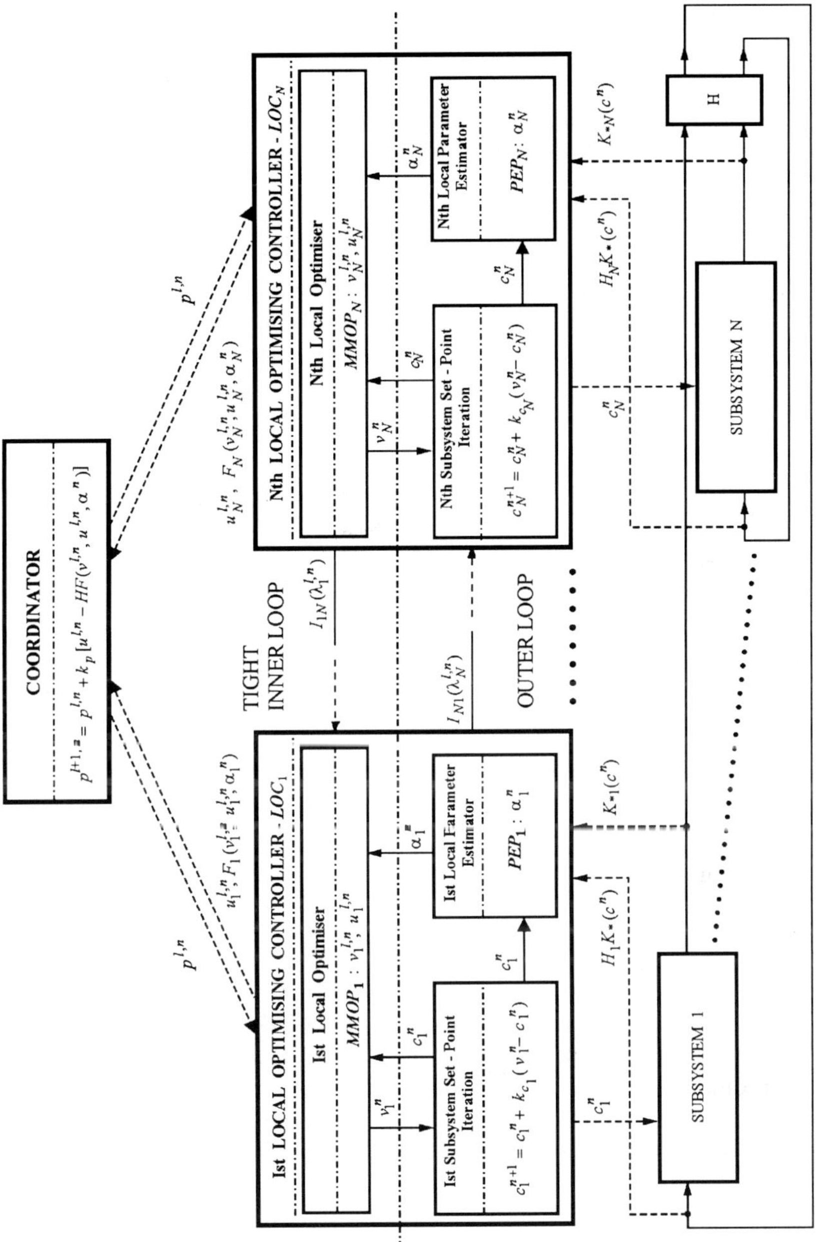

Fig. 7.12 Information structure of AISOPE model-based double-loop iterative strategy with input-output measurements and with tight inner-loop.

- If a difference between the two exists then the control inputs at hand c_n are adjusted by the outer loop algorithm and applied to the system in order to gather new information. This new measurement information is used to update the model parameters and the modifier for the Model-Based Modified Optimization Problem (7.98).

The inner and outer loop algorithmic mappings Ψ_{c_i} and Ψ_p, respectively need to be designed so that the loop convergence is achieved. As for the double-loop system-based algorithm a relaxation algorithm with a scalar relaxation gain is chosen for the outer loop. Hence, the corresponding outer loop local algorithmic mappings are:

$$\Psi_{c_i} : \mathbb{R}^{n_{c_i}} \mapsto \mathbb{R}^{n_{c_i}}$$
$$\Psi_{c_i}(c_i, v_i) = c_i + k_{c_i}(\nu_i - c_i), \quad i \in \overline{1:N} \tag{7.127}$$

where $k_{c_i} > 0$ is the ith outer loop relaxation mapping scalar gain. The corresponding outer loop decentralized algorithm reads:

$$c_i^n = c_i^n + k_{c_i}(\nu_i^n - c_i^n), \quad i \in \overline{1:N}. \tag{7.128}$$

Regarding a design of the inner loop algorithm let us consider the following optimization problem:

$$\begin{aligned}
&\text{minimize}_{\nu,u}\{q(\nu,u,\hat{\alpha}(c)) + \rho_c\|\nu - c\|^2 + \rho_u\|u - HK_*(c)\|^2 + \\
&\qquad\qquad -\lambda^T(c,\tilde{p},\hat{\alpha}(c))\nu &&(7.129\text{a})\\
&\text{subject to:} \quad u = HF(\nu,u,\hat{\alpha}(c) &&(7.129\text{b})\\
&\qquad\qquad g(\nu,u) \le 0 &&(7.129\text{c})
\end{aligned}$$

where c and $\tilde{p}$ are is fixed and $\lambda(c,\tilde{p},\hat{\alpha}(c))$ is calculated according to (7.94) and (7.90) for $\alpha = \hat{\alpha}(c)$.

Comparison of the performance function (7.129a) and the constraints (7.129b) and (7.129c) of this optimization problem with the performance and function and the constraints of the MMOP (see (7.98)) and one might be tempted to apply the hierarchical IBM algorithm to solve the inner loop problem. Indeed, the IBM lower level problem would then be:

$$\begin{aligned}
&\text{minimize}_{\nu,u}\{q(\nu,u,\hat{\alpha}(c)) + \rho_c\|\nu - c\|^2 + \rho_u\|u - HK_*(c)\|^2 + \\
&\qquad -\lambda^T(c,\tilde{p},\hat{\alpha}(c))\nu + p^T(u - HF(\nu,u,\hat{\alpha}(c))\\
&\text{subject to:} \quad g(c,\nu) \le 0. &&(7.130)
\end{aligned}$$

The IBM coordinator would iterate Lagrange price p in order to achieve the interaction balance. If the interaction balance is achieved at $\hat{p} = \tilde{p}$ then $\hat{p}$ solves the inner loop problem (7.118). However, if $\hat{p} \neq \tilde{p}$ then the function minimized in (7.130) is different at $p = \hat{p}$ from the function that is minimized in (7.98) as the modifier λ in (7.130) is calculated at $\tilde{p}$ but not at $\hat{p}$. Hence, the interaction balance is achieved but not at the solution of MMOP (see (7.98)) and consequently $\hat{p}$ is not solution of the inner loop problem (7.129a). One would have to adjust $\tilde{p}$ and repeat solving (7.130). As the modifier has changed the interaction balance would be achieved at different than $\hat{p}$ value of p. Thus, bringing these to price values towards the desired equality $\hat{p} = \tilde{p}$ require additional iterations of $\tilde{p}$.

It is an open problem if such equality can be achieved at all. It now becomes clear that an existence of the inner loop solution for sufficiently large range of control inputs c can be assumed as granted. An experience indicates that some specific assumptions needs to be made in order to force the inner loop problem solution exists and this is main limitation of an applicability of the double-loop algorithm with input-output measurements and tight inner loop.

Nevertheless, being motivated by the same arguments as used when designing an outer loop algorithmic mapping for the double-loop system-based algorithm we shall choose here the inner loop algorithmic mapping as $\Psi_p : \mathbb{R}^{n_u} \times \mathbb{R}^{n_u} \mapsto \mathbb{R}^{n_u}$,

$$\Psi_p(p, u - HF(\nu, u, \alpha)) = p + k_p[u - HF(\nu, u, \alpha)]. \tag{7.131}$$

With this inner loop algorithmic mapping the corresponding inner loop algorithm reads:

$$p^{l+1,n} = p^{l,n} + k_p[u^{l,n} - HF(\nu^{l,n}, u^{l,n}, \alpha^n)] \tag{7.132}$$

where $k_p > 0$ is a step coefficient of the inner loop gradient type of algorithm.

We shall analyze the convergence properties of the overall double-loop algorithm for a linear-quadratic case. That is

$$Q(x, y) = \frac{1}{2}(x - d)^T M(x - d) \tag{7.133}$$

where $x \triangleq (c^T, u^T)^T$, M is symmetric matrix, d is constant vector, and linear input-output relationships for the system and its point-parametric model are

$$F_*(x) = D_{*1}c + D_{*2}u + d_* \quad (7.134)$$
$$F(x, \alpha) = D_1 c + D_2 u + P(\alpha). \quad (7.135)$$

The matrices D_1, D_2 and vector function $P(\cdot)$ are chosen such that the model is point-parametric on the constraint set $CU = C \times U = \mathbb{R}^{n_c} \times \mathbb{R}^{n_c}$ Without loss of generality, the performance index is assumed not output dependent in order to simplify notation.

It is assumed that the inverses $(HD_{*2} - I_u)^{-1}$ and $(HD_2 - I_u)^{-1}$, where I_u is identity matrix of dimension $n_u \times n_u$, exists. Consequently, the interaction input mapping $HK_*(\cdot)$ and $HK(\cdot)$ are well defined.

Let us define the following matrices:

$$B \triangleq [-HD_1, I_u - HD_2] \quad (7.136)$$
$$B_* \triangleq [(HD_{*2} - I_u)^{-1} HD_{*1}, I_c] \quad (7.137)$$
$$\widetilde{B} \triangleq (I_u - HD_2)B_* \quad (7.138)$$

where I_c is identity matrix of dimension $n_c \times n_c$.

It is straightforward to show that the constraint $u - HK_*(c) = 0$ in $SOCP$ can be written as $B_*x - b_* = 0$, where $b_* = -[I_u - HD_{*2}]^{-1}Hd_*$ and its model-based counterpart $u - HF(c, u, \alpha) = 0$ as $Bx - HP(\alpha) = 0$.

We shall also assume that the matrix B_*has full rank, and that the second order sufficient conditions for optimality hold for the SOCP, i.e.,

$$x^T M x > 0 \ \text{ for every } x \in C \times U, x \neq 0 \text{ such that } \ B_* x = 0. \quad (7.139)$$

Note that this implies that there is a unique solution $c^{opt}$to the SOCP and a unique corresponding price vector p^{opt} associated with the constraints $u - HK_*(c) = 0$. Consequently, there is a unique solution of the outer loop problem c^{opt} (see (7.119)) and a unique solution of the corresponding inner loop problem $\overline{p}$ (see (7.118)), where

$$\overline{p} = [(I_u - HD_2)^{-1}]^T p^{opt} \quad (7.140)$$

Let us note that a non-convex case is not excluded from the considerations. It is assumed, without loss on generality, that $\rho_c = \rho_u = \rho$ in (7.121), and that the value of ρ is chosen so that

$$M_\rho \triangleq M + \rho I_x > 0 \quad (7.141)$$

where I_x is identity matrix of dimension $(n_c + n_u) \times (n_c + n_u)$.

Finally, let us define the matrices

$$L \triangleq \begin{bmatrix} I_c \\ (I_u - HD_2)^{-1} HD_1 \end{bmatrix} \tag{7.142}$$

and

$$L_* \triangleq \begin{bmatrix} I_c \\ (I_u - HD_{*2})^{-1} HD_{*1} \end{bmatrix}. \tag{7.143}$$

The *convergence conditions* can now be formulated for the double iterative loop algorithm with a tight inner loop.

Theorem 7.1
Assume that

(i) $BM_\rho^{-1}\widetilde{B}^T > 0$

(ii) $L_*^T M_\rho L > 0$

Then

(a) The inner loop problem is well defined.

(b) There exists such number $\overline{k_p} > 0$ *that for any* $k_p \in (0, \overline{k_p})$ *the inner loop iterative scheme is convergent for every* $c \in C$.

(c) If the relaxation gain k_c *is such that*

$$L_*^T M_\rho L - \frac{k_c}{2} L_*^T M L_* > 0 \tag{7.144}$$

then the outer loop iterative scheme is convergent to c^{opt} *while the corresponding sequence of prices is convergent to* $\overline{p}$ *given by (7.140).*

Proof of this theorem is given in Appendix B.1.

It follows from the proof that the condition (i) is only required in order to preserve the convergence of the inner loop iterative algorithm while the condition (ii) implies the outer loop algorithm convergence.provided that k_c satisfies (7.144). The latter can be easily achieved by selecting the relaxation gain to be sufficiently small. Indeed, due to (i) the condition (7.144) is satisfied for sufficiently small values of k_c. It can be seen from (7.144) that due to the convexifying terms in (7.121), $\rho > 0$ may allow larger values of the relaxation gain k_c to be employed which makes the outer loop iterative scheme more efficient and, consequently, reduce a number of iterations performed on a real system. This was fully confirmed by numerical simulations. The conditions (i) and (ii) are entirely imposed by the model-reality differences. Indeed, when the model is perfect, then $\widetilde{B} = B$ (see (7.136), (7.137) and (7.138)) and $\widetilde{L} = L$ (see (7.142) and (7.143)) and these conditions are satisfied in that particular situation. Therefore, the only reason for restrictive applicability of the algorithm in the linear-quadratic case is a model-reality difference.

7.3.4.2 *Model-based double-loop algorithm with relaxed inner loop*

The method was first proposed in (Brdyś *et al.*, 1989) for standard ISOPE algorithm and then generalized in (Brdyś *et al.*, 1990a) to produce the AISOPE structure and algorithm presented below. A reduced number of iterations to be performed on the controlled system is an extremely important advantage of a model-based double-loop algorithm over the system based one. A limited applicability due to a stiffness of its inner loop problem implying that solution of the inner loop problem may not exist for the required range of control inputs is a drawback. Our problem now is to think of such modification of the algorithm structure that the advantage remains while the drawback is removed. The solution comes in a straightforward manner from the analysis of the inner loop problem that was performed in Section 7.3.4.1. Indeed, all we need to do is to distinguish between the price variable entering an expression for the modifier λ and the price variable that is responsible for model-based interaction balance condition. This leads to *relaxed model-based modified optimization problem* (RMMOP) defined as:

$$\begin{aligned}
&\text{minimize}_{\nu,u}\{q(\nu,u,\hat{\alpha}(c)) + \rho_c\|\nu - c\|^2 + \rho_u\|u - HK_*(c)\|^2 - \\
&\qquad\qquad +\lambda^T(c,p_2,\hat{\alpha}(c))\nu + p_1^T(u - HF(\nu,u,\hat{\alpha}(c))\} \\
&\text{subject to:}\quad g(c,\nu) \le 0 \qquad\qquad (7.145)
\end{aligned}$$

where c, p_1, p_2 are the RMMOP inputs.
Let us denote the solution by

$$\hat{\nu}(c,p_1,p_2) \quad \text{and} \quad \hat{u}(c,p_1,p_2) \qquad (7.146)$$

The core conditions defining a solution of a double-loop iterative strategy (7.100) and (7.101) can now be written in equivalent but relaxed manner as:

$$\hat{\nu}(c, p_1, p_2) = c \tag{7.147a}$$

$$\hat{u}(c, p_1, p_2) = HF(\hat{\nu}(c, p_1, p_2), \hat{u}(c, p_1, p_2), \hat{\alpha}(c)) \tag{7.147b}$$

$$p_1 = p_2 \tag{7.147c}$$

Given c and p_2, the *relaxed inner loop* iterates p_1 in order to meet the interaction balance condition (7.147b). It is done by solving the *RMMOP* with the prescribed values of c and p_2 yielding:

price: $\hat{p}_1(c, p_2)$ (7.148a)

predicted control inputs: $\hat{\nu}(c, p_2) \triangleq \hat{\nu}(c, \hat{p}_1(c, p_2), p_2)$ (7.148b)

predicted interaction inputs: $\hat{u}(c, p_2) \triangleq \hat{u}(c, \hat{p}_1(c, p_2), p_2)$ (7.148c)

The outer loop iterates the control input and price variables c and p_2 in order to satisfy the conditions (7.147a) and (7.147c). Hence, for prescribed by the outer loop c and p_2 *the inner loop task* is to find such value of $p_1$that

$$\hat{u}(c, p_2) = HF(\hat{\nu}(c, p_2), \hat{u}(c, p_2), \hat{\alpha}(c)) \tag{7.149}$$

while the *outer loop task* is to find such values of p_2 and c that

$$\hat{\nu}(c, p_2) = c \tag{7.150}$$

$$\hat{p}_1(c, p_2) = p_2. \tag{7.151}$$

The *model-based double-loop algorithm with relaxed inner loop* can now be formulated:

Start. Given initial points $c^0, p^{0,0}$, inner and outer loop algorithmic mappings Ψ_{p_1} and $\{\Psi_{c_i}\}_{i\in\overline{1:N}}$, $\{\Psi_{p_2,i}\}$,respectively, convexifying coefficients ρ_c and ρ_u and the loop solution accuracies $\varepsilon_c = [\varepsilon_{c_1}, ...\varepsilon_{c_N}]$, $\varepsilon_{c_i} > 0$, $\varepsilon_{p_2} = [\varepsilon_{p_2,1,...,}\varepsilon_{p_2,N}]$, $\varepsilon_{p_2,i} > 0$, $\varepsilon_{p1} > 0$. Set $n := 0$ and $l = 0$; n and l denote the outer and inner loop iteration numbers, respectively.

Step 1. Apply c^n to the controlled plant and measure the outputs $y^n = K_*(c^n)$, $y^{on} = K_*^o(c^n)$ and interaction inputs $u^n = HK_*(c^n)$. Calculate the plant output mapping derivatives $K_*^{'}(c^n), K_*^{o'}(c^n)$ by using additional perturbation of c^n or by employing previous output measurements as it is done by ISOPED.

Step 2. Solve the *parameter estimation problem* PEP defined by (7.97a) and (7.97b). The PEP decomposes into N independent *local parameter estimation problems* $\text{PEP}_i, i \in \overline{1,N}$ yielding the parameter estimates $\alpha_i^n = \hat{\alpha}_i(c^n)$ satisfying

$$F_i^o(c_i^n, H_i K_*(c^n), \alpha_i^n) = K_{*i}^o(c^n) \qquad (7.152a)$$
$$F_i(c_i^n, H_i K_*(c^n), \alpha_i^n) = K_{*i}(c^n). \qquad (7.152b)$$

Step 3. For each subsystem calculate modifier $\lambda_i^{l,n} = \lambda_i(c^n, \alpha^n, p_2^n)$, $i \in \overline{1:N}$ by applying formulae (7.90) and (7.94). This can not be done independently for each subsystem due to interconnections between subsystems. Hence, an *information exchange* between *local subsystem controllers* is needed in order to perform this step.

Step 4. For $c := c^n$, $p_2 := p_2^n$ $p_1 := p_1^{l,n}$ solve the *relaxed modified model-based optimization problem RMMOP* defined by (7.145) with $\lambda := \lambda^{l,n}$ and $\alpha := \alpha^n$. The RMMOP decomposes into N independent *local relaxed modified model-based optimization problems RMMOP$_i$*, $i \in \overline{1,N}$ (see (7.145) and (7.4)

$$\begin{aligned} &\text{minimize}_{\nu_,,u_i}\{q_i(\nu_i, u_i, \alpha_i^n) + \rho_c\|\nu_i - c_i^n\|^2 + \\ &\quad +\rho_u\|u_i - H_i K_*(c^n)\|^2 - (\lambda_i^{l,n})^T \nu_i + (p_1^{l,n})_i^T u_i + \\ &\quad - \sum_{j=1}^{N} (p_1^{l,n})_j^T H_{ji} F_i(\nu_i, u_i, \alpha_i^n)\} \\ &\text{subject to: } g_i(\nu_i, u_i) \le 0. \end{aligned} \qquad (7.153)$$

Solving the RMMOP_i $i \in \overline{1:N}$ produces $\nu_i^{l,n} = \hat{\nu}_i(c_i^n, p_1^{l,n}, p_2^n)$ and $u_i^{l,n} = \hat{u}_i(c_i^n, p_1^{l,n}, p_2^n)$ (see (7.146)). If for $i \in \overline{1:N}$

$$\| u_i^{l,n} - H_i F(\nu^{l,n}, u^{l,n}, \alpha^n) \| \le \varepsilon_{p_1,i} \qquad (7.154)$$

then terminate the inner loop iterations.
The inner loop problem solution $p_1^n = [(p_1^n)_1^T, .., (p_1^n)_1^T] = \hat{p}_1(c^n, p_2^n)$ has now been found ((7.149) is satisfied). Continue from Step 6.
The condition (7.154) can not be verified at the subsystem level without information exchange between local controllers. Typically, an *inner loop coordinator* is introduced to carry out this operation.

Step 5. Set

$$p_1^{l+1,n} := \Psi_{p_1}(p_1^{l,n}, u^{l,n} - HF(\nu^{l,n}, u^{l,n}, \alpha^n)) \qquad (7.155)$$

set $l := l + 1$ and continue from Step 4.

This inner loop price iteration can not be carried out at the subsystem level that is in a decentralized manner and it is performed by the inner loop coordinator.

Step 6. The price, interaction inputs and auxiliary variable determined at the inner loop problem solutions are (see (7.148a), (7.148b) and (7.148c)):

$$p_1^n = \hat{p}_1(c^n, p_2^n) \tag{7.156a}$$

$$u^n = \hat{u}(c^n, p_2^n) \tag{7.156b}$$

$$\nu^n = \hat{\nu}(c^n, p_2^n) \tag{7.156c}$$

If

$$\| \; c_i^n - \nu_i^n \; \| \leq \varepsilon_{c_i} \tag{7.157}$$

$$\| \; p_{1,i}^n - p_{2,i}^n \; \| \leq \varepsilon_{p_{2},i} \tag{7.158}$$

then terminate the outer loop iterations.

The outer loop problem solution has now been found ((7.147a) and (7.147c) are satisfied).

Step 7. Set

$$c_i^{n+1} \; : = \Psi_{c_i}(c_i^n, \nu_i^n), \quad i \in \overline{1:N} \tag{7.159}$$

$$p_{2,i}^{n+1} \; : = \Psi_{p_2,i}(p_{2,i}^n, p_{1,i}^n), \quad i \in \overline{1:N} \tag{7.160}$$

set $n := n + 1$ and continue from Step 1.

The control input and the price iterations according to (7.159) and (7.160) are completely decentralized. Hence they are performed at the subsystem level and this is a key benefit of decentralization. However, as the modifier λ_i depends on all the components of the vectors c and p_2 (see (7.90) then so do the mappings $\hat{\nu}_i(\cdot, \cdot)$ and $\hat{p}_{2,i}(\cdot, \cdot)$ (see (7.145). Hence, applying a completely decentralized algorithmic mapping at the outer loop in order to solve (7.150) and (7.151) limits convergence of the iterations (7.159) and (7.160).

An information structure of the *double-loop model-based with relaxed inner loop algorithm* is a two-level hierarchical with information exchange between local subsystem controllers and it is illustrated in Figure 7.13.

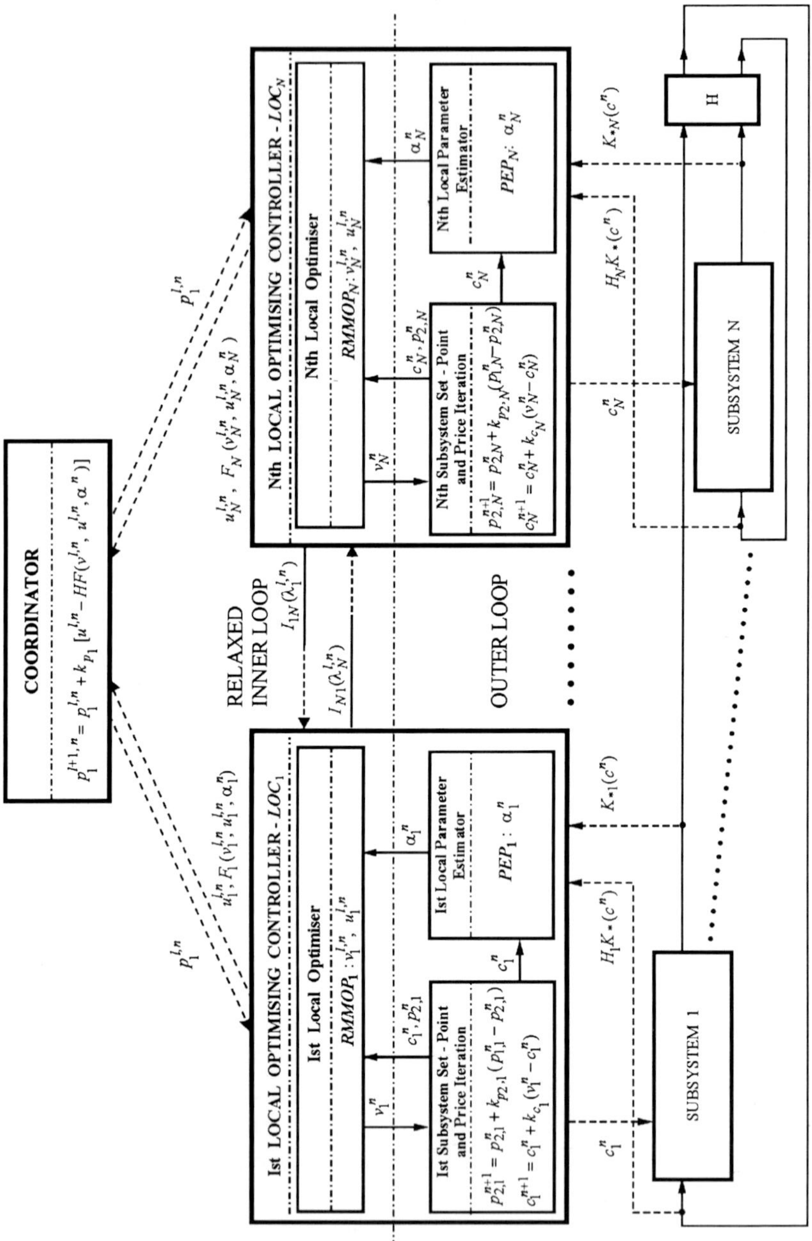

Fig. 7.13 Information structure of AISOPE model-based double-loop iterative strategy with input-output measurements and with relaxed inner loop.

Comparison of the Figure 7.13 with the Figure 7.12 illustrating the double-loop model-based with tight inner loop algorithm shows that exactly the same units are present. However, the data regarding price vectors are now different and they are differently processed. Both the local controllers iterate the price vectors but the outer loop ensures that at the solution the prices are the same. As a number of price variables has increased twice the computing and the communication load have also increased. Otherwise the structure operation is the same.

The inner and outer loop algorithmic mappings Ψ_{p_1} and $\{\Psi_{c_i}\}_{i\in\overline{1:N}}$, $\{\Psi_{p_{2,i}}\}$ respectively, need to be designed so that the loop convergence is achieved. As for the double-loop model-based with tight inner loop algorithm a relaxation algorithm with a scalar relaxation gain is chosen for the outer loop. Hence, for the control inputs the corresponding outer loop local algorithmic mappings are:

$$\begin{aligned}\Psi_{c_i} &: CU_i \mapsto \mathbb{R}^{n_{c_i}} \\ \Psi_{c_i}(c_i, v_i) &= c_i + k_{c_i}(\nu_i - c_i), \quad i \in \overline{1:N}\end{aligned} \tag{7.161}$$

where $k_{c_i} > 0$ is the ith outer loop control relaxation mapping scalar gain. The corresponding outer loop decentralized algorithm for iterating the control inputs reads:

$$c_i^{n+1} = c_i^n + k_{c_i}(\nu_i^n - c_i^n), \quad i \in \overline{1:N}. \tag{7.162}$$

Similarly, the prices at the outer loop are iterated according to:

$$\begin{aligned}\Psi_{p_{2,i}} &: \mathbb{R}^{n_{u_i}} \mapsto \mathbb{R}^{n_{u_i}} \\ \Psi_{p_{2,i}}(p_{2,i}, p_{1,i}) &= p_{2,i} + k_{p_{2,i}}(p_{1,i} - p_{2,i}), \quad i \in \overline{1:N}\end{aligned} \tag{7.163}$$

where $k_{p_{2,i}} > 0$ is the ith outer loop price relaxation mapping scalar gain.

The corresponding outer loop decentralized algorithm for iterating the prices influencing the modifier λ reads:

$$p_{2,i}^{n+1} = p_{2,i}^n + k_{p_{2,i}}(p_{1,i}^n - p_{2,i}^n). \tag{7.164}$$

Thanks to relaxation of the inner loop a design of the inner loop algorithmic mapping is straightforward:

$$\begin{aligned}\Psi_{p_1} &: \mathbb{R}^{n_u} \mapsto \mathbb{R}^{n_u} \\ \Psi_{p_1}(p_1, u - HF(\nu, u, \alpha)) &= p_1 + k_{p_1}[u - HF(\nu, u, \alpha)].\end{aligned} \tag{7.165}$$

The corresponding inner loop algorithm for iterating the prices that forces the interaction balance reads:

$$p_1^{l+1,n} = p_1^{l,n} + k_{p_1}[u^{l,n} - HF(\nu^{l,n}, u^{l,n}, \alpha^n)]. \tag{7.166}$$

As previously, we shall analyze the convergence properties of the double-loop model-based with relaxed inner loop algorithm in a linear-quadratic case. We shall not repeat here the assumptions, notation and definitions as they remain the same as introduced and considered for the case with tight inner loop.
Let us define matrix M_γ as

$$M_\gamma \triangleq M + \gamma \widetilde{B}^T B \tag{7.167}$$

where the matrices M, B and $\widetilde{B}$ are defined by (7.133), (7.136) and (7.138). and where $\gamma > 0$ is chosen such that $M_\gamma > 0$. Note that according to (7.137) and due to assumption (7.139) such a choice of γ always exists (Luenberger, 1984). It is assumed additionally that the matrix B has full rank.

The *convergence conditions* can now be formulated for the double iterative loop algorithm with a relaxed inner loop.

Theorem 7.2
Assume that

$$\begin{bmatrix} \frac{1}{k_c} L_*^T M_\rho L - \frac{1}{2} L_*^T M_\gamma L_* & \frac{1}{2k_c} \widetilde{B} M_\gamma^{-1} M_\rho L + \frac{1}{2k_p} B L_* \\ \frac{1}{2k_c} \widetilde{B} M_\gamma^{-1} M_\rho L + \frac{1}{2k_p} B L_* & \frac{1}{k_p} \widetilde{B} M_\gamma^{-1} B^T - \frac{1}{2} \widetilde{B} M_\gamma^{-1} \widetilde{B}^T \end{bmatrix} > 0 \tag{7.168}$$

Then

(i) The relaxed inner loop problem is well defined.
(ii) There exists a number $\overline{k}_{p_1} > 0$ such that for any $k_{p_1} \in (0, \overline{k}_{p_1})$ the inner loop iterative scheme is convergent for every $c \in C$ and $p_2 \in U$.
(iii) The outer loop iterative scheme is well defined and generates a sequence of $\{(c^n, p_2^n)\}$ which is convergent to $(c^{opt}, \overline{p})$.

Proof of this theorem is given in Appendix B.2.

Let us note that condition (7.168) implies that condition *(ii)* of Theorem (7.1) is satisfied and condition *(ii)* of the Theorem (7.1) in which matrix M_ρ^{-1} is replaced by matrix M_γ^{-1}, is also satisfied. A proof of the Theorem (7.2) shows that the condition (7.168) is only required to preserve a convergence of the outer loop iterative scheme while there are no restrictions implied by the inner loop. Therefore, as expected, although the convergence properties of the relaxed inner loop are better, the outer loop imposes conditions such that the overall convergence conditions of the second technique seem to be not less restrictive than the technique with tight inner loop, in the linear quadratic case.

Clearly, the maximum ranges of feasible values of the algorithm relaxation gains may be significantly different and so may the technique efficiency. It is expected that in some more general nonlinear situations the double-loop method with a relaxed inner loop will be more efficient due to unquestionable better efficiency of the inner loop.

Let us now examine the conditions (i) and (ii) of the Theorem(7.1) and the condition (7.168).in a convex case, that is if $M > 0$. Let us also assume that $k_c = k_p = \chi$. The values $\rho_c = \rho_u = \rho = 0$ are then allowed and the conditions (i) and (ii) take the form $BM_\rho^{-1}\widetilde{B}^T > 0$ and $L_*^T M_\rho L > 0$ while the condition (7.168) reads

$$\begin{bmatrix} \frac{1}{\chi}L_*^T ML - \frac{1}{2}L_*^T ML_* & \frac{1}{2\chi}(\widetilde{B}L + BL_*) \\ \frac{1}{2\chi}(\widetilde{B}L + BL_*) & \frac{1}{\chi}\widetilde{B}MB^T - \frac{1}{2}\widetilde{B}M\widetilde{B}^T \end{bmatrix} > 0. \tag{7.169}$$

The above conditions are exactly the same as those obtained separately for the convex case by (Brdys *et al.*,1989). Moreover, it can be easily verified that $\widetilde{B}L + BL_* = 0$. Therefore, the condition (7.168) is equivalent to the conditions (i) and (ii) in the convex case.

7.3.5 *Simulation studies*

Example 7.3

A simple system consisting of two interconnected subsystems structured as depicted in Fig. 7.14 will first be considered in order to illustrate the convergence conditions of the theorems 7.1 and 7.2 in a convex case. Also, the impact of the time varying disturbance and measurement noise will be investigated. Hence, the interconnection structure matrix reads

$$H = \begin{bmatrix} 0 & 1 & 0 \\ 1 & 0 & 0 \end{bmatrix}. \tag{7.170}$$

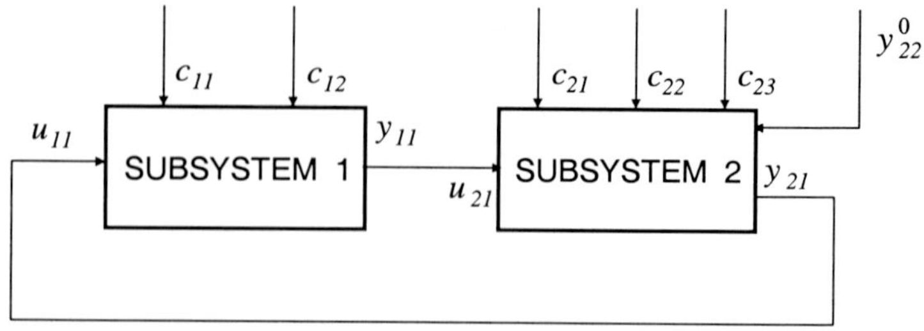

Fig. 7.14 Structure of the system in the example problem, Ex. 7.3.

The model and reality equations are

$$\begin{aligned}
y_1 &= y_{11} = F_{11}(c_1, u_1) = (c_{11} - c_{12} + 2u_{11} + \alpha_{11} && (7.171a)\\
y_{21} &= F_{21}(c_2, u_2) = c_{21} - c_{22} + u_{21} + \alpha_{21} && (7.171b)\\
y_{22} &= F_{22}(c_2, u_2) = 2c_{22} - c_{23} - u_{21} + \alpha_{22} && (7.171c)
\end{aligned}$$

and

$$\begin{aligned}
y_{11} &= F_{*11}(c_1, u_1) = 1.4c_{11} - 0.6c_{12} + 1.8u_{11} && (7.172a)\\
y_{21} &= F_{*21}(c_2, u_2) = 1.3c_{21} - 1.1c_{22} + 1.1u_{21} && (7.172b)\\
y_{22} &= F_{*22}(c_2, u_2) = 2.3c_{22} - 0.7c_{23} - u_{21}. && (7.172c)
\end{aligned}$$

There are constraints imposed on both subsystem variables

$$CU_1 = \{(c_1, u_1) : \ 2c_{12} + 0.6u_{11} - 0.8 \leq 0; \ -1 \leq c_{11} \leq 1; \\ -1 \leq c_{12} \leq 1\} \quad (7.173a)$$

$$CU_2 = \{(c_2, u_2) : \ -1 \leq c_{21} \leq 1; \ -1 \leq c_{21} \leq 1; \ -1 \leq c_{21} \leq 1\}. \quad (7.173b)$$

Initially, the output independent performance functions are considered

$$\begin{aligned}
Q_1(c_1, u_1, y_1) &= (c_{11})^2 + (c_{12})^2 + 2(u_{11} - 2)^2 && (7.174a)\\
Q_2(c_2, u_2, y_2) &= (c_{21})^2 + (c_{22})^2 + (c_{23})^2 + (u_{21} - 1)^2. && (7.174b)
\end{aligned}$$

Using (7.136), (7.137), (7.138), (7.142) and (7.143) with moderate matrix calculations enables us to calculate

$$\widetilde{B}M^{-1}\widetilde{B}^T = \begin{bmatrix} 1.727 & -0.882 \\ -0.882 & 3.295 \end{bmatrix} \tag{7.175a}$$

$$BM^{-1}\widetilde{B}^T = \begin{bmatrix} 1.73 & -0.756 \\ -1.101 & 2.723 \end{bmatrix} \tag{7.175b}$$

and

$$L_*^T M L_* = \begin{bmatrix} 15.94 & -5.97 & 15.14 & -12.83 & 0.0 \\ -5.97 & 4.58 & -6.49 & 5.5 & 0.0 \\ 15.14 & 4.58 & -6.49 & 5.5 & 0.0 \\ 15.14 & -6.49 & 20.43 & -15.61 & 0.0 \\ 0.0 & 0.0 & 0.0 & 0.0 & 2.0 \end{bmatrix} \tag{7.176a}$$

$$L_*^T M L = \begin{bmatrix} 11.14 & -9.14 & 12.0 & -12.0 & 0.0 \\ -3.92 & 5.92 & -5.15 & 5.15 & 0.0 \\ 10.07 & -10.07 & 16.85 & -14.85 & 0.0 \\ -8.54 & 8.54 & -12.85 & 14.58 & 0.0 \\ 0.0 & 0.0 & 0.0 & 0.0 & 2.0 \end{bmatrix}. \tag{7.176b}$$

It can be easily verified that $BM^{-1}\widetilde{B}^T > 0$ and $L_*^T ML > 0$, so that the conditions $(i), (ii)$ of Theorem 7.1 and the condition (7.168) of Theorem 7.2 are satisfied under the existing model-reality differences. Consequently, both iterative algorithms are convergent if their step coefficients are appropriately chosen. We assume that the technique with relaxed inner loop is to be implemented with $k_c = k_p = \chi$. The condition (7.168) yields a feasible range for χ as $0 < \chi < 1.52$. It can be easily verified that the condition (7.144) is not satisfied for $\chi = 1.5$. The limit on a maximum allowed step coefficient of the algorithm with tight inner loop when compared with the relaxed inner loop algorithm can then be noted and better efficiency of the later one can be expected. However, it can be verified that the condition (7.144) is satisfied with $\chi = 1.0$. Therefore, $k_p = 1.0$ and $k_c = 1.0$ constitute a proper choice for the step coefficients and the above analysis shows that the convergence of the iterative algorithms can be readily achieved.

The simulation studies have been performed to investigate the performance of the algorithm with tight inner loop but in the case where the performance function is output dependent

$$Q_1(c_1, u_1, y_1) = (c_{11})^2 + (c_{12})^2 + (y_{11} - 1)^2 \tag{7.177}$$

$$Q_2(c_2, u_2, y_2) = (c_{21})^2 + (c_{22})^2 + (c_{23})^2 + \\ +2(y_{21} - 2)^2 + (y_{22}^0 - 3)^2. \quad (7.178)$$

The optimal values, in terms of the convergence efficiency, of the step coefficients were found as $k_c = 1.0$ and $k_p = 0.8$ and the simulations were carried out with these values. A convergence of the performance function to the true optimal value is shown in Fig. 7.15. The double-loop algorithm needs only 4 outer loop iterations, hence three applications of the set-points, to achieve a performance within 1.5% of the true system minimum. With a total 143 inner loop iterations that are are performed entirely on the system model and do not involve changing the controller set-points in the system, the algorithm requires only nine set point applications to converge to the true optimum with the accuracies $\varepsilon_p < 0.0001$ and $\varepsilon_c < 0.0001$ of solving the inner-loop and the outer-loop tasks (see (7.122) and (7.124), respectively). Because the inner loop iterations are solely model based, the computational time required to perform the 143 by the inner loop is negligible comparing to the time required for applying the nine set-points with associated times for obtaining the required steady-state measurements when the algorithm is used to control a typically slow industrial process.

The algorithm has been found to respond efficiently to various disturbances applied to the system and simulated by resetting the starting point after convergence from the previous point. This is illustrated in Fig. 7.16, which demonstrates the self-adaptive nature of the method. Because the calculation of the modifier vector λ by using the formulae (7.90) and (7.94) requires evaluation of real process derivatives, which in practice are obtained by applying perturbations to the controller set-points and using finite difference formulae, it is expected that the behavior of the method could deteriorate when faced with noisy measurements. This indeed occurs, as illustrated in Fig. 7.17, which gives the performance under the measurement noise with a standard deviation of 0.033, representing a noise-to-signal power ratio of 10%.However, this important problem can be overcome using simple digital filter techniques. This is illustrated by the results shown in Fig. 7.18, which demonstrate that by using an averaging filter on each measurement (which in this particular situation took an average of 50 rapidly taken measurements) adequate performance is achieved even in the presence of noise.

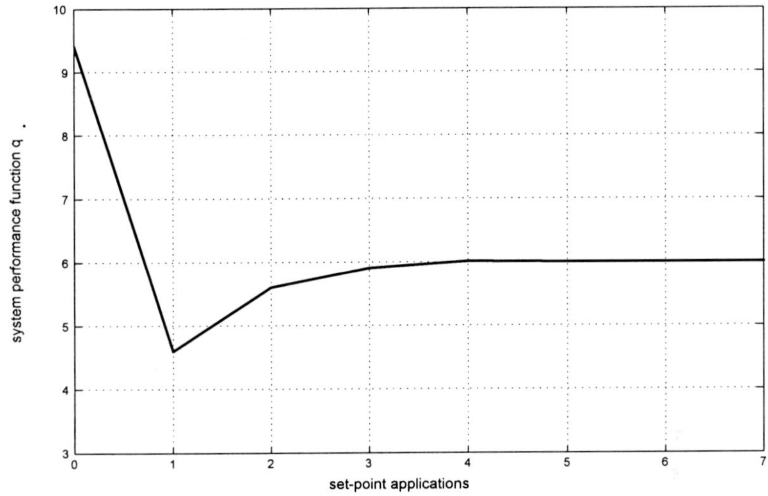

Fig. 7.15 System performance function trajectory generated by model-based double-loop algorithm with tight inner loop and with input-output measurements for the example problem, Ex.7.3.

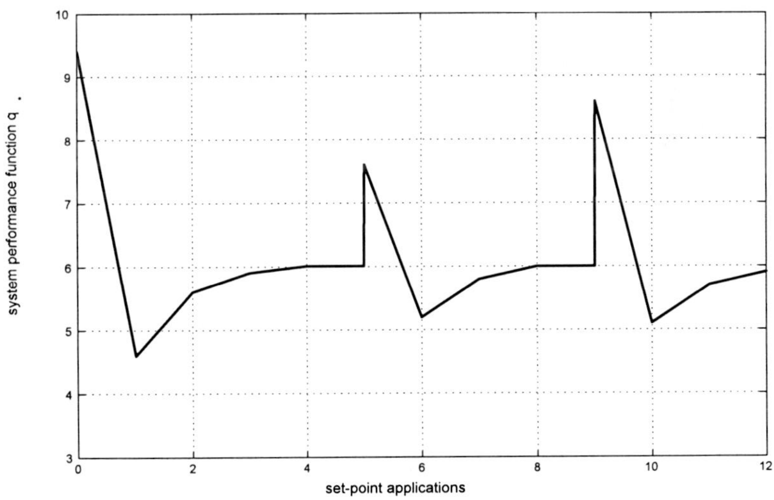

Fig. 7.16 Response to different disturbances in the example problem, Ex. 7.3.

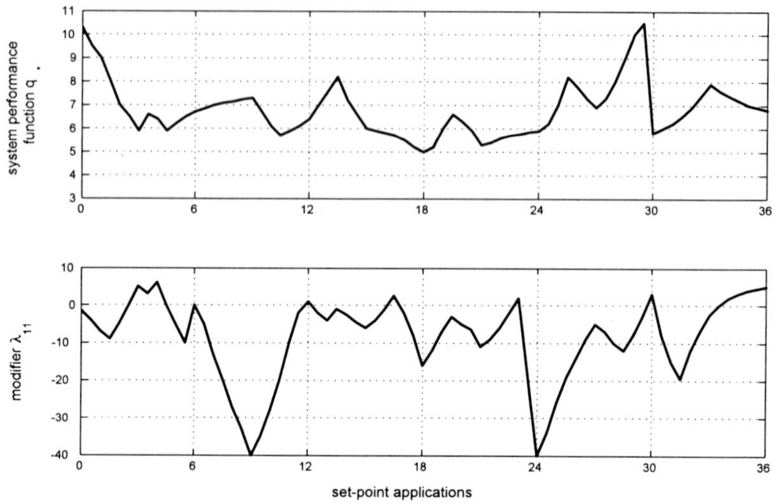

Fig. 7.17 Deterioration in performance when subject to measurement noise for the example problem, Ex. 7.3.

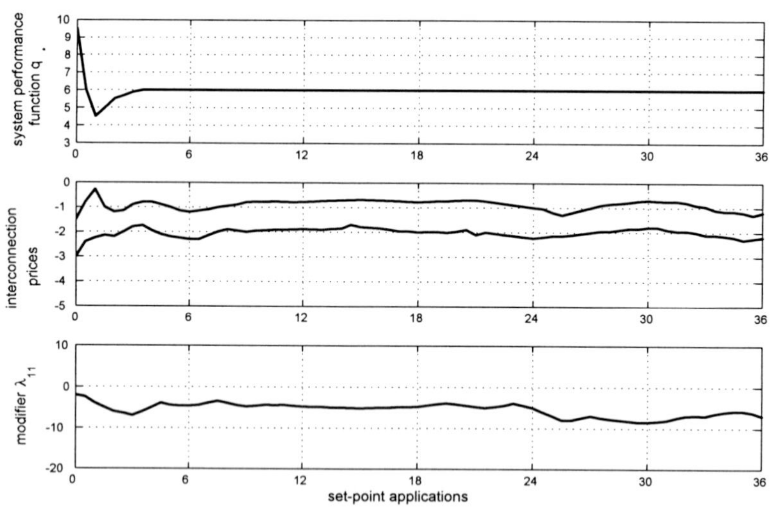

Fig. 7.18 Application of averaging filter to overcome the measurement noise for the example problem, Ex. 7.3.

Example 7.4

A simple system consisting of two interconnected subsystems structured as depicted in Fig. 7.19 will be considered in order to compare the convergence properties of the methods with tight inner loop and relaxed inner loop in convex and nonconvex cases.

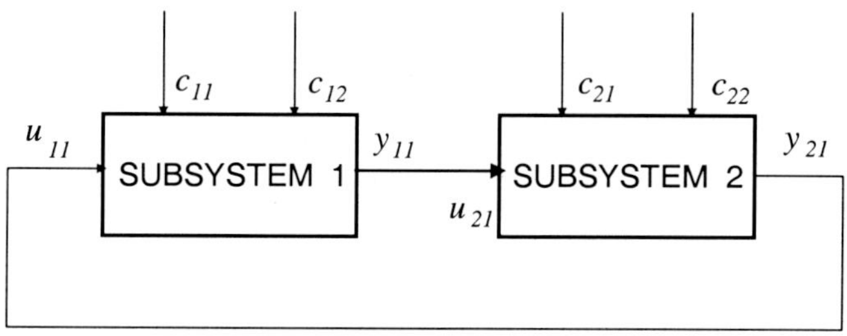

Fig. 7.19 Structure of the system in the example problem, Ex. 7.4.

Case 1 - convex

The model and reality equations are:

$$y_1 - F_1(c_1, u_1) - 1.7c_{11} - 0.5c_{12} + u_1 + \alpha_1 \qquad (7.179a)$$
$$y_2 = F_2(c_2, u_2) = 2.2c_{21} - 0.2c_{22} + 0.5u_2 + \alpha_2 \qquad (7.179b)$$

and

$$y_1 = F_{*1}(c_1, u_1) = 2.0c_{11} - c_{12} + 0.6u_1 + 0.15u_1c_{11} \qquad (7.180a)$$
$$y_2 = F_{*2}(c_2, u_2) = 2.2c_{21} - 0.2c_{22} + 0.9u_2 + 0.1c_{21}c_{22}. \qquad (7.180b)$$

The constraints on the interaction and control inputs (local constraints) are:

$$CU_1 = \{(c_1, u_1) : (c_{11})^2 + (c_{12})^2 \leq 1; \; -1 \leq c_{11} \leq 1; \; -1 \leq c_{12} \leq 1; \; 0 \leq u_1 \leq 0.5\} \qquad (7.181a)$$

$$CU_2 = \{(c_2, u_2) : 0.5c_{21} + c_{22} \leq 1; \; -1 \leq c_{21} \leq 1; \; -1 \leq c_{22} \leq 1; \; -1 \leq u_2 \leq 1\}. \qquad (7.181b)$$

The subsystem performance functions are output independent

$$Q_1(c_1, u_1, y_1) = 0.5(c_{11})^2 + 0.5(c_{12})^2 + 0.5(u_1 - 1)^2 \quad (7.182a)$$
$$Q_2(c_2, u_2, y_2) = 0.5(c_{21})^2 + (c_{22})^2 + 0.5(u_{11} - 2)^2. \quad (7.182b)$$

Case 2 -nonconvex

The subsystem performance functions are chosen nonconvex

$$Q_1(c_1, u_1, y_1) = -(c_{11})^2 - (c_{12})^2 + 0.5(u_1 - 1)^2 \quad (7.183)$$
$$Q_2(c_2, u_2, y_2) = (c_{21})^2 + (c_{22})^2 - 0.5(u_2 - 2)^2. \quad (7.184)$$

The input-output relations, their models and local constraints are the same as in case 1.

Initially, simulation was performed to determine suitable values of the convexifying parameters ρ_u, ρ_c (see (7.121)) and ρ in (7.141), γ in (7.167), step coefficients (relaxation gains) k_c, k_p ((7.128) and (7.132)) for the method with tight inner loop and k_c, k_{p_1}, k_{p_2} (see (7.162), (7.165) and (7.164)), for the method with relaxed inner loop. In order to simplify the simulations trials and and the same time without major loss on a generality, it was assumed that $\rho_u = \rho_c = \rho = \gamma$. Also, the same values of k_c was assumed for both methods. Finally, $k_p = k_{p_1}$ was set. The results are summarized in Table 7.2 regarding the optimal, in terms of the convergence efficiency, values of these for parameters and gains.

	ρ	k_c	k_{p_1}	k_{p_2}	set point iterations	total iter.	optimal perform.
Case 1-convex							
tight inner loop	0.0	0.7	0.3	n/a	30	228	0.6876
	0.2	0.7	0.3	n/a	34	157	0.6878
	4.0	0.7	1.0	n/a	39	218	0.6875
rel. inner loop	0.2	0.7	0.3	0.7	24	113	0.6873
	4.0	0.7	0.1	0.7	30	164	0.6872
Case 2-nonconvex							
tight inner loop	5.0	0.5	1.0	1.0	57	290	−5.9025
rel. inner loop	5.0	0.5	1.0	0.7	55	280	−5.9017

Table 7.2 Convergence properties of model-based double iterative loop algorithms with input-output measurements for the example problem.

The results results from Table 7.2 show that when $\rho = 0$ and there is no convexifying term in the MMOP, an algorithm with tight inner loop achieves solution in the Case 1 with 30 set-point changes (iterations performed on the system) and with 228 total iterations. The latter represents a measure of an overall communication effort due to a total information exchange within the control system that is needed to determine the optimal system set-points. In order to optimize the algorithm efficiency the largest allowable values of the relaxation gains (step-sizes) at all the iterative loops were applied, that is $k_c = 0.7$ and $k_p = k_{p,1} = 0.3$. With these values of the gains the MMOP augmentation was applied with $\rho = 0.2$. For the algorithm with tight inner loop a total number of set-point changes increases to 34 but a total information exchange reduces to 157 overall iterations. The algorithm with relaxed inner loop further manages to reduces the set-point changes number to 240 and also to reduce the total number of iterations to 113.

The augmentation strength was then further increased to $\rho = 4.0$ allowing to increase the inner loop gain k_{p_1} to 1.0. This, however, did not improve, as expected, the algorithm efficiency. Both the total number of set point changes and the overall iteration number increased to 39 and 218, respectively for the tight inner loop algorithm and to 30 and 164, respectively for the relaxed loop algorithm (see Table 7.2)). The real system performance function trajectories produced by these algorithms are illustrated in Fig. 7.20 and Fig. 7.21. In this example, the increased step-size in the inner possible as the result of the increased augmentation strength, does not however, imply an increase of the outer loop efficiency. Hence, an overall iterative loop efficiency gets worse.

The two iterative strategies with input-output measurements were also applied to the Case 2 example that is nonconvex. The simulation results are illustrated in Fig. 7.22 and the quantitative information is given in Table 7.2. The MMOP augmentation is necessary in this example as the problem is not convex and $\rho = 5.0$ was applied. The algorithm with relaxed inner loop is slightly better than the one with tight inner loop and achieves the optimal solution with 55 set-point changes and with 290 iterations overall.

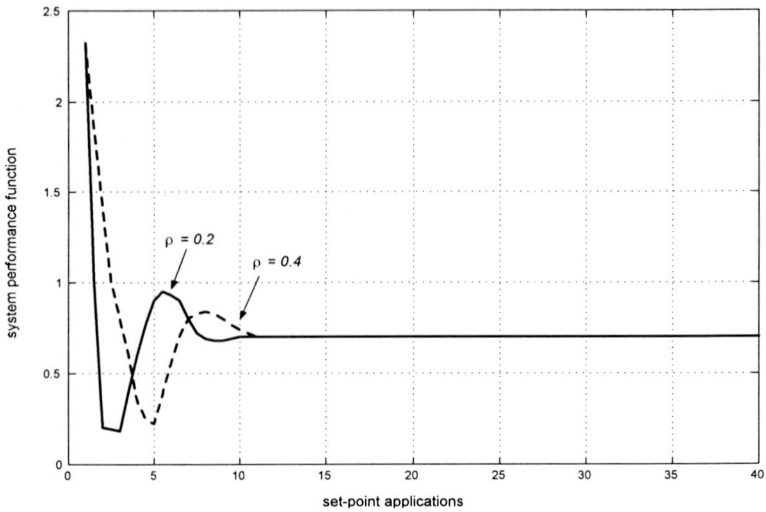

Fig. 7.20 The performance function trajectory produced by the double-loop algorithm with relaxed inner loop and with input-output measurements in the convex case, Ex. 7.4.

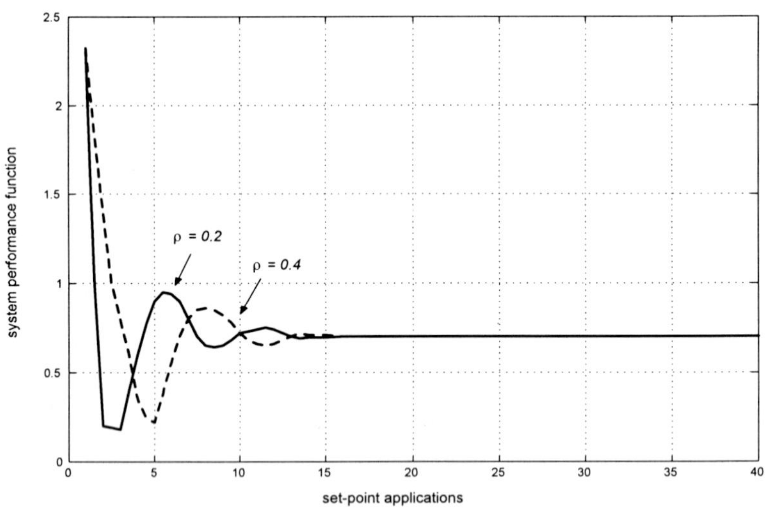

Fig. 7.21 The performance function trajectory produced by the double-loop algorithm with tight inner loop and with input-output measurements in the convex case, Ex. 7.4.

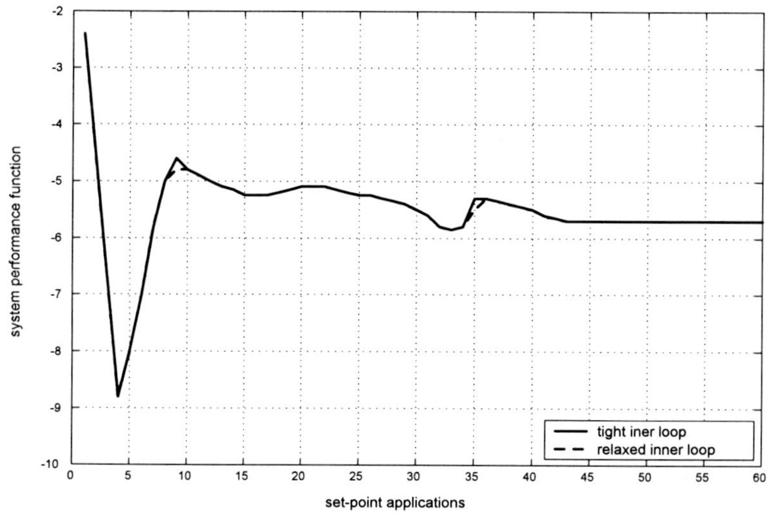

Fig. 7.22 The performance function trajectory produced by the model-based algorithms with the tight inner loop and relaxed inner loop and with input-output measurements in the nonconvex case, Ex. 7.4.

7.4 ISOPE Double-Loop Iterative Algorithms with Output Measurements

As it was already pointed out in Section 7.3, using the interaction input measurements for ISOPE type of control might be doubtful if the measurement noise level is high. This is due to a necessity of obtaining the interaction input derivatives. Regardless of how these derivatives are produced its manufacturing means an operation on the noisy data. Clearly, the interaction inputs can be estimated by using the output measurements and the interconnection equations (7.4) but this still produces noisy data. Alternatively, we would think of exploiting models and design the control algorithm in such a way that the interaction values needed would be provided by the models. In this subsection we shall derive such control architecture and also similarly as in Section 7.3 double loop control algorithms that do not require measurements of the interaction inputs.

7.4.1 *Structure*

The structure development combines work presented by (Brdyś and Roberts, 1986; Brdyś *et al.*, 1989; Brdyś *et al.*, 1990b). We shall start with transforming the SOCP1 form described by (7.80a), (7.80b), (7.80c), (7.80d), (7.80e), (7.80f) and (7.80g) of the SOCP (see (7.79)) into its another representation SOCP4 as:

$$\text{minimize}_{c,\upsilon,u,w,\alpha} \left\{ q(\upsilon, w, \alpha) + \rho_c \|\upsilon - c\|^2 + \rho_u \|w - u\|^2 \right\} \tag{7.185a}$$

subject to:

$$\overline{F}(c, u, \alpha) = \overline{K}_*(c) \tag{7.185b}$$
$$w = HF(\upsilon, w, \alpha) \tag{7.185c}$$
$$g(\upsilon, w) \leq 0 \tag{7.185d}$$
$$\alpha \in A \tag{7.185e}$$
$$\upsilon = c \tag{7.185f}$$
$$w = u \tag{7.185g}$$

where

$$\overline{F}(c, u, \alpha)^T \triangleq [F^0(c, u, \alpha)^T, F(c, u, \alpha)^T] \tag{7.185h}$$

and the mapping $\overline{K}_*(\cdot)$ is defined by (7.82b).

The above representation of the SOCP is analogous to the SOCP3 representation derived in Section 7.3 for a purpose of deriving the double-loop iterative control algorithms with input-output measurements (see (7.84a), (7.84b), (7.84c), (7.84d), (7.84e) and (7.84f)). Let us notice that indeed the SOC4 representation of SOCP does not require the interaction input measurements as the term $HK_*(c)$ in (7.84b) was replaced in (7.185a) and (7.185b) by free interaction variable u. In addition to the auxiliary variable ν another auxiliary variable w was introduced in (7.185c) and (7.185d) in order to separate a parameter estimation from a modified model optimization.

The Lagrange function for the SOCP4 reads:

$$\begin{aligned} L(c, \nu, u, w, \alpha, \lambda, t, p, \xi, \mu) \;=\; & q(\nu, u, \alpha) + \rho_c \|\nu - c\|^2 + \rho_u \|w - u\|^2 + \\ & + \lambda^T (c - \nu) + t^T (u - w) + \\ & + p^T (u - HF(\nu, w, \alpha) + \mu^T g(\nu, w) + \\ & + \xi^T (\overline{F}(c, u, \alpha) - \overline{K}_*(c)) \end{aligned} \tag{7.186}$$

where λ, t, p, ξ, μ are the Lagrange multipliers associated with the constraints (7.185f), (7.185g), (7.185c), (7.185b) and (7.185d), respectively.

Let us consider the Kuhn-Tucker necessary optimality conditions for SOCP4:

$$L'_{[\nu,w]}(c,\nu,u,w,\alpha,\lambda,t,p,\xi,\mu)^T = q'_{[\nu,w]}(\nu,w,\alpha)^T + 2\rho_c \begin{bmatrix} \nu - c \\ 0 \end{bmatrix} +$$

$$+2\rho_u \begin{bmatrix} 0 \\ w-u \end{bmatrix} - \begin{bmatrix} \lambda \\ 0 \end{bmatrix} - \begin{bmatrix} 0 \\ t \end{bmatrix} + \{(u - HF(\nu,w,\alpha))'_{[\nu,w]}\}^T p + g'(\nu,w)^T \mu = 0$$

$$g(\nu,w) \leq 0, \ \mu \geqslant 0, \ \mu^T g(\nu,w) = 0 \qquad (7.187a)$$

$$L'_c(c,\nu,u,w,\alpha,\lambda,t,p,\xi,\mu)^T = 2\rho_c(c-\nu) + \lambda + [\overline{F}'_c(c,u,\alpha)^T - \overline{K}'_*(c)^T]\xi = 0 \qquad (7.187b)$$

$$L'_u(c,\nu,u,w,\alpha,\lambda,t,p,\xi,\mu)^T = 2\rho_u(u-w) + t + \overline{F}'_u(c,u,\alpha)^T\xi = 0 \qquad (7.187c)$$

$$L'_\alpha(c,\nu,u,w,\alpha,\lambda,t,p,\xi,\mu)^T = q'_\alpha(\nu,w,\alpha)^T - F'_\alpha(\nu,w,\alpha)^T H^T p +$$
$$+ \overline{F}'_\alpha(c,u,\alpha)^T\xi = 0 \qquad (7.187d)$$

$$L'_\lambda(c,\nu,u,w,\alpha,\lambda,t,p,\xi,\mu)^T = \nu - c = 0 \qquad (7.187e)$$

$$L'_t(c,\nu,u,w,\alpha,\lambda,t,p,\xi,\mu)^T = w - u = 0 \qquad (7.187f)$$

$$L'_p(c,\nu,u,w,\alpha,\lambda,t,p,\xi,\mu)^T = w - HF(\nu,w,\alpha) = 0 \qquad (7.187g)$$

$$L'_\xi(c,\nu,u,w,\alpha,\lambda,t,p,\xi,\mu)^T = \overline{F}(c,u,\alpha) - \overline{K}_*(c) = 0. \qquad (7.187h)$$

The double-loop ISOPE control algorithm with output measurements will be derived as a double-loop structured iterative method for solving the stated above necessary optimality conditions corresponding to SOCP4. Hence, we shall now proceed analogously as in Section 7.3. From (7.187d) the multiplier ξ can be calculated as:

$$\xi(c,\nu,u,w,\alpha,p) = [\overline{F}'_\alpha(c,u,\alpha)\overline{F}'_\alpha(c,u,\alpha)^T]^{-1}\overline{F}'_\alpha(c,u,\alpha) \times$$
$$\times[F'_\alpha(\nu,w,\alpha)^T H^T p - q'_\alpha(\nu,w,\alpha)^T]. \qquad (7.188)$$

The inverse in (7.188) exists for reasonably parameterized models. Substituting the expression for ξ into the equations (7.187b) and (7.187c) yields the expressions for the multipliers λ and t in terms of c, ν, u, w, α and p as

$$\lambda(c,\nu,u,w,\alpha,p) = 2\rho_c(\nu - c) + [\overline{K}_*'(c)^T - \overline{F}_c'(c,u,\alpha)^T]\xi \tag{7.189}$$

$$t(c,\nu,u,w,\alpha,p) = 2\rho_c(w - u) - \overline{F}_u'(c,u,\alpha)^T\xi \tag{7.190}$$

where ξ is given by (7.188).

It follows from (7.187e) and (7.187f) that at a solution of the optimality conditions $c = \nu$ and $u = w$. Hence, at the solution the expressions (7.189) and (7.190) for λ and t, respectively simplify to:

$$\lambda(c,u,\alpha,p) \triangleq \lambda(c,c,u,u,\alpha,p) = [\overline{K}_*'(c)^T - \overline{F}_c'(c,u,\alpha)^T]\xi(c,u,\alpha,p) \tag{7.191}$$

$$t(c,u,\alpha,p) \triangleq t(c,c,u,u,\alpha,p) = -\overline{F}_u'(c,u,\alpha)^T\xi(c,u,\alpha,p). \tag{7.192}$$

Substituting now (7.93) into (7.188) more suitable expression for calculating ξ is obtained:

$$\begin{aligned}\xi(c,u,\alpha,p) \triangleq\ &\xi(c,u,\alpha,p) = -\overline{Q}_{\overline{y}}(c,u,\overline{F}(c,u,\alpha))^T + \\ &+ [\overline{F}_\alpha'(c,u,\alpha)\overline{F}_\alpha'(c,u,\alpha)^T]^{-1}\overline{F}_\alpha'(c,u,\alpha)F_\alpha'(c,u,\alpha)^T H^T p.\end{aligned} \tag{7.193a}$$

Let us notice that in case of a single plant the variables u, p, H disappear and also $\overline{F}(\cdot) = \overline{F}^o(\cdot) = F(\cdot)$. The formula (7.193a) becomes then the same as the formula (4.7) in Chapter 4. Also, substituting the multiplier ξ into (7.191) gives the expression for λ that is exactly the same as derived in Chapter 4 (see (4.10)).

The second term in (7.193a) looks complicated. However, this is only superficial. Indeed, suppose that all the outputs are of the interconnection type. Then $y = y^o$. The second term in (7.193a) becomes then equal to $H^T p$ and the expression (7.193a) reads

$$\xi(c,u,\alpha,p) = -Q_y'(c,u,F(c,u,\alpha))^T + H^T p \tag{7.194}$$

yielding the expressions for λ and t (see (7.191) and (7.192) as

$$\lambda(c,u,\alpha,p) = [F_c'(c,u,\alpha)^T - K_*'(c)^T][Q_y'(c,u,F(c,u,\alpha))^T - H^T p] \tag{7.195}$$

$$t(c,u,\alpha,p) = F_u^{'}(c,u,\alpha)^T[Q_y^{'}(c,u,F(c,u,\alpha))^T - H^T p]. \tag{7.196}$$

It will be shown that the multipliers λ and t are used to modify a model-based optimization problem. As it follows from the formula (7.195) and (7.196) both the multipliers depend on the model parameters that are updated on line based on the output measurements and also on the price variable p that is responsible for meeting the interconnection structure equations, or in other words, for achieving the interaction balance (see first term in the third line of (7.186). Thus the multipliers cater for model reality differences and for meeting the subsystem interconnections.

So far we have managed to solve three out of eight necessary optimality conditions that are (7.187b), (7.187c) and (7.187d) and the conditions (7.187e) and (7.187f) have been imbedded in the obtained solutions. Under prescribed values of of the control and interaction inputs c and u respectively, the condition (7.187g) allows for an estimation of the model parameter vector α given as the sytem output response to the control inputs $\overline{K}_*(c)$ is known from the output measurements.

The *parameter estimation problem* (PEP) is defined as of finding for given control input c and interaction input u such parameter values $\hat{\alpha}(c,u) \in A$ that

$$\begin{aligned} F^o(c,u,\hat{\alpha}(c,u)) &= K_*^o(c) \\ F(c,u,\hat{\alpha}(c,u)) &= K_*(c). \end{aligned} \tag{7.197a}$$

Different methods of solving PEP were explained in Section 7.3 that use suitable models of uncertainty. It was also pointed out that a nondeterministic formulation of ISOPE method still remains a challenge.

Finally, the remaining conditions (7.187a), (7.187g) and still (7.187e) and (7.187f) are met as follows. First a *modified model-based optimization problem (*MMOP) is defined for prescribed vectors c, u and p as:

$$\begin{aligned} &\text{minimize}_{\nu,w}\{q(\nu,w,\alpha) + \rho_c\|\nu - c\|^2 + \rho_u\|w-u\|^2 + \\ &\qquad -\lambda^T\nu - t^T w + p^T(w - HF(\nu,w,\alpha))\} \\ &\text{subject to: } g(\nu,w) \le 0 \end{aligned} \tag{7.198}$$

where $\alpha = \hat{\alpha}(c,u)$ is defined as solution of PEP under given c and u; the multipliers $\lambda = \lambda(c,u,\hat{\alpha}(c,u),p)$ and $t = t(c,u,\hat{\alpha}(c,u),p)$ are calculated from (7.193a), (7.191) and (7.192) under $\alpha = \hat{\alpha}(c)$.

The MMOP is defined for prescribed values of the control and interaction inputs c, u and prices p that are the MMOP inputs. The multipliers λ and t are calculated from known formulae based on the MMOP input data. Let us denote the MMOP solution by

$$\hat{\nu}(c,u,p) \text{ and } \hat{w}(c,u,p). \tag{7.199a}$$

Second, meeting the conditions (7.187a) and (7.187g) and completing the meeting of the conditions (7.187e) and (7.187f) is achieved by solving the following set of equations:

$$\hat{\nu}(c,u,p) = c \tag{7.200a}$$
$$\hat{w}(c,u,p) = u \tag{7.200b}$$
$$\hat{w}(c,u,p) = HF(\hat{\nu}(c,u,p), \hat{w}(c,u,p), \hat{\alpha}(c,u)). \tag{7.200c}$$

Indeed, writing down Kuhn-Tucker necessary conditions for optimality for MMOP yields the condition (7.187a) with

$$c, \nu = \hat{\nu}(c,u,p), w = \hat{w}(c,u,p), \alpha = \hat{\alpha}(c,u) \tag{7.201}$$

and λ as applied in the MMOP. Moreover, the equations (7.200a), (7.200b) and (7.200c) represent nothing else but the conditions (7.187e), (7.187f) and (7.187g)), respectively with c, u, α as given by (7.201).

The equations (7.200a), (7.200b) and (7.200c) constitute a base for deriving the ISOPE double-loop iterative strategies utilizing output measurements. As in Subsection 7.3 two possible ways of double-loop structuring of these equations will be considered. As the result the *system-based double-loop algorithm* and the *model-based double-loop algorithm* will be derived. We shall start with an investigation of the optimality properties of the solutions produced.

7.4.2 *Optimality*

Let Ω denotes the solution set. It follows immediately from presented previously derivation of the ISOPE strategy with output measurements that: if $(\bar{c}, \bar{u}, p) \in \Omega$ and in a point $(\bar{c}, \bar{u}) \in CU$ regularity conditions are satisfied then the point $(\bar{c}, \bar{u})$ satisfies Kuhn-Tucker necessary conditions for optimality (Lasdon, 1970) of SOCP. Hence a quality of the solutions provided by the method is high. It is however, extremely important to assess the method strength as a tool to solving the SOCP. Hence, a fundamental question to be answered is this: if $(\bar{c}, K_*(\bar{c}))$ is known to be solution of SOCP can it be found by applying the method? In other words does a price value $\bar{p}$ exist such that $(\bar{c}, K_*(\bar{c}), \bar{p})$ belong to the solution set Ω ? It will be shown that under standard assumptions the answer is positive.

We shall begin with transferring the expression for the modifiers λ and t into forms that are more suitable for this analysis. In order to simplify notation but without loss of a generality a case when all the system outputs are at the same time the interconnection outputs is to be investigated. In this case the modifiers are expressed by formulae (7.195) and (7.196), respectively. First, let us consider λ. The eq. (7.195) with $\hat{\alpha} = \hat{\alpha}(c, u)$ yields:

$$\begin{aligned} \lambda(c, u, \hat{\alpha}, p)^T &= Q_y'(c, u, F(c, u, \hat{\alpha}))F_c'(c, u, \hat{\alpha}) - Q_y'(c, u, F(c, u, \hat{\alpha}))K_*'(c) \\ &\quad -p^T HF_c'(c, u, \hat{\alpha}) + p^T HK_*'(c). \end{aligned} \tag{7.202}$$

Let us define:

$$q_*(c, u) \triangleq Q(c, u, K_*(c)) \tag{7.203}$$

$$q(c, u, \alpha) \triangleq Q(c, u, F(c, u, \alpha)) \tag{7.204}$$

$$G_*(c, u) \triangleq u - HK_*(c) \tag{7.205}$$

$$G(c, u, \alpha) \triangleq u - HF(c, u, \alpha). \tag{7.206}$$

Note that the function $q_*(\cdot)$ is the interconnected system performance function while the function $q(\cdot)$ is the system performance function model. Similarly, the mapping $G_*(\cdot, \cdot)$ describes the system interconnections while the mapping represents its model.

As $\hat{\alpha}$ solves PEP with c an u then from (7.204), (7.203), (7.206) and (7.205) the following hold:

$$Q_y'(c, u, F(c, u, \hat{\alpha}))F_c'(c, u, \hat{\alpha}) = q_c'(c, u, \hat{\alpha}) - Q_c'(c, u, F(c, u, \hat{\alpha})) \tag{7.207a}$$

$$\begin{aligned} -Q_y'(c, u, F(c, u, \hat{\alpha}))K_*'(c) &= -q_{*c}'(c, u) + Q_c'(c, u, K_*(c)) \\ &= -q_{*c}'(c, u) + Q_c'(c, u, F(c, u, \hat{\alpha})) \end{aligned} \tag{7.207b}$$

$$-p^T HF_c(c, u, \hat{\alpha}) = p^T G_c'(c, u, \hat{\alpha}) \tag{7.207c}$$

$$p^T HK_*'(c) = -p^T G_{*c}(c, u). \tag{7.207d}$$

Substituting (7.207a), (7.207b), (7.207c) and (7.207d) in (7.202) yields:

$$\begin{aligned}\lambda(c,u,\hat{\alpha},p)^T &= q_c^{'}(c,u,\hat{\alpha}) - Q_c^{'}(c,u,F(c,u,\hat{\alpha})) + \\ &\quad -q_{*c}^{'}(c,u) + Q_c^{'}(c,u,F(c,u,\hat{\alpha})) + \\ &\quad +p^T G_c^{'}(c,u,\hat{\alpha}) - p^T G_{*c}(c,u) \\ &= q_c^{'}(c,u,\hat{\alpha}) - q_{*c}^{'}(c,u) + \\ &\quad +p^T G_c^{'}(c,u,\hat{\alpha}) - p^T G_{*c}(c,u). \end{aligned} \tag{7.208}$$

Similarly

$$t(c,u,\hat{\alpha},p)^T = q_u^{'}(c,u,\hat{\alpha}) - q_{*u}^{'}(c,u)p^T + G_u^{'}(c,u,\hat{\alpha}) - p^T G_{*u}(c,u). \tag{7.209}$$

Finally, using (7.208) and (7.209) an overall modifier that is used in MMOP (see (7.198)) can be expressed as:

$$\begin{aligned}[\lambda(c,u,\hat{\alpha},p)^T, t(c,u,\hat{\alpha},p)^T] &= q_{[c,u]}^{'}(c,u,\hat{\alpha}) + p^T G_{[c,u]}^{'}(c,u) + \\ &\quad -q_{*[c,u]}^{'}(c,u) - p^T G_{*[c,u]}(c,u). \end{aligned} \tag{7.210}$$

Let $\bar{c}$ be the solution of SOCP and $\bar{u} = HK_*(\bar{c})$ the corresponding system interaction input. Notice, that the SOCP can now be written in an equivalent augmented form ASOCP as:

$$\begin{aligned} &\text{minimize}_{c,u}\{q_*(c,u) + \rho_u \parallel \bar{c} - c \parallel^2 + \rho_u \parallel \bar{u} - u \parallel^2\} \\ &\text{subject to: } G_*(c,u) = 0 \\ &\qquad (c,u) \in CU. \end{aligned} \tag{7.211}$$

Indeed, for any $(c,u) \in CU$ a contribution of the added terms is not negative and only at $(\bar{c},\bar{u})$ these terms are equal to zero. Hence, no new minimum points are introduced introduced and the solution of OCP remains the AOCP solution.

The ASOCP Lagrange function reads:

$$L_*^a(c,u,p) = q_*(c,u) + \rho_u \parallel \bar{c} - c \parallel^2 + \rho_u \parallel \bar{u} - u \parallel^2 + p^T G_*(c,u) \tag{7.212}$$

If the feasible set CU is compact in $\mathbb{R}^{n_c} \times \mathbb{R}^{n_u}$ then there exist $\overline{\rho}_u > 0$ and $\overline{\rho}_c > 0$ such that for any $p \in \mathbb{R}^{n_u}$ the function $L_*^a(\cdot,\cdot,p)$ is convex on CU if $\rho_c \geq \overline{\rho}_c$ and $\rho_u \geq \overline{\rho}_u$ (see Brdyś *et al.*, 1986). Assume in addition that the set CU is convex and regularity conditions are satisfied at $(\bar{c}, \bar{u})$. Hence, there exists such a price $\overline{p} \in \mathbb{R}^{n_u}$ that (Lasdon, 1970)

$$(\bar{c}, \bar{u}) = \text{Arg}\ \min_{(c,u)\in CU} L^a_*(c, u, \overline{p}). \tag{7.213}$$

Hence, as the set CU is convex the following inequality holds (Lasdon, 1970)

$$L^a_{*[c,u]}(\bar{c}, \bar{u}, \bar{p})' \begin{bmatrix} \nu - \bar{c} \\ w - \bar{u} \end{bmatrix} \geq 0$$
$$\text{for any } (\nu, w) \in CU \tag{7.214}$$

or (see (7.212))

$$\{q'_{*[c,u]}(\bar{c}, \bar{u}) + 2[\rho_c(\bar{c} - \bar{c})^T, \rho_u(\bar{u} - \bar{u})^T] + p^T G'_{*[c,u]}(\bar{c}, \bar{u})\} \begin{bmatrix} \nu - \bar{c} \\ w - \bar{u} \end{bmatrix} \geq 0 \tag{7.215a}$$

or finally

$$\{q'_{*[c,u]}(\bar{c}, \bar{u}) + p^T G'_{*[c,u]}(\bar{c}, \bar{u})\} \begin{bmatrix} \nu - \bar{c} \\ w - \bar{u} \end{bmatrix} \geq 0. \tag{7.215b}$$

Consider now the MMOP (see (7.198)) with the input data $\bar{c}, \bar{u}$ and $\bar{p}$. The MMOP reads:

$$\begin{aligned} \text{minimize}_{\nu,w} \{ & q(\nu, w, \hat{\alpha}(\bar{c}, \bar{u})) + \rho_c\|\nu - \bar{c}\|^2 + \rho_u\|w - \bar{u}\|^2 + \\ & -[\lambda(\bar{c}, \bar{u}, \hat{\alpha}(\bar{c}, \bar{u}), \bar{p})^T, t(\bar{c}, \bar{u}, \hat{\alpha}(\bar{c}, \bar{u}), \bar{p})^T] \begin{bmatrix} \nu \\ w \end{bmatrix} + \\ & +\bar{p}^T(w - HF(\nu, w, \hat{\alpha}(\bar{c}, \bar{u})))\} \\ \text{subject to: } & (c, u) \in CU. \end{aligned} \tag{7.216}$$

The term

$$-[\lambda(\bar{c}, \bar{u}, \hat{\alpha}(\bar{c}, \bar{u}), \bar{p})^T, t(\bar{c}, \bar{u}, \hat{\alpha}(\bar{c}, \bar{u}), \bar{p})^T \begin{bmatrix} \nu \\ w \end{bmatrix}$$

in (7.216) is a linear function of (ν, w).Moreover, as it has been shown above the remaining terms are nothing else but $L^a_*(\nu, w, \bar{p})$ that was already proved convex on CU. Hence, the function in the (7.216) that is minimized with respect to (v, w) is convex on CU. Let us denote this function as $\phi(\nu, w)$.The function $\phi(\cdot)$ is parameterized by variables c, u and p but this fact is omitted in the notation for the sake of a notational simplicity. Hence, a sufficient condition for optimality a point of the MMOP at a point $(\bar{\nu}, \bar{w})$ is:

$$\phi'_{[v,w]}(\bar{\nu}, \bar{w}) \begin{bmatrix} \nu - \bar{\nu} \\ w - \bar{w} \end{bmatrix} \geq 0 \tag{7.217}$$

where

$$\phi'_{[v,w]}(\bar{\nu}, \bar{w}) = q'_{[\nu,w]}(\bar{\nu}, \bar{w}, \hat{\alpha}(\bar{c}, \bar{u}) + 2[\rho_c(\bar{\nu} - \bar{c})^T, \rho_u(\bar{w} - \bar{u})^T] + \\ -[\lambda(\bar{c}, \bar{u}, \hat{\alpha}(\bar{c}, \bar{u}), \bar{p})^T, t(\bar{c}, \bar{u}, \hat{\alpha}(\bar{c}, \bar{u}), \bar{p})^T] + \bar{p}^T G'_{[\nu,w]}(\bar{\nu}, \bar{w}, \hat{\alpha}(\bar{c}, \bar{u})). \tag{7.218}$$

Applying (7.210) to (7.218) yields:

$$\begin{aligned} \phi'_{[v,w]}(\bar{\nu}, \bar{w}) &= q'_{[\nu,w]}(\bar{\nu}, \bar{w}, \hat{\alpha}(\bar{c}, \bar{u}) + 2[\rho_c(\bar{\nu} - \bar{c})^T, \rho_u(\bar{w} - \bar{u})^T] + \\ &\quad -q'_{[c,u]}(\bar{c}, \bar{u}, \hat{\alpha}(\bar{c}, \bar{u})) - \bar{p}^T G'_{[c,u]}(\bar{c}, \bar{u}) + \\ &\quad +q'_{*[c,u]}(\bar{c}, \bar{u}) + \bar{p}^T G_{*[c,u]}(\bar{c}, \bar{u}) + \\ &\quad +\bar{p}^T G'_{[\nu,w]}(\bar{\nu}, \bar{w}, \hat{\alpha}(\bar{c}, \bar{u})). \end{aligned} \tag{7.219}$$

Consider the sufficient condition for MMOP (see (7.216)) optimality at $\bar{\nu} = \bar{c}$ and $\bar{w} = \bar{u}$ and apply this data to (7.219). Then

$$\begin{aligned} \phi'_{[v,w]}(\bar{c}, \bar{u}) &= q'_{[\nu,w]}(\bar{c}, \bar{u}, \hat{\alpha}(\bar{c}, \bar{u}) + 2[\rho_c(\bar{c} - \bar{c})^T, \rho_u(\bar{u} - \bar{u})^T] + \\ &\quad -q'_{[c,u]}(\bar{c}, \bar{u}, \hat{\alpha}(\bar{c}, \bar{u})) - \bar{p}^T G'_{[c,u]}(\bar{c}, \bar{u}) + \\ &\quad +q'_{*[c,u]}(\bar{c}, \bar{u}) + \bar{p}^T G_{*[c,u]}(\bar{c}, \bar{u}) + \\ &\quad +\bar{p}^T G'_{[\nu,w]}(\bar{c}, \bar{u}, \hat{\alpha}(\bar{c}, \bar{u})) \end{aligned} \tag{7.220}$$

and finally

$$\phi'_{[v,w]}(\bar{c}, \bar{u}) = q'_{*[c,u]}(\bar{c}, \bar{u}) + \bar{p}^T G_{*[c,u]}(\bar{c}, \bar{u}) \tag{7.221}$$

Hence, the condition (7.217) becomes

$$\phi'_{[v,w]}(\bar{c}, \bar{u}) \begin{bmatrix} \nu - \bar{c} \\ w - \bar{w} \end{bmatrix} \geq 0. \tag{7.222}$$

The inequality is exactly the same as the inequality (7.215b). The latter has been already shown to be satisfied. Thus, the following has been shown:

- for the optimal control input (solution of SOCP) $\bar{c}$ and the system interaction input $\bar{u} = HK_*(\bar{c})$ there exists price vector $\bar{p}$ such that with the input data $\bar{c}, \bar{u}, \bar{p}$ the MMOP solution satisfies:

$$\hat{\nu}(\bar{c}, \bar{u}, \bar{p}) = \bar{c} \tag{7.223}$$

$$\hat{w}(\bar{c}, \bar{u}, \bar{p}) = \bar{u}. \tag{7.224}$$

Hence, the first two core conditions of the method, namely (7.200a) and (7.200b) are satisfied at $(\bar{c}, \bar{u}, \bar{p})$. To complete the proof it is sufficient to show that the third core condition (7.200c) is also met. This is straightforward. Indeed, as the set CU is convex and compact and the Lagrange function was shown convex on CU then the triple $(\bar{c}, \bar{u}, \bar{p})$ is a saddle point of the function $L_*^a(\cdot, \cdot, \cdot)$ on $CU \times \mathbb{R}^{n_u}$ and consequently (Lasdon, 1970)

$$G_*(c, u) = 0 \tag{7.225}$$

or (see (7.205))

$$\hat{w}(\bar{c}, \bar{u}, \bar{p}) = HF(\hat{\nu}(\bar{c}, \bar{u}, \bar{p}), \hat{w}(\bar{c}, \bar{u}, \bar{p}), \hat{\alpha}(\bar{c}, \bar{u})). \tag{7.226}$$

The terms convexifying the model-based performance function in MMOP can make the resulting function $\psi(\nu, w)$ not only convex but also strictly convex. Hence, for convex CU the MMOP solution is unique and therefore, it is guaranteed that the optimal control inputs $\bar{c}$ but other point will be found after the MMOP has been solved. This is very important property of the augmented hierarchical ISOPE with output measurements. The same can be shown for the augmented ISOPE hierarchical methods with input-output measurements. As the reasoning is very analogous to the presented here it is omitted.

Let us summarize our findings in a form of a Lemma.

Lemma 7.1

Assume that

(i) a solution of the SOCP exists and it is denoted by $\bar{c}$,

(ii) the regularity conditions at the point $(\bar{c}, HK_*(\bar{c})$ *are satisfied,*

(iii) set CU is convex and compact in $\mathbb{R}^{n_c}\mathbb{R}^{n_u}$,

(iv) the coefficients ρ_c *and* ρ_u *in MMOP are selected sufficiently large to strictly convexify the Lagrange function of the Augmented System Optimizing Control Problem-ASOCP on CU.*

Then

(a) there exists such price vector $\bar{p}$ *that the triple* $(\bar{c}, HK_*(\bar{c}), \bar{p})$ *is a saddle point of the ASOCP Lagrange function* $L_*(\cdot, \cdot, \cdot)$ *on* $CU \times \mathbb{R}^{n_u}$.

(b) $(\bar{c}, HK_*(\bar{c}), \bar{p}) \in \Omega$ *thus the optimal control inputs can be found by hierarchical AISOPE strategy with output measurements. Moreover, the optimal control input vector is a unique solution of the MMOP.*

7.4.3 *System-based double-loop algorithm*

7.4.3.1 *Algorithm*

We shall now generalize the ISOPE algorithm that was developed by (Brdyś *et al.*, 1990b) to produce the augmented ISOPE (AISOPE) algorithm. The system based double-loop algorithm with output measurements is made up of two nested iterative loops, each having different frequency of intervention: an *inner loop* and an *outer loop*. The outer loop task is to solve equation (7.200c) by iterating price variable p while the inner loop task is to solve equations (7.200a) and (7.200b) by iterating the control input and interaction inputs c and u, respectively under p prescribed by the outer loop. Let us denote the inner loop solution as $\hat{c}(p) \triangleq \hat{\nu}(\hat{c}(p), \hat{u}(p), p)$ and $\hat{u}(p) \triangleq \hat{w}(\hat{c}(p), \hat{u}(p), p)$. The corresponding parameters are: $\hat{\alpha}(p) \triangleq \hat{\alpha}(\hat{c}(p), \hat{u}(p))$.

The outer-loop task (see (7.200c) can now be stated as of finding such value of p that

$$\hat{u}(p) = HF(\hat{c}(p), \hat{u}(p), \hat{\alpha}(p)). \tag{7.227}$$

The strategy is called system-based because the inner loop utilizes both the system model and the measurements (see (7.197a)) while the outer loop does not directly utilize the measurements.

The *system-based double-loop algorithm with output measurements* can now be formulated:

Start. Given initial points $c^{0,0}, p^0$, inner and outer loop algorithmic mappings $\{\Psi_{c_i}\}_{i \in \overline{1:N}}$, $\{\Psi_{u_i}\}_{i \in \overline{1:N}}$ and Ψ_p, respectively, convexifying coefficients ρ_c and ρ_u and the loop solution accuracies $\varepsilon_c = [\varepsilon_{c_1}, ... \varepsilon_{c_N}], \varepsilon_{c_i} > 0$, $\varepsilon_u = [\varepsilon_{u_1}, ... \varepsilon_{u_N}]$, $\varepsilon_{c_i} > 0$, $\varepsilon_{u_i} > 0$ and $\varepsilon_p > 0$ and Set $n := 0$ and $l = 0$; n and l denote the outer and inner loop iteration numbers, respectively.

Step 1. Apply $c^{l,n}$ to the controlled plant and measure the outputs $y^{l,n} =$

$K_*(c^{i,n})$, $y^{ol,n} = K_*^o(c^{l,n})$. Calculate the plant output mapping derivatives $K_*^{'}(c^{l,n}), K_*^{o'}(c^{l,n})$ by using additional perturbation of $c^{l,n}$ or by employing previous output measurements as it is done by ISOPED.

Step 2. Solve the *parameter estimation problem* PEP defined by (7.197a). The PEP decomposes into N independent *local parameter estimation problems* $\text{PEP}_i, i \in \overline{1,N}$ yielding the parameter estimates $\alpha_i^{l,n} = \hat{\alpha}_i(c^{l,n})$ satisfying

$$F_i^o(c_i^{l,n}, u_i^{l,n}, \alpha_i^{l,n}) = K_*^o(c^{l,n}) \quad (7.228a)$$

$$F_i(c_i^{l,n}, u_i^{l,n}, \alpha_i^{l,n}) = K_{*i}(c^{l,n}). \quad (7.228b)$$

Step 3. For each subsystem calculate modifiers $\lambda_i^{l,n} = \lambda_i(c^{l,n}, u^{l,n}\alpha^{l,n}, p^n)$ and $t_i^{l,n} = t_i(c^{l,n}, u^{l,n}\alpha^{l,n}, p^n)$ $i \in \overline{1:N}$ by applying formulae (7.191) and (7.192), respectively. This can not be done independently for each subsystem due to the interconnections. Hence, an *information exchange* between *local subsystem controllers* is needed in order to perform this step.

Step 4. For $c = c^{l,n}$, $u = u^{l,n}$, $p = p^n$ solve the *modified model-based optimization problem* MMOP defined by (7.198) with $\lambda = \lambda^{l,n}$ and $t = t^{l,n}$ and $\alpha = \alpha^{l,n}$. The MMOP decomposes into N independent *local modified model-based optimization problems* MMOP_i, $i \in \overline{1,N}$ (see (7.198) and (7.4):

$$\begin{aligned} \text{minimize}_{\nu_i, w_i} \{ & q_i(\nu_i, w_i, \alpha_i^{l,n}) + \rho_c \|\nu_i - c_i^{l,n}\|^2 + \\ & + \rho_u \|w_i - u_i\|^2 - (\lambda_i^{l,n})^T \nu_i - (t_i^{l,n})^T w_i + \\ & + (p_i^{l,n})^T u_i + - \sum_{j=1}^{N} (p_j^{l,n})^T H_{ji} F_i(\nu_i, w_i, \alpha_i^n) \} \\ \text{subject to: } & g_i(\nu_i, w_i) \leq 0. \end{aligned} \quad (7.229)$$

Solving the MMOP_i $i \in \overline{1:N}$ produces $\nu_i^{l,n} = \hat{\nu}_i(c_i^{l,n}, u^{l,n}, p^n)$ and $w_i^{l,n} = \hat{w}_i(c_i^{l,n}, u^{l,n}, p^n)$ (see (7.199a)).
If for $i \in \overline{1:N}$

$$\| c_i^{l,n} - \nu_i^{l,n} \| \leq \varepsilon_{c_i} \quad (7.230)$$

and

$$\| u_i^{l,n} - w_i^{l,n} \| \leq \varepsilon_{u_i} \quad (7.231)$$

then terminate the inner loop iterations.

The inner loop problem solution $c^n = [(c_1^n)^T, .., (c_N^n)^T] = \hat{c}(p^n)$ and $u^n = [(u_1^n)^T, .., (u_N^n)^T] = \hat{u}(p^n)$ has now been found ((7.200a) and (7.200b are satisfied). Continue from Step 6.

Step 5. Set

$$c_i^{l+1,n} := \Psi_{c_i}(c_i^{l,n}, \nu_i^{l,n}) \tag{7.232}$$

and

$$u_i^{l+1,n} := \Psi_{u_i}(u_i^{l,n}, w_i^{l,n}) \tag{7.233}$$

set $l := l + 1$ and continue from Step 1.
The control input iteration according to (7.232) and (7.233) are completely decentralized. Hence they are performed at the subsystem level. However, as the modifiers λ and t depend on all the components of vectors c and u (see (7.191) and (7.192)) then so do the mappings $\hat{\nu}_i(\cdot)$ and $\hat{w}(\cdot)$ (see (7.199a) and (7.198)). Hence, applying a completely decentralized algorithmic mapping at the inner loop in order to solve (7.200a) and (7.200b) or in other words to find a fixed point of the mappings:

$$\begin{aligned} \{\hat{\nu}, \hat{w}\} \quad : \quad & \mathbb{R}^{n_c} \times \mathbb{R}^{n_u} \longmapsto \mathbb{R}^{n_c} \times \mathbb{R}^{n_u} \\ & \hat{w}(c, u, p) = u \\ & \hat{\nu}(c, u, p) = c. \end{aligned} \tag{7.234}$$

Step 6. The control and interaction inputs, and model parameters determined at the inner loop problem solutions are : $c^n = \hat{\nu}(c^n, u^n, p^n)$, $u^n = \hat{w}(c^n, u^n, p^n)$ and $\alpha^n = \hat{\alpha}(c^n, u^n)$.
If

$$\| u^n - HF(c^n, u^n, \alpha^n) \| \leq \varepsilon_p \tag{7.235}$$

then terminate the outer loop iterations.
The outer loop problem solution has now been found (the eq.(7.227) is satisfied).

Step 7. Set

$$p^{n+1} := \Psi_p(p^n, u^n - HF(c^n, u^n, \alpha^n)) \tag{7.236}$$

set $n := n + 1$ and continue from Step 1.

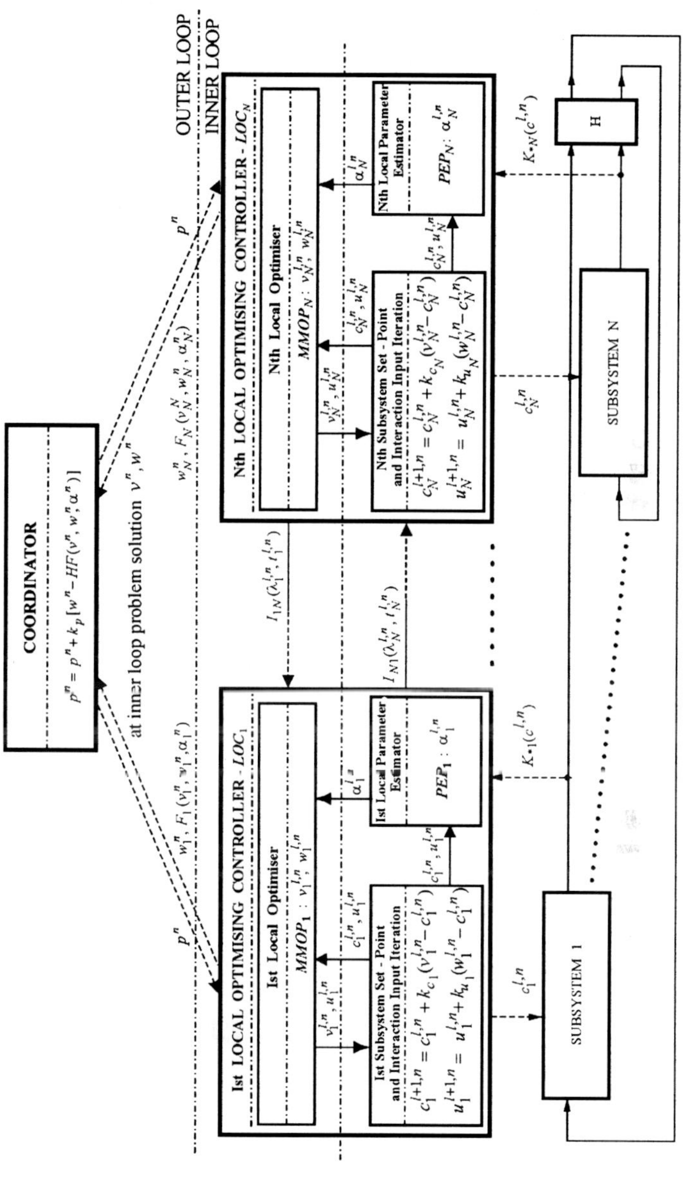

Fig. 7.23 Information structure of AISOPE system-based double-loop iterative strategy with output measurements.

An information structure of the double-loop system-based algorithm is two-level hierarchical with information exchange between local subsystem controllers and it is illustrated in Fig. 7.23.

The outer loop task is solved by the *coordinator*. The inner loop task is solved in parallel by the subsystem *local optimizing controllers* (LOC_i). The structure is very similar to the system-based double-loop algorithm with input-output measurements structure (see Fig. 7.11). The only difference is that now the interaction inputs are not gathered from the measurement data but they are predicted from the models utilized by distributed $\mathrm{LOC}_i, i \in \overline{1:N}$ to carry out their iterative processes.

The inner and outer loop algorithmic mappings Ψ_{c_i}, Ψ_{u_i}, $i \in \overline{1:N}$ and Ψ_p, respectively, need to be design so that the loop convergence is achieved. The inner loop problem is a classic problem of finding a fix point of a mapping while the outer loop problems is to find such price vector that the interaction balance is achieved. The mappings design and the algorithm convergence is presented in the following subsection.

7.4.3.2 *Algorithmic mappings and convergence*

In order to design the loop algorithmic mappings we shall gain an insight into important properties of the inner loop.

Let $\bar{c}, \bar{u}$ and $\bar{p}$ constitute input data to the inner loop. We shall assume that the set CU is convex and compact and that the coefficients ρ_c, ρ_u are large enough so that the MMOP performance function $\phi(\nu, w)$ is strictly convex on CU. Hence, the solution of MMOP exists and is unique. Let us denote the solution by $\hat{\nu}(\bar{c}, \bar{u}, p)$, $\hat{w}(\bar{c}, \bar{u}, p)$. It follows from the optimality properties of a convex function on a convex set that:

$$\phi'_{[v,w]}(\hat{\nu}(\bar{c}, \bar{u}, p), \hat{w}(\bar{c}, \bar{u}, p)) \begin{bmatrix} \nu - \hat{\nu}(\bar{c}, \bar{u}, p) \\ w - \hat{w}(\bar{c}, \bar{u}, p) \end{bmatrix} \geq 0 \tag{7.237}$$

for any $(\nu, w) \in CU$, where (see (7.220))

$$\begin{aligned} \phi'_{[v,w]}(\hat{\nu}(\bar{c}, \bar{u}, p), \hat{w}(\bar{c}, \bar{u}, p)) \quad = \quad & q'_{[\nu,w]}(\hat{\nu}(\bar{c}, \bar{u}, p), \hat{w}(\bar{c}, \bar{u}, p), \hat{\alpha}(\bar{c}, \bar{u})) + \\ & +2[\rho_c(\hat{\nu}(\bar{c}, \bar{u}, p) - \bar{c})^T, \rho_u(\hat{w}(\bar{c}, \bar{u}, p) - \bar{u})^T] + \\ & -q'_{[c,u]}(\bar{c}, \bar{u}, \hat{\alpha}(\bar{c}, \bar{u})) - p^T G'_{[c,u]}(\bar{c}, \bar{u}) + \\ & +q'_{*[c,u]}(\bar{c}, \bar{u}) + p^T G_{*[c,u]}(\bar{c}, \bar{u}) + \\ & +p^T G'_{[\nu,w]}(\bar{\nu}, \bar{w}, \hat{\alpha}(\bar{c}, \bar{u})). \end{aligned} \tag{7.238a}$$

If in addition, $\hat{\nu}(\bar{c}, \bar{u}, p)$, $\hat{w}(\bar{c}, \bar{u}, p)$ is a solution of the inner loop problem (see (7.200a) and (7.200b)) then $\hat{\nu}(\bar{c}, \bar{u}, p) = c$ and $\hat{w}(\bar{c}, \bar{u}, p) = \bar{u}$. Then the (7.238a) becomes:

$$\begin{aligned}
\phi'_{[v,w]}(\bar{c}, \bar{u}) \quad = \quad & q'_{[\nu,w]}(\bar{c}, \bar{u}, \hat{\alpha}(\bar{c}, \bar{u})) + \\
& +2[\rho_c(\bar{c} - \bar{c})^T, \rho_u(\bar{u} - \bar{u})^T] + \\
& -q'_{[c,u]}(\bar{c}, \bar{u}, \hat{\alpha}(\bar{c}, \bar{u})) - p^T G'_{[c,u]}(\bar{c}, \bar{u}, \hat{\alpha}(\bar{c}, \bar{u})) + \\
& +q'_{*[c,u]}(\bar{c}, \bar{u}) + p^T G_{*[c,u]}(\bar{c}, \bar{u}) + \\
& +p^T G'_{[\nu,w]}(\bar{c}, \bar{u}, \hat{\alpha}(\bar{c}, \bar{u}))
\end{aligned} \tag{7.238b}$$

and, finally

$$\phi'_{[v,w]}(\bar{c}, \bar{u}) = q'_{*[\nu,w]}(\bar{c}, \bar{u}) + p^T G_{*[c,u]}(\bar{c}, \bar{u}). \tag{7.238c}$$

Let us write down an *augmented system optimizing control problem* (ASOCP) as (see (7.203), (7.205):

$$\begin{aligned}
& \text{minimize}_{c,v,u,w}\{q_*(c,u) + \rho_c \parallel \nu - c \parallel^2 + \rho_u \parallel w - u \parallel^2\} \\
& \text{subject to:} \quad G_*(c,u) = 0 \\
& \qquad\qquad\quad (c,u) \in CU \\
& \qquad\qquad\quad \nu = c \\
& \qquad\qquad\quad w = u.
\end{aligned} \tag{7.239}$$

The ASOCP Lagrange function reads :

$$L^A_*(c,u,\nu,w,p) \quad = \quad q_*(c,u) + \rho_c \parallel \nu - c \parallel^2 + \rho_u \parallel w - u \parallel^2 + p^T G_*(c,u) \tag{7.240}$$

and its derivative with respect to (c,ν, u, w) equals to:

$$\begin{aligned}
L^A_{*[c,u,\nu,w]}(c,u,\nu,w,p)' \quad = \quad & [q'_{*[c,u]}(c,u) + (2\rho_c(c-\nu)^T, 2\rho_u(w-u)^T) + \\
& p^T G'_{*[c,u]}(c,u),\ 2\rho_c(\nu - c)^T,\ 2\rho_u(w-u)^T].
\end{aligned} \tag{7.241}$$

Comparing (7.238c) with (7.241) and (7.237) yield that for any p the following inequality holds at a solution $\hat{c}(p), \hat{u}(p)$ of the inner loop problem:

$$L^A_{*[c,u,\nu,w]}(\hat{c}(p), \hat{u}(p), \hat{c}(p), \hat{u}(p), p) \begin{bmatrix} c - \hat{c}(p) \\ u - \hat{u}(p) \\ \nu - \hat{c}(p) \\ w - \hat{u}(p) \end{bmatrix} \geq 0. \tag{7.242}$$

As the Lagrange function $L^A_*(\cdot,\cdot,\cdot,\cdot,p)$ is strictly convex on the set

$$\begin{aligned} CUVW \triangleq \{(c,u,\nu,w) \in \mathbb{R}^{n_c} \times \mathbb{R}^{n_u} \times \mathbb{R}^{n_c} \times \mathbb{R}^{n_u} : G_*(c,u) = 0, \\ (c,u) \in CU, \quad \nu = c, \quad w = u\} \end{aligned} \tag{7.243}$$

then the augmented inner loop solution $(\hat{c}(p), \hat{u}(p), \hat{c}(p), \hat{u}(p))$ minimizes $L^A_*(\cdot,\cdot,\cdot,\cdot,p)$ on the set $CUVW$.

Notice, that the Lagrange function for SOCP reads:

$$L_*(c,u,p) = q_*(c,u) + p^T G_*(c,u). \tag{7.244}$$

Comparison of (7.240) with (7.244) implies that $(\hat{c}(p), \hat{u}(p))$ minimizes on CU.

- In summary, it has now been shown that for any $p \in \mathbb{R}^{n_u}$ the inner loop solution minimizes Lagrange function of SOCP.

Clearly, the optimization problem

$$\begin{aligned} &\text{minimize}_{c,u} L_*(c,u,p) \\ &\text{subject to:} \quad G_*(c,u) = 0 \\ &\qquad\qquad\qquad (c,u) \in CU \end{aligned} \tag{7.245}$$

is under uncertainty as the system input-output mappings $F_*(c,u)$ and $F^o_*(c,u)$ needed to define the mappings $q_*(c,u)$ (see (7.203)) and $G_*(c,u)$ (7.205) are not known. Hence, one can think of applying the ISOPE method to this problem. It is not difficult to verify just by comparing problem (7.245) with the MMOP (see (7.198)) and the multiplier formula (7.210) that what the inner loop is doing is nothing else but applying ISOPE method to solving the optimization problem (7.245). Indeed, the expression for the overall multiplier can be written as:

$$\begin{aligned} [\lambda(c,u,\hat{\alpha},p)^T, t(c,u,\hat{\alpha},p)^T] &= q'_{[c,u]}(c,u,\hat{\alpha}) + p^T G'_{[c,u]}(c,u,\hat{\alpha}) + \\ &\quad - q'_{*[c,u]}(c,u) - p^T G_{*[c,u]}(c,u) \end{aligned} \tag{7.246}$$

and

$$[\lambda(c,u,\hat{\alpha},p)^T, t(c,u,\hat{\alpha},p)^T] = L'_{[c,u]}(c,u,\hat{\alpha}(c,u),p) - L'_{*[c,u]}(c,u,p) \quad (7.247)$$

where

$$L(c,u,p,\alpha) \triangleq q(c,u,\alpha) + p^T G(c,u,\alpha). \quad (7.248)$$

The model-based Lagrange function $L(c,u,\alpha)$ represents nothing else but point-parametric model of the function $L_*(c,u,p)$.

The MMOP can now be written as:

$$\begin{aligned} &\min\text{imize}_{v,w}\{L(\nu,w,\hat{\alpha}(c,u)) + \rho_c \parallel \nu - c \parallel^2 + \rho_u \parallel w - u \parallel^2 + \\ &\qquad - [\lambda(c,u,\hat{\alpha}(c,u),p)^T, t(c,u,\hat{\alpha}(c,u),p)^T] \begin{bmatrix} \nu \\ w \end{bmatrix} \} \\ &\text{subject to: } g(\nu,w) \le 0. \end{aligned} \quad (7.249)$$

Hence, the "identification" of the inner loop activities has now been completed.

The inner loop algorithmic mapping follows the immediately from the above observation. It can be a standard relaxation mapping in a decentralized form:

$$\begin{aligned} \Psi_{c_i} &: \mathbb{R}^{n_{c_i}} \mapsto \mathbb{R}^{n_{c_i}} \\ \Psi_{c_i}(c_i,\nu_i) &= c_i + k_{c_i}(\nu_i - c_i), \quad i \in \overline{1:N} \end{aligned} \quad (7.250a)$$

and

$$\begin{aligned} \Psi_{u_i} &: \mathbb{R}^{n_{u_i}} \mapsto \mathbb{R}^{n_{u_i}} \\ \Psi_{u_i}(u_i,w_i) &= u_i + k_{u_i}(w_i - u_i), \quad i \in \overline{1:N} \end{aligned} \quad (7.250b)$$

leading to the inner loop iterative algorithm

$$\begin{aligned} c_i^{l+1,n} &= c_i^{l,n} + k_{c_i}(\nu_i^{l,n} - c_i^{l,n}), \quad i \in \overline{1:N} \\ u_i^{l+1,n} &= u_i^{l,n} + k_{u_i}(w_i^{l,n} - u_i^{l,n}), \quad i \in \overline{1:N}. \end{aligned} \quad (7.250c)$$

Let us now consider the outer loop algorithmic mapping. Given p, the inner loop solves the optimization problem (7.245) by employing an augmented ISOPE method. The value $L_*(\widehat{c}(p),\widehat{u}(p),p)$ is a value at p of a dual function of $L_*(\cdot,\cdot,p)$ on CU.A gradient of the dual function at p equals to $G_*(\widehat{c}(p), \widehat{u}(p))$ (Lasdon, 1970). Clearly the following holds (see (7.205), (7.206) and (7.197a)):

$$G_*(\widehat{c}(p), \widehat{u}(p)) = G(\widehat{c}(p), \widehat{u}(p), \widehat{\alpha}(p)). \tag{7.251}$$

A gradient algorithm that maximizes the dual function with respect to p will bring its gradient to zero. Due to (7.251) this algorithm will also find such value of the price vector that the outer loop condition (7.200c) will be met. Hence, the outer loop algorithmic mapping is chosen as:

$$\begin{aligned} \Psi_p \quad &: \quad \mathbb{R}^{n_u} \mapsto \mathbb{R}^{n_u} \\ \Psi_p(p, u - HF(c, u, \alpha)) \quad &= \quad p + k_p[u - HF(c, u, \alpha)]. \end{aligned} \tag{7.252a}$$

With this outer loop algorithmic mapping the corresponding outer loop iterative algorithm reads:

$$p^{n+1} = p^n + k_p[u^n - HF(c^n, u^n, \alpha^n)]. \tag{7.252b}$$

where $k_p > 0$ is the outer loop gradient algorithm step coefficient.

7.4.4 *Model-based double-loop algorithms*

The model-based algorithm will be introduced for ISOPE double-loop method with output measurements similarly as for the method with input-output measurements described in section 7.3.4. Also two double-loop structures and algorithms will be defined: with tight inner loop (see Section 7.3.4.1) and with relaxed inner loop (see Section 7.3.4.2). The interaction input is now not taken from the measurements but remains a free vector variable and the previously presented double-loop algorithms need to be modified accordingly. The algorithms derivation principles and the motivation remains exactly the same as previously.

7.4.4.1 *Model-based double-loop algorithm with tight inner loop*

We shall now generalize the ISOPE algorithm that was developed by (Brdyś *et al.*, 1990b) to produce the augmented ISOPE (AISOPE) algorithm. The model-based double-loop algorithm consists of two nested iterative loops, each having different frequency of interventions: an *inner loop* and an *outer loop*. The inner loop task is to solve the equation (7.200c) by iterating prices p under the control input and interaction inputs c and u, respectively that are prescribed by the outer loop. The outer loop task is to solve the equations (7.200a) and (7.200b) by iterating the control and interaction inputs c and u, respectively. Let us denote the inner loop problem solution as $\hat{p}(c, u)$. The corresponding auxiliary variables w and v are:

$$\widehat{w}(c, u) \triangleq \widehat{w}(c, u, \hat{p}(c, u)) \tag{7.253}$$

and

$$\hat{\nu}(c, u) \triangleq \hat{\nu}(c, u, \hat{p}(c, u)) \tag{7.254}$$

respectively. The inner loop task is then to find such value of p that

$$\widehat{w}(c, u) = HF(\hat{\nu}(c, u), \widehat{w}(c, u), \hat{\alpha}(c, u)) \tag{7.255}$$

while the outer loop tasks is to find such values of c and u that

$$\hat{\nu}(c, u) = c \tag{7.256}$$

and

$$\widehat{w}(c, u) = u. \tag{7.257}$$

The strategy is called model-based because the inner loop entirely utilizes the system model. The model-based double-loop algorithm with tight inner loop can now be formulated:

Start. Given initial points $c^0, u^0, p^{0,0}$, inner and outer loop algorithmic mappings $\Psi_p, \{\Psi_{c_i}\}, \{\Psi_{u_i}\}_{i \in \overline{1 \cdot N}}$, respectively, convexifying coefficients ρ_c and ρ_u and the loop solution accuracies $\varepsilon_c = [\varepsilon_{c_1}, \ldots \varepsilon_{c_N}], \varepsilon_{c_i} > 0, \varepsilon_u = [\varepsilon_{u_1}, \ldots \varepsilon_{u_N}], \varepsilon_{u_i} > 0, \varepsilon_p > 0$. Set $n := 0$ and $l = 0$; n and l denote the outer and inner loop iteration numbers, respectively.

Step 1. Apply c^n, u^n to the controlled plant and measure the outputs $y^n = K_*(c^n)$, $y^{on} = K_*^o(c^n)$. Calculate the plant output mapping derivatives $K_*^{'}(c^n), K_*^{o'}(c^n)$ by using additional perturbation of c^n or by employing previous output measurements as it is done by ISOPED.

Step 2. Solve the *parameter estimation problem* PEP defined by (7.197a). The PEP decomposes into N independent *local parameter estimation problems* $\text{PEP}_i, i \in \overline{1, N}$ yielding the parameter estimates $\alpha_i^n = \hat{\alpha}_i(c^n, u^n)$ satisfying

$$F_i^o(c_i^n, u_i^n, \alpha_i^n) = K_{*i}^o(c^n) \tag{7.258a}$$

$$F_i(c_i^n, u_i^n, \alpha_i^n) = K_{*i}(c^n) \tag{7.258b}$$

Step 3. For each subsystem calculate modifiers $\lambda_i^{l,n} = \lambda_i(c^n, u^n, \alpha^n, p^{l,n})$, $t_i^{l,n} = t_i(c^n, u^n, \alpha^n, p^{l,n}), i \in \overline{1:N}$ by applying formulae (7.191) and (7.192). This can not be done independently for each subsystem due to interconnections between subsystems. Hence, an *information exchange* between *local subsystem controllers* is needed in order to perform this step.

Step 4. For $c := c^n$, $u := u^n$, $p := p^{l,n}$ solve the *modified model-based optimization problem* MMOP defined by (7.198) with $\lambda := \lambda^{l,n}, t := t^{l,n}$ and $\alpha = \alpha^n$. The MMOP decomposes into N independent *local modified model-based optimization problems* MMOP_i, $i \in \overline{1,N}$ (see (7.198) and (7.4)

$$\begin{aligned} \text{minimize}_{\nu,,w_i} \{ & q_i(\nu_i, w_i, \alpha_i^n) + \rho_c \|\nu_i - c_i^n\|^2 + \rho_u \|w_i - u_i^n\|^2 + \\ & -(\lambda_i^{l,n})^T \nu_i - (t_i^{l,n})^T w_i + (p_i^{l,n})^T w_i + \\ & - \sum_{j=1}^{N} (p_j^{l,n})^T H_{ji} F_i(\nu_i, w_i, \alpha_i^n) \} \\ \text{subject to: } & g_i(\nu_i, w_i) \le 0. \end{aligned} \tag{7.259}$$

Solving the MMOP_i $i \in \overline{1:N}$ produces $\nu_i^{l,n} = \hat{\nu}_i(c_i^n, u_i^n, p^{l,n})$ and $w_i^{l,n} = \widehat{w}_i(c_i^n, u_i^n, p^{l,n})$ (see (7.199a)). If for $i \in \overline{1:N}$

$$\| w_i^{l,n} - H_i F(\nu^{l,n}, w^{l,n}, \alpha^n) \| \le \varepsilon_p \tag{7.260}$$

then terminate the inner loop iterations.

The inner loop problem solution $p^n = [(p_1^n)^T, .., (p_N^n)^T] = p(c^n, u^n)$ has now been found (the eq.(7.255) is satisfied). Continue from Step 6.

The condition (7.260) can not be verified at the subsystem level without information exchange between local controllers. Typically, an *inner loop coordinator* is introduced to carry out this operation.

Step 5. Set

$$p^{l+1,n} := \Psi_p(p^{l,n}, w^{l,n} - HF(\nu^{l,n}, w^{l,n}, \alpha^n)) \tag{7.261}$$

set $l := l + 1$ and continue from Step 4.

This inner loop price iteration can not be carried out at the subsystem level that is in a decentralized manner and it is performed by the inner loop coordinator.

Step 6. The auxiliary variables determined at the inner loop problem solutions are (see (7.253) and (7.254)): $w^n = \widehat{w}(c^n, u^n, p^n)$ and $\nu^n = \hat{\nu}(c^n, u^n, p^n)$. If

$$\| c_i^n - \nu_i^n \| \leq \varepsilon_{c_i} \tag{7.262}$$

and

$$\| u_i^{l,n} - w_i^{l,n} \| \leq \varepsilon_{u_i} \tag{7.263}$$

then terminate the outer loop iterations.
The outer loop problem solution has now been found ((7.256) and (7.257) are satisfied).

Step 7. Set

$$c_i^{n+1} := \Psi_{c_i}(c_i^n, \nu_i^n), \quad i \in \overline{1:N} \tag{7.264}$$

and

$$u_i^{n+1} := \Psi_{u_i}(u_i^n, w_i^n), \quad i \in \overline{1:N} \tag{7.265}$$

set $n := n + 1$ and continue from Step 1.

An information structure of the *double-loop model-based with tight inner loop algorithm* is two-level hierarchical with information exchange between local subsystem controllers and it is illustrated in Fig. 7.24. Comparison of Figure 7.24 with Figure 7.23 illustrating the *double-loop system-based algorithm* shows that exactly the same units are present and exactly the same data are processed in exactly the same manner. The only difference is in the unit intervention frequency that implies different location of the inner and outer loops in the Figures.

Clearly, the structure in Fig. 7.24 is very similar to the structure of the model-based double-loop algorithm with input-output measurements and with tight inner loop (see Fig. 7.12). The only difference is that now the interaction inputs are not gathered from the measurement data but they are predicted from the models utilised by distributed $\text{LOC}_i, i \in \overline{1:N}$ to carry out their iterative processes.

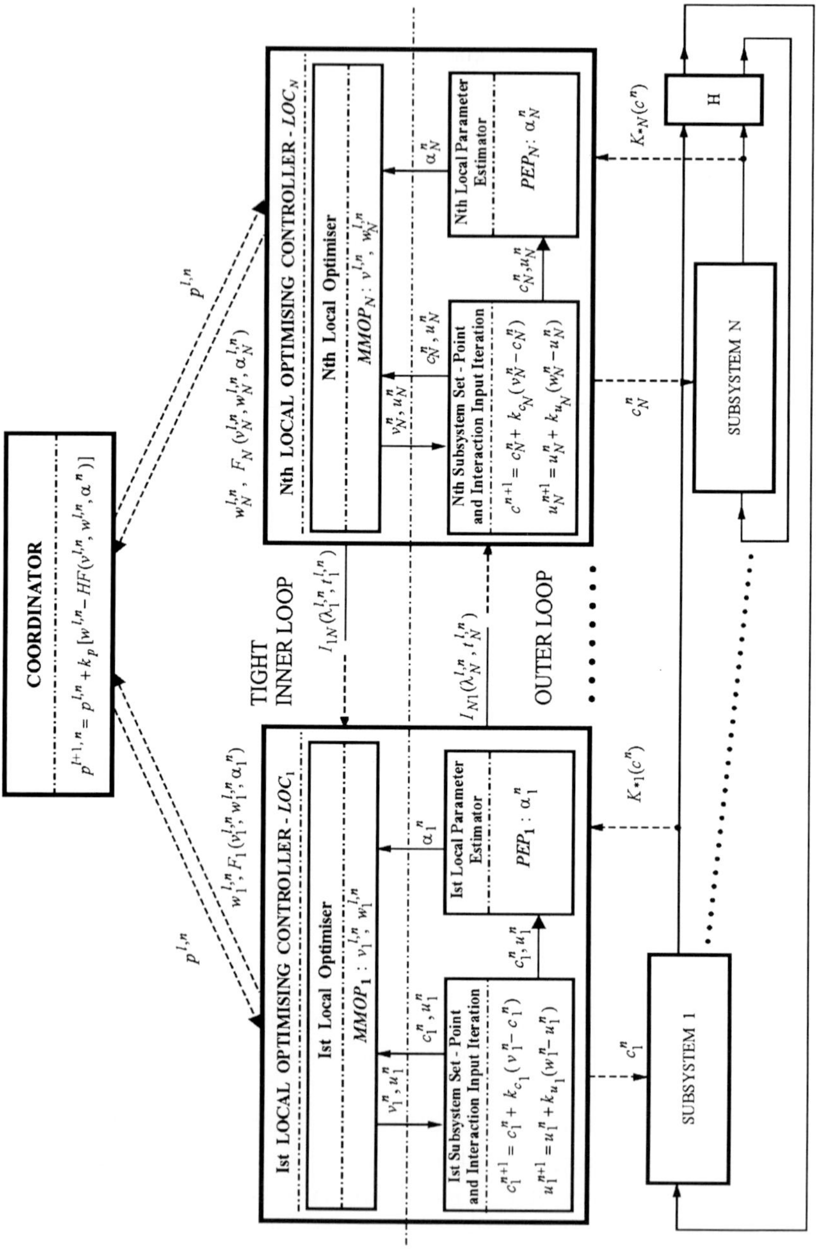

Fig. 7.24 Information structure of AISOPE model-based double-loop iterative strategy with output measurements and with tight inner loop.

The inner and outer loop algorithmic mappings Ψ_{c_i}, Ψ_{u_i} and Ψ_p, respectively need to be designed so that the loop convergence is achieved. As previously, the relaxation algorithms with the scalar relaxation gains are chosen for the outer loop. Hence, the corresponding outer loop local algorithmic mappings are:

$$\begin{aligned} \Psi_{c_i} &: \mathbb{R}^{n_{c_i}} \times \mathbb{R}^{n_{c_i}} \mapsto \mathbb{R}^{n_{c_i}} \\ \Psi_{c_i}(c_i, v_i) &= c_i + k_{c_i}(\nu_i - c_i), \quad i \in \overline{1:N} \end{aligned} \tag{7.266a}$$

and

$$\begin{aligned} \Psi_{u_i} &: \mathbb{R}^{n_{u_i}} \times \mathbb{R}^{n_{u_i}} \mapsto \mathbb{R}^{n_{u_i}} \\ \Psi_{u_i}(u_i, w_i) &= u_i + k_{u_i}(w_i - c_i), \; i \in \overline{1:N} \end{aligned} \tag{7.266b}$$

where $k_{c_i} > 0$ and $k_{u_i} > 0$ are the ith outer loop relaxation mapping scalar gains. The corresponding outer loop decentralized algorithms read:

$$\begin{aligned} c_i^{n+1} &= c_i^n + k_{c_i}(\nu_i^n - c_i^n), \quad i \in \overline{1:N} \\ u_i^{n+1} &= u_i^n + k_{u_i}(w_i^n - u_i^n), \quad i \in \overline{1:N}. \end{aligned} \tag{7.266c}$$

The inner loop algorithmic mappings and the corresponding algorithm with information exchange are (see (7.131) and (7.132)):

$$\begin{aligned} \Psi_p &: \mathbb{R}^{n_u} \times \mathbb{R}^{n_u} \mapsto \mathbb{R}^{n_u} \\ \Psi_p(p, w - HF(\nu, w, \alpha)) &= p + k_p[w - HF(\nu, w, \alpha)]. \end{aligned} \tag{7.267a}$$

With this inner loop algorithmic mapping the corresponding inner loop algorithm reads:

$$p^{l+1,n} = p^{l,n} + k_p[w^{l,n} - HF(\nu^{l,n}, w^{l,n}, \alpha^n)] \tag{7.267b}$$

where $k_p > 0$ is a step coefficient of the inner loop gradient type of algorithm.

7.4.4.2 *Model-based double-loop algorithm with relaxed inner loop*

We shall now generalize the ISOPE algorithm that was developed by (Brdyś *et al.*, 1990b) to produce the AISOPE algorithm. A reduced number of iterations to be performed on the controlled system is an extremely important advantage of a model-based double-loop algorithm over the system based one. A limited applicability due to a stiffness of its inner loop problem implying that solution of the inner loop problem may not exist for the required range of control inputs is a drawback. Our problem now is to think of such modification of the algorithm structure that the advantage remains while the drawback is removed. The solution comes in a straightforward manner from the analysis of the inner loop problem that was performed in Section 7.3.4.2. Indeed, all we need to do is to distinguish between the price variable entering an expression for the modifiers λ and t and the price variable that is responsible for model-based interaction balance condition. This leads to *relaxed modified model based optimization problem* (RMMOP) defined as (see (7.198)):

$$\begin{aligned} &\text{minimize}_{\nu,w}\{q(\nu, w, \hat{\alpha}(c,u)) + \rho_c\|\nu - c\|^2 + \rho_u\|w - u\|^2 + \\ &\qquad -\lambda(c,u,p_2,\hat{\alpha}(c,u))^T\nu - t(c,u,p_2,\hat{\alpha}(c,u))^T w + \\ &\qquad\qquad + p_1^T(w - HF(\nu, w, \hat{\alpha}(c)))\} \\ &\text{subject to} : g(\nu, w) \leq 0 \end{aligned} \tag{7.268}$$

where $\alpha = \hat{\alpha}(c,u)$ is defined as solution of PEP (see (7.197a)) under given c and u, the multipliers $\lambda = \lambda(c,u,p_2,\hat{\alpha}(c,u))$ and $t = t(c,u,p_2,\hat{\alpha}(c,u))$ are calculated from (7.193a), (7.191) and (7.192) under $\alpha = \hat{\alpha}(c,u)$ and $p = p_2$. The inputs to the RMMOP are $c, u, p_1$and p_2.

Let us denote the RMMOP solution by

$$\begin{aligned} &\hat{\nu}(c,u,p_1,p_2) \\ &\widehat{w}(c,u,p_1,p_2). \end{aligned} \tag{7.269}$$

The core conditions definining a solution of a double-loop iterative strategy (7.200a), (7.200b) and (7.200c) can now be written in a relaxed but equivalent way as:

$$\hat{\nu}(c,u,p_1,p_2) = c \tag{7.270a}$$

$$\widehat{w}(c,u,p_1,p_2) = u \tag{7.270b}$$

$$\widehat{w}(c,u,p_1,p_2) = HF(\hat{\nu}(c,u,p_1,p_2), \widehat{w}(c,u,p_1,p_2), \hat{\alpha}(c,u)) \tag{7.270c}$$

$$p_1 = p_2. \tag{7.270d}$$

Given c, u and p_2,the *relaxed inner loop* iterates p_1 in order to meet the interaction balance condition (7.270c). It is done by solving the RMMOP with the prescribed values of c, u and p_2 yielding:

$$\text{price:} \quad \hat{p}_1(c, u, p_2) \tag{7.271}$$

predicted control inputs:

$$\hat{\nu}(c, u, p_2) \overset{\Delta}{=} \hat{\nu}(c, u, \hat{p}_1(c, u, p_2), p_2) \tag{7.272}$$

predicted interaction inputs:

$$\hat{w}(c, u, p_2) \overset{\Delta}{=} \widehat{w}(c, u, \hat{p}_1(c, u, p_2), p_2). \tag{7.273}$$

The *outer loop* iterates the control input, interaction input and price variables c, u and p_2 in order to satisfy the conditions (7.270a), (7.270b) and (7.270d).

Hence, for prescribed by the outer loop c, u and p_2 the *inner loop task* is to find such value of p_1 that

$$\widehat{w}(c, u, p_2) = HF(\hat{\nu}(c, u, p_2), \widehat{w}(c, u, p_2), \hat{\alpha}(c, u)) \tag{7.274}$$

while the *outer loop task* is to find such values of p_2, c and u that

$$\hat{\nu}(c, u, p_2) = c \tag{7.275}$$

$$\widehat{w}(c, u, p_2) = u \tag{7.276}$$

$$\hat{p}_1(c, u, p_2) = p_2 \tag{7.277}$$

The *model-based double-loop algorithm with relaxed inner loop* can now be formulated for double-loop ISOPE with the output measurements:

Start. Given initial points $c^0, u^0, p^{0,0}$, inner and outer loop algorithmic mappings Ψ_{p_1} and $\{\Psi_{c_i}\}_{i \in \overline{1:N}}$, $\{\Psi_{u_i}\}_{i \in \overline{1:N}}$, $\{\Psi_{p_2,i}\}$,respectively, convexifying coefficients ρ_c and ρ_u and the loop solution accuracies $\varepsilon_c = [\varepsilon_{c_1}, \ldots \varepsilon_{c_N}]$, $\varepsilon_{c_i} > 0$, $\varepsilon_u = [\varepsilon_{u_1}, \ldots \varepsilon_{u_N}]$, $\varepsilon_{u_i} > 0$, $\varepsilon_{p_2} = [\varepsilon_{p_2,1}, \ldots, \varepsilon_{p_2,N}]$, $\varepsilon_{p_2,i} > 0$, $\varepsilon_{p1} > 0$. Set $n := 0$ and $l = 0$; n and l denote the outer and inner loop iteration numbers, respectively.

Step 1. Apply c^n to the controlled plant and measure the outputs $y^n = K_*(c^n)$, $y^{on} = K_*^o(c^n)$. Calculate the plant output mapping derivatives $K_*^{'}(c^n), K_*^{o'}(c^n)$ by using additional perturbation of c^n or by employing previous output measurements as it is done by DISOPE.

Step 2. Solve the *parameter estimation problem* PEP defined by (7.197a). The PEP decomposes into N independent *local parameter estimation problems* $\text{PEP}_i, i \in \overline{1, N}$ yielding the parameter estimates

$\alpha_i^n = \hat{\alpha}_i(c^n, u^n)$ satisfying

$$F_i^o(c_i^n, u_i^n, \alpha_i^n) = K_{*i}^o(c^n) \qquad (7.278a)$$
$$F_i(c_i^n, u_i^n, \alpha_i^n) = K_{*i}(c^n). \qquad (7.278b)$$

Step 3. For each subsystem calculate modifiers $\lambda_i^{l,n} = \lambda_i(c^n, u^n, \alpha^n, p_2^{l,n})$, $t_i^{l,n} = t_i(c^n, u^n, \alpha^n, p_2^{l,n}), i \in \overline{1:N}$ by applying formulae (7.191) and (7.192). This can not be done independently for each subsystem due to interconnections between subsystems. Hence, an *information exchange* between *local subsystem controllers* is needed in order to perform this step.

Step 4. For $c := c^n$, $u := u^n$, $p_2 := p_2^n$, $p_1 := p_1^{l,n}$ solve the *relaxed modified model-based optimization problem* RMMOP defined by (7.268) with $\lambda := \lambda^{l,n}, t := t^{l,n}$ and $\alpha := \alpha^n$. The RMMOP decomposes into N independent *local relaxed modified model-based optimization problems* RMMOP$_i$, $i \in \overline{1, N}$ (see (7.145) and (7.4))

$$\begin{aligned} &\text{minimize}_{\nu_i, w_i} \{q_i(\nu_i, w_i, \alpha_i^n) + \rho_c \|\nu_i - c_i^n\|^2 + \rho_u \|w_i - u_i^n\|^2 + \\ &\qquad -(\lambda_i^{l,n})^T \nu_i - (t_i^{l,n})^T w_i + (p_{1,i}^{l,n})^T w_i + \\ &\qquad - \sum_{j=1}^{N} (p_{1,j}^{l,n})^T H_{ji} F_i(\nu_i, w_i, \alpha_i^n)\} \\ &\text{subject to: } g_i(\nu_i, w_i) \leq 0. \end{aligned} \qquad (7.279)$$

Solving the RMMOP$_i$ $i \in \overline{1:N}$ produces $\nu_i^{l,n} = \hat{\nu}_i(c_i^n, u_i^n, p_1^{l,n}, p_2^n)$ and $w_i^{l,n} = \widehat{w}_i(c_i^n, u_i^n, p_1^{l,n}, p_2^n)$(see (7.269)).
If for $i \in \overline{1:N}$

$$\| w_i^{l,n} - H_i F(\nu^{l,n}, w^{l,n}, \alpha^n) \| \leq \varepsilon_{p_1,i} \qquad (7.280)$$

then terminate the inner loop iterations. The inner loop problem solution $p_1^n = [(p_1^n)_1^T, .., (p_1^n)_1^T] = \hat{p}_1(c^n, u^n, p_2^n)$ has now been found (see (7.274)). Continue from Step 6.
The condition (7.280) cannot be verified at the subsystem level without information exchange between local controllers. Typically, an *inner loop coordinator* is introduced to carry out this operation.

Step 5. Set

$$p_1^{l+1,n} := \Psi_{p_1}(p_1^{l,n}, w^{l,n} - HF(\nu^{l,n}, w^{l,n}, \alpha^n)) \qquad (7.281)$$

set $l := l + 1$ and continue from Step 4.

This inner loop price iteration can not be carried out at the subsystem level that is in a decentralized manner and it is performed by the inner loop coordinator.

Step 6. The price, predictions of the control and interaction inputs determined at the inner loop problem solutions are : $p_1^n = \hat{p}_1(c^n, u^n, p_2^n)$, $\nu^n = \hat{\nu}(c^n, u^n, p_2^n)$ and $w^n = \widehat{w}(c^n, u^n, p_2^n)$ (see (7.271), (7.272), (7.273)).
If for $i \in \overline{1:N}$ the following hold:

$$\| \quad c_i^n - \nu_i^n \quad \| \leq \varepsilon_{c_i} \tag{7.282}$$

$$\| \quad u_i^n - w_i^n \quad \| \leq \varepsilon_{c_i} \tag{7.283}$$

$$\| \quad p_{1,i}^n - p_{2,i}^n \quad \| \leq \varepsilon_{p_2,i} \tag{7.284}$$

then terminate the outer loop iterations.
The outer loop problem solution has now been found ((7.275), (7.276), (7.277) are satisfied).

Step 7. Set

$$c_i^{n+1} \quad : \quad = \Psi_{c_i}(c_i^n, \nu_i^n), \quad i \in \overline{1:N} \tag{7.285}$$

$$u_i^{n+1} \quad : \quad = \Psi_{u_i}(u_i^n, \nu_i^n), \quad i \in \overline{1:N} \tag{7.286}$$

$$p_{2,i}^{n+1} \quad : \quad = \Psi_{p_2,i}(p_{2,i}^n, p_{1,i}^n), \quad i \in \overline{1:N} \tag{7.287}$$

set $n := n + 1$ and continue from Step 1.

The control and interaction inputs and the price iterations according to (7.285), (7.286) and (7.287) are completely decentralized. Hence they are performed at the subsystem level and this is a key benefit of decentralization. However, as the modifiers λ_i and t_i depend on all the components of the vectors c, u and p_2 (see (7.191), (7.192) and (7.193a) then so do the mappings $\hat{\nu}_i(\cdot,\cdot)$ and $\hat{p}_{2,i}(\cdot,\cdot)$ (see (7.268)). Hence, applying a completely decentralized algorithmic mapping at the outer loop in order to solve (7.270a), (7.270b) and (7.270d) limits convergence of the iterations (7.285), (7.286) and (7.287).

An information structure of the *double-loop model-based with relaxed inner loop algorithm* is two-level hierarchical with information exchange between local subsystem controllers and it is presented in Fig. 7.25.

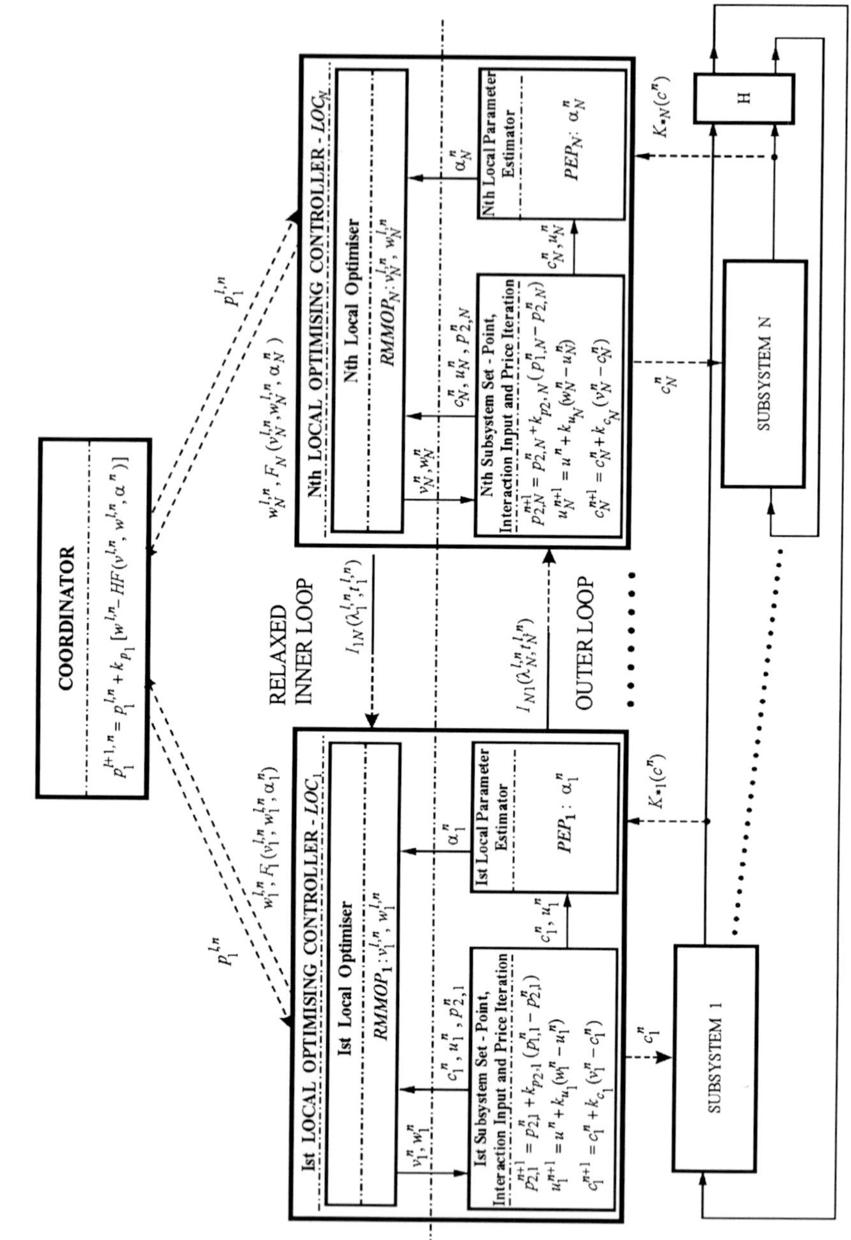

Fig. 7.25 Information structure of AISOPE model-based double-loop iterative strategy with output measurements and with relaxed inner-loop.

Comparison of Fig. 7.25 with Fig. 7.24 illustrating the *double-loop model-based with tight inner loop algorithms* shows that exactly the same units are present. However, the data regarding price vectors are now different and they are differently processed. Both the local controllers iterate the price vectors but the outer loop ensures that at the solution the prices are the same. As the number of price variables has increased twice the computing and the communication load have also increased. The structure is also very similar to the structure of the model-based double-loop algorithm with input-output measurements and with relaxed inner-loop (see Fig. 7.13)). The only difference is that now the interaction inputs are not gathered from the measurement data but they are predicted from the models utilized by distributed $\mathrm{LOC}_i, i \in \overline{1:N}$ to carry out their iterative processes.

The inner and outer loop algorithmic mappings Ψ_{p_1} and $\{\Psi_{c_i}\}_{i\in\overline{1:N}}$, $\{\Psi_{u_i}\}_{i\in\overline{1:N}}$, $\{\Psi_{p_2,i}\}$ respectively, need to be designed so that the loop convergence is achieved. As for the double-loop model-based with tight inner loop algorithm a relaxation algorithm with a scalar relaxation gain is chosen for the outer loop. Hence, for the control and interaction inputs the corresponding outer loop local algorithmic mappings are:

$$\begin{aligned} \Psi_{c_i} &: \mathbb{R}^{n_{c_i}} \mapsto \mathbb{R}^{n_{c_i}} \\ \Psi_{c_i}(c_i, v_i) &= c_i + k_{c_i}(\nu_i - c_i), \quad i \in \overline{1:N} \qquad (7.288a) \\ \Psi_{u_i} &: \mathbb{R}^{n_{u_i}} \mapsto \mathbb{R}^{n_{u_i}} \\ \Psi_{u_i}(u_i, w_i) &= u_i + k_{u_i}(w_i - u_i), \quad i \in \overline{1:N} \qquad (7.288b) \end{aligned}$$

where $k_{c_i} > 0$, $k_{u_i} > 0$ are the ith outer loop control and interaction relaxation mapping scalar gains.

The corresponding outer loop decentralized algorithm for iterating the control and interaction inputs read:

$$\begin{aligned} c_i^{n+1} &= c_i^n + k_{c_i}(\nu_i^n - c_i^n), \quad i \in \overline{1:N} \qquad (7.289a) \\ u_i^{n+1} &= u_i^n + k_{u_i}(w_i^n - u_i^n), \quad i \in \overline{1:N}. \qquad (7.289b) \end{aligned}$$

Similarly, the prices at the outer loop are iterated according to:

$$\begin{aligned} \Psi_{p_2,i} &: \mathbb{R}^{n_{u_i}} \mapsto \mathbb{R}^{n_{u_i}} \\ \Psi_{p_2,i}(p_{2,i}, p_{1,i}) &= p_{2,i} + k_{p_2,i}(p_{1,i} - p_{2,i}), \quad i \in \overline{1:N} \qquad (7.290) \end{aligned}$$

where $k_{p_{2,i}} > 0$ is the ith outer loop control relaxation mapping scalar gain. The corresponding outer loop decentralized algorithm for iterating the prices influencing the modifiers λ and t read:

$$p_{2,i}^{n+1} = p_{2,i} + k_{p_2,i}(p_{1,i}^n - p_{2,i}^n). \tag{7.291}$$

Thanks to a relaxation of the inner loop a design of the inner loop algorithmic mapping is straithforward:

$$\begin{aligned} \Psi_{p_1} \quad &: \quad \mathbb{R}^{n_u} \mapsto \mathbb{R}^{n_u} \\ \Psi_{p_1}(p_1, w - HF(\nu, w, \alpha) \quad &= \quad p_1 + k_{p_1}[w - HF(\nu, w, \alpha)]. \end{aligned} \tag{7.292}$$

The corresponding inner loop algorithm for iterating the prices that forces the interaction balance reads:

$$p_1^{l+1,n} = p_1^{l,n} + k_{p_1}[w^{l,n} - HF(\nu^{l,n}, w^{l,n}, \alpha^n)]. \tag{7.293}$$

As previously, we shall analyze the convergence properties of the double-loop model-based with relaxed inner loop algorithm for a linear-quadratic convex case. As the corresponding optimization problem is strictly convex then in order to simplify formulae it is assumed that $\rho_c = \rho_u = 0$, hence the convexifying terms disappear.

We shall now consider more general outer loop algorithm than the algorithm described above by (7.289a), (7.289b) and (7.291), that envolves an information exchange during performing the outer loop iterations. Thus, it has a wider applicability than the completely decentralized algorithm described by (7.289a), (7.289b) and (7.291). The possibly not sparse matrix gains are to be used in order to improve the outer loop convergence properties at a cost of the increased information exchange. Also, the full autonomy of the decision units is now lost. The generalized outer loop iterative scheme reads as:

$$\begin{bmatrix} c^{n+1} \\ u^{n+1} \end{bmatrix} = \begin{bmatrix} c^n \\ u^n \end{bmatrix} + R_x \begin{bmatrix} v^n - c^n \\ w^n - u^n \end{bmatrix} \tag{7.294a}$$

and

$$p_2^{n+1} = p_2^n + R_p(p_1^n - p_2^n) \tag{7.294b}$$

where the matrices $R_x$and R_p have inverses R_x^{-1} and R_p^{-1} respectively and are suitably chosen to preserve convergence, and where

$$v^n = \widehat{v}(c^n, u^n, \widehat{p}_1(c^n, u^n, p_2^n), p_2^n) \quad (7.295a)$$
$$w^n = \widehat{w}(c^n, u^n, \widehat{p}_1(c^n, u^n, p_2^n), p_2^n) \quad (7.295b)$$
$$p_1^n = \widehat{p}_1^n(c^n, u^n, p_2^n) \quad (7.295c)$$

is the inner loop solution for $c^n, u^n, p_2^n, \alpha^n$.

Regarding iterations of the variables c_i, u_i and $p_{2,i}$ for $i \in \overline{1:N}$, let us note that if all the scalar relaxation gains k_{c_i}, k_{u_i} and $k_{p_{2,i}}$ in the decentralized algorithm are equal to k_c, k_u, k_{p_2}, respectively, then

$$R_x = \begin{bmatrix} k_c I_c & 0 \\ 0 & k_u I_u \end{bmatrix} \quad (7.296a)$$

where matrices I_c, I_u are identity matrices of the dimensions $n_c \times n_c$ and $n_u \times n_u$, respectively, and

$$R_p = k_{p_2} I_p \quad (7.296b)$$

where I_p is identity matrix of dimension $n_u \times n_u$.

Finally, if $k_c = k_u = k_x$ and denoting $x \triangleq (c^T, u^T)^T$ the equation (7.296a) takes the form of $R_x = k_x I_x$, where I_x is identity matrix of dimension $(n_c + n_u) \times (n_c + n_u)$.

Let us assume a quadratic performance index as

$$Q(x, y) = \frac{1}{2}(x - d)^T M (x - d) + \frac{1}{2}(y - e)^T E (y - e) \quad (7.297)$$

where M and E are symmetric and positive definite matrices, d and e are constant vectors, and linear input-output relationships for the system and its point-parametric model as

$$F_*(x) = D_{*1}c + D_{*2}u + d_* \quad (7.298a)$$
$$F(x, \alpha) = D_1 c + D_2 u + P(\alpha). \quad (7.298b)$$

Let us define the following matrices:

$$B \triangleq [-HD_1, I_u - HD_2] \quad (7.299)$$
$$B_* \triangleq [(HD_{*2} - I_u)^{-1} HD_{*1}, I_u] \quad (7.300)$$
$$\widetilde{M} \triangleq M + D^T E D \quad (7.301)$$
$$M_* \triangleq M + D_*^T E D_* \quad (7.302)$$

where $D_* = [D_{*1}, D_{*2}]$ and $D = [D_1, D_2]$. The iterations are terminated when $(c^n, u^n, p^n) \in \Omega$, where Ω denotes set of solutions of (7.200a), (7.200b) and (7.200c). The convergence conditions can now be formulated.

Theorem 7.3

Assume that

(*i*) *the equations* $u - HD_1c - HD_2u = 0$ *and* $u - HK_*(c, u) = 0$ *are both linearly independent.*

(*ii*) $R_p^{-T}B = BR_x^{-1}$

(*iii*) $\widetilde{M}R_x^{-1} = R_x^{-1}\widetilde{M}$

(*iv*) $\widetilde{M}R_x^{-1} - \frac{1}{2}M_* > 0$

(*v*) $$B_*M_*^{-1}B^TR_p^{-1} + (R_p^{-1})^TBM_*^{-1}B_*^T + \\ -B_*M_*^{-1}(\widetilde{M}R_x^{-1} + M_*)M_*^{-1}B_*^T > 0$$

Then

(*a*) *There exists a unique solution* $(\overline{c}, \overline{u})$ *of the SOCP. The algorithm solution set* Ω *contains a single point* $(\overline{c}, \overline{u}, \overline{p})$.

(*b*) *The algorithm either stops at* Ω *or generates a sequence* $\{(c^n, u^n, p_2^n)\}_{n\in\mathbb{N}}$ *,,that is convergent to the point* $(\overline{c}, \overline{u}, \overline{p})$.

A proof of the Theorem 7.3 is presented in Appendix B.3.

The assumption (*i*) is always satisfied in realistic practical situations. The assumption (*i*) is purely technical and it is made only to express the assumption (*v*) a more readable form. A simplest choice of the matrices R_p and R_x is according to (7.296a) and (7.296b) with $k_{p_2} = k_c = k_u = \chi > 0$. Clearly, the assumptions (*ii*) and (*iii*) are then satisfied, while the assumption (*iv*) is also satisfied in this case if χ is sufficiently small. As it has been pointed out before such structure of the matrix gains R_p and R_x in the outer loop algorithms leads to a completely decentralized structure of the outer loop that is most attractive. Now investigating assumption (*v*), however, we can see that it imposes certain structural requirements that may not be satisfied for every value of χ. This is why a general matrix gain rather than the scalar gain type of iterative strategy was proposed for solving the outer loop equations (7.275), (7.276) and (7.277). The assumption (*v*) is very technical and it follows from a proof of the Theorem 7.3.

7.4.5 *Simulation studies*

Example 7.5

The double-loop system based algorithm and the double-loop model-based algorithm with relaxed inner loop are applied to three example systems and their applicability and convergence properties are investigated by simulation.

Example 7.5 a

The system structure is illustrated in Fig.7.26. Hence, the interconnection structure matrix reads

$$H = \begin{bmatrix} 0 & 1 & 0 & 0 \\ 1 & 0 & 0 & 0 \\ 0 & 0 & 0 & 1 \\ 0 & 0 & 1 & 0 \end{bmatrix}. \tag{7.303}$$

The subsystem input-output relations are

$$y_1 = y_{11} = F_{*1}(c_1, u_1) = 1.3c_{11} - c_{12} + 2u_{11} + 0.15u_{11}c_{11} \tag{7.304a}$$

$$y_{21} = F_{*21}(c_2, u_2) = c_{21} - c_{22} + 1.2u_{21} - 3u_{22} + 0.1(c_{22})^2 \tag{7.304b}$$

$$y_{22} = F_{*22}(c_2, u_2) = 2c_{21} - 1.25c_{23} - u_{21} + u_{22} + +0.25c_{22}c_{23} + 0.1 \tag{7.304c}$$

$$y_3 = y_{31} = F_{*3}(c_3, u_3) = 0.8c_{31} + 2.5c_{32} - 4.2u_{31}. \tag{7.304d}$$

Note that the system is described by nonlinear equations.

The real system nonlinear input-output subsystem mappings are mod-

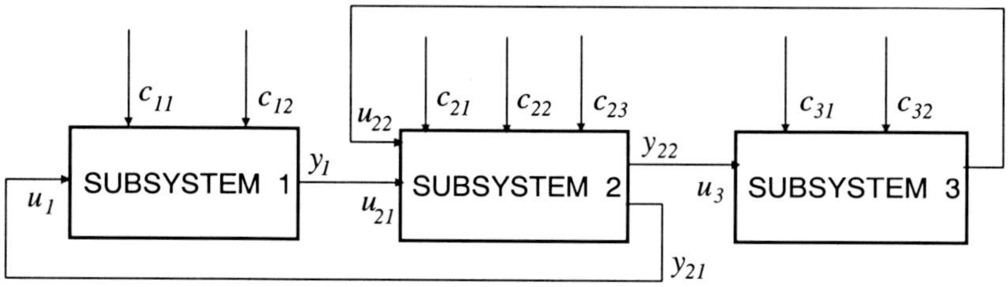

Fig. 7.26 Structure of the system in the example problem 7.5 a.

eled by point-parametric parameterized linear mappings as follows:

$$y_1 = y_{11} = F_1(c_1, u_1) = c_{21} - c_{12} + 2u_{11} + \alpha_{11} \quad (7.305a)$$

$$y_{21} = F_{21}(c_2, u_2) = c_{21} - c_{22} + u_{21} - 3u_{22} + \alpha_{21} \quad (7.305b)$$

$$y_{22} = F_{22}(c_2, u_{2)} = 2c_{21} - c_{23} - u_{21} + u_{22} + \alpha_{22} \quad (7.305c)$$

$$y_3 = y_{31} = F_3(c_3, u_3) = c_{31} + 2.5c_{32} - 4u_{31} + \alpha_{31}. \quad (7.305d)$$

There are two different situations in which such a model may arrive. First, even if the real system mappings are exactly known we may prefer to use their linear approximations in order to be able to efficiently solve a nonlinear and hence, nonconvex optimization task of a large scale. However, the solution might be very suboptimal due to the linear approximation of a nonlinear reality. The ISOPE technology can be then viewed as a method that achieves the nonlinear optimization problem solution by solving a number of linearized optimization tasks, each of them corresponding to different parameter values. It is entirely due to the ISOPE internal mechanism to update the subsequent parameter values in order to achieve convergence to the true optimum. In summary, the ISOPE is a method for solving nonlinear optimization tasks.

The second possible situation is that our *a priori* knowledge about uncertainty in the system model is crude. Hence, the point-parametric modeling approach is employed in order to reduce the modeling uncertainty. For example, we may apply a linear model structure in order to model a nonlinear reality if there is no better choice. As our *a priori* knowledge regarding the nonlinearities is crude the model parameters need to be estimated. However, due to a large structural modeling error resulting from the linearization even the best constant parameter estimates may not achieve required model accuracy. In order to improve the modeling accuracy the parameters must be made input dependent. In other words, given input that implies the system output, the best parameter values are used to explain the system output by using a model fed by the same input. The point-parametric modeling introduced first by (Brdyś, 1983) was recently further developed in general terms in (Chang *et al.*, 2004). Successful applications to drinking water networks and wastewater systems were reported in (Brdyś and Chang, 2002; Chang, *et al.*, 2003; Rutkowski *et al.*, 2004). It is pointed out that ISOPE technology is capable of handling this severe uncertainty and achieve an exact solution of the system optimizing control problem.

The subsystem local constraints read:

$$CU_1 = \{(c_1, u_1) \in \mathbb{R}^3 : \ (c_{11})^2 + (c_{12})^2 \le 1; \ 0 \le u_{11} \le 0.5\} \quad (7.306a)$$

$$\begin{aligned} CU_2 = \{(c_2, u_2) \in \mathbb{R}^5 : \ & 0.5c_{21} + c_{22} + c_{23} \le 1; \ 4(c_{21})^2 + \\ & +2c_{21}u_{21} + 0.4u_{21} + c_{21}c_{23} + 0.5(c_{23})^2 + (u_{21})^2 \le 4\} \end{aligned} \quad (7.306b)$$

$$CU_3 = \{(c_3, u_3) \in \mathbb{R}^3 : \ -0.5 \le c_{31} + u_{31}; \ 0 \le c_{32} \le 1\}. \quad (7.306c)$$

The subsystem performance functions are

$$Q_1(c_1, u_1) = (u_{11} - 1)^4 + 5(c_{11} + c_{12} - 1)^2 \quad (7.307a)$$

$$Q_2(c_2, u_2) = 4(u_{21})^2 + (u_{22})^2 + 2(c_{21} - 2)^2 + +(c_{22})^2 + 3(c_{23})^2 \quad (7.307b)$$

$$Q_3(c_3, u_3) = (u_{31} - 1)^2 + (c_{31} + 1)^2 + 2.5(c_{32})^2. \quad (7.307c)$$

Note, that the constraints functions are also nonlinear but they are not linearized in order to avoid feasibility problems.

Example 7.5 b
The subsystem performance functions are now output dependent:

$$Q_1(c_1, u_1, y_1) = (u_{11} - 1)^4 + 5(c_{11} + c_{12} - 1)^2 + +0.1(y_{11} - 1)^2 \quad (7.308a)$$

$$Q_2(c_2, u_2) = 4(u_{21})^2 + (u_{22})^2 + 2(c_{21} - 2)^2 + +(c_{22})^2 + 3(c_{23})^2 \quad (7.308b)$$

$$Q_3(c_3, u_3, y_3) = (u_{31} - 1)^2 + (c_{31} + 1)^2 + 2.5(c_{32})^2 + +0.2(y_{31} - 2)^2. \quad (7.308c)$$

The subsystem mappings, local constraints and the structure equations are the same as in the Example 7.5 a.

Example 7.5 c
The system structure is illustrated in Fig. 7.27. Hence, the interconnection equations are as follows:

$$\begin{bmatrix} u_{11} \\ u_{21} \end{bmatrix} = \begin{bmatrix} 0 & 1 \\ 1 & 0 \end{bmatrix} \begin{bmatrix} y_{11} \\ y_{21} \end{bmatrix}. \quad (7.309)$$

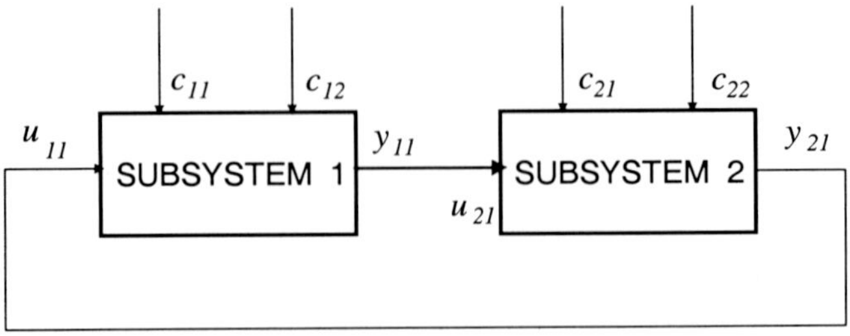

Fig. 7.27 Structure of the system in the example problem 7.5 c.

The reality and its model equations are

$$y_1 = y_{11} = F_{*1}(c_1, u_1) = 2c_{11} - c_{12} + 0.5u_{11} + 0.15u_{11}c_{11} \quad (7.310a)$$
$$y_1 = y_{11} = F_1(c_1, u_1) = 1.7c_{11} - 0.5c_{12} + u_{11} + \alpha_{11} \quad (7.310b)$$

and

$$y_2 = y_{21} = F_{*2}(c_1, u_1) = 2.3c_{21} - 0.5c_{22} + u_{21} + 0.1(c_{21})^2 + 0.1 \quad (7.311a)$$
$$y_2 = y_{21} = F_2(c_1, u_1) = 2.2c_{21} - 0.2c_{22} + 0.5u_{21} + \alpha_{21}. \quad (7.311b)$$

The subsystem local constraints read

$$CU_1 = \{(c_1, u_1) \in \mathbb{R}^3 : \ (c_{11})^2 + (c_{12})^2 \leq 1; \ 0 \leq u_{11} \leq 0.5\} \quad (7.312a)$$

$$CU_2 = \{(c_2, u_2) \in \mathbb{R}^4 : \ 0.5c_{21} + c_{22} + 2c_{23} \leq 1; \ 4(c_{21})^2 + 2c_{21}u_{21} + 0.4u_{21} + c_{21}c_{23} + 0.5(c_{23})^2 + (u_{21})^2 \leq 4\} \quad (7.312b)$$

while the performance functions are

$$Q_1(c_1, u_1) = 0.5(u_{11} - 1)^2 + 0.5(c_{11})^2 + 0.5(c_{12})^2 \quad (7.313a)$$
$$Q_2(c_2, u_2) = 0.5(u_{21})^2 + 0.5(c_{21} - 2)^2 + 0.5(c_{22})^2 + 0.5(c_{23})^2. \quad (7.313b)$$

The optimal solutions were obtained with high accuracy of $10^{-4}[\%]$ regarding the performance function optimal values.

Example 7.5 a:

$$c^{opt} = [0.48126, 0.87657, -0.17921, 0.03615. - 0.51997, 0.34258]^T$$
$$u^{opt} = [0.13645, 0.03130, 0.35662\ 0.01997]^T$$
$$Q^{opt} = 6.32593$$

Example 7.5 b:

$$
\begin{aligned}
c^{opt} &= [0.49606, 0.86828, 0.98270, -0.19647, 0.00943, -0.51299, 0.33443]^T \\
u^{opt} &= [0.13331, 0.05289, 0.37112, 0.01299]^T \\
Q^{opt} &= 6.95338.
\end{aligned}
$$

Example 7.5 c:

$$
\begin{aligned}
c^{opt} &= [-0.29488, 0.144224, 0.45437, 0.32317, 0.00000]^T \\
u^{opt} &= 0.50000, -0.50412]^T \\
Q^{opt} &= 1.55240.
\end{aligned}
$$

Simulations were carried out for the system-based double-loop algorithm and model-based double-loop algorithm with relaxed inner loop without convexifying terms in the MMOP (see.(7.198). Hence, $\rho_u = \rho_c = 0$. Initially, simulation were performed to determine suitable values of step-sizes (relaxation gains) k_c, k_u and k_p ((7.250c) and (7.252b)) for the system-based method, and k_c, k_u, k_{p_1} and k_{p_2} (see (7.288a), (7.288b), (7.293) and (7.291)), for the relaxed inner loop method. In order to simplify the simulations trials and and without major loss on a generality, the same values of k_c and k_u were assumed for the both methods. Also, $k_p = k_{p_1}$ was set. The results are summarized in Table 7.3 regarding the optimal, in terms of the convergence efficiency, values of these for parameters and gains.

	k_c, k_u	k_p	k_{p2}	set-point iterations	total iterations	real perf. achieved
Example 7.5a						
system-based	1.0	0.4	n/a	37	37	6.2662
model-based	*no convergence*					
Example 7.5b						
system-based	0.9	0.6	n/a	47	47	6.8718
model-based	*no convergence*					
Example 7.5c						
model-based rel. inn. loop	0.3	0.5	0.2	8	51	1.5353

Table 7.3 Convergence properties of the system-based and model-based algorithms with output measurements.

The algorithms terminated when the real system performance achieved its optimal value with a relative accuracy not greater than 2 [%]. The results from Table 7.3 show wider applicability of the system based algorithm. However, it requires more set-point changes to achieve the optimum. Rate of convergence of the algorithm for the systems in Example 7.5a and Example 7.5b is illustrated in Fig. 7.28 and in Fig. 7.29, respectively.

Rate of convergence of the model-based algorithm for the system in Example 7.5c is illustrated in Fig. 7.30. Trajectories of the multipliers generated by the model-based algorithm are shown in Fig. 7.31. The price trajectories in the inner loop and in the outer loop of the algorithm are illustrated in Fig. 7.32.

Finally, a number of iterations of the inner loop required by one iteration of the outer loop is illustrated for model-based algorithm in Fig. 7.33.

Let us notice that the inner loop iteration per one outer loop iteration is quite high but what is most important, a total outer loop iterations required to achieve the optimal set-points is small.

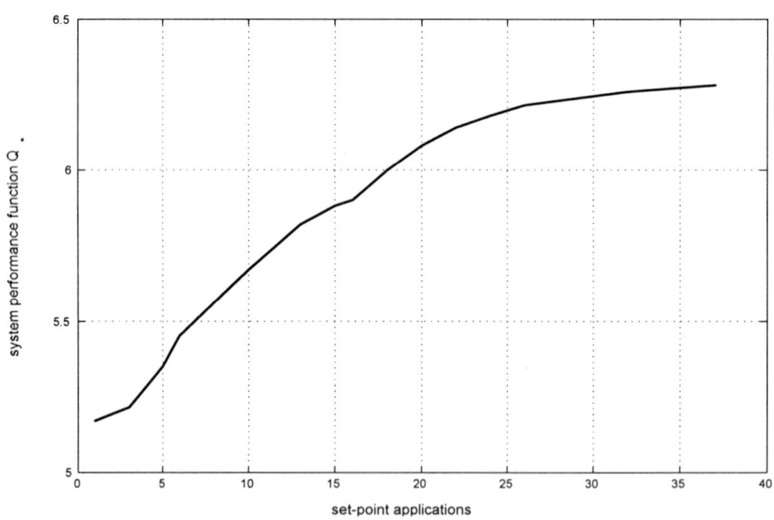

Fig. 7.28 Convergence rate of system-based double-loop algorithm with output measurements in the example problem 7.5a.

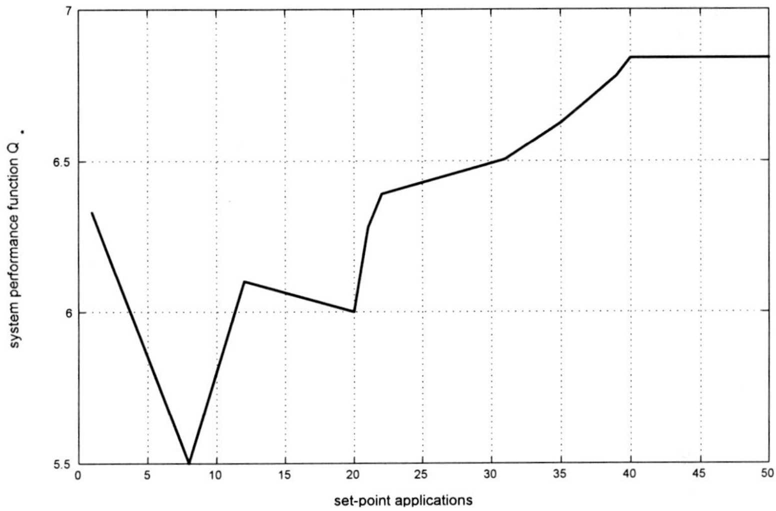

Fig. 7.29 Convergence rate of system-based double-loop algorithm with output measurement in the example problem 7.5b.

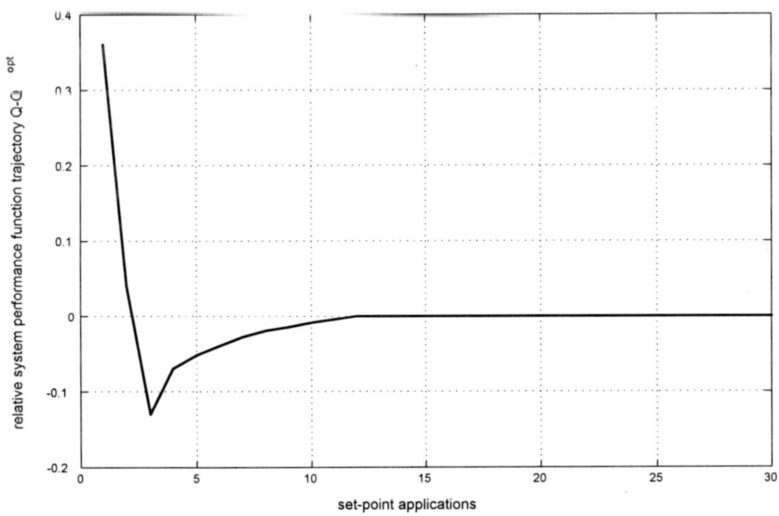

Fig. 7.30 Convergence rate of model-based double-loop algorithm with output measurements the example problem 7.5c

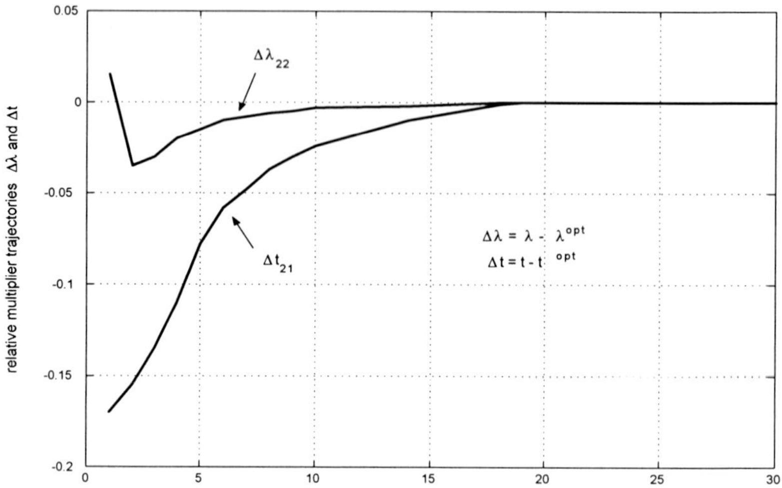

Fig. 7.31 Trajectories of multipliers generated by model-based algorithm with output measurement in the example problem 7.5c.

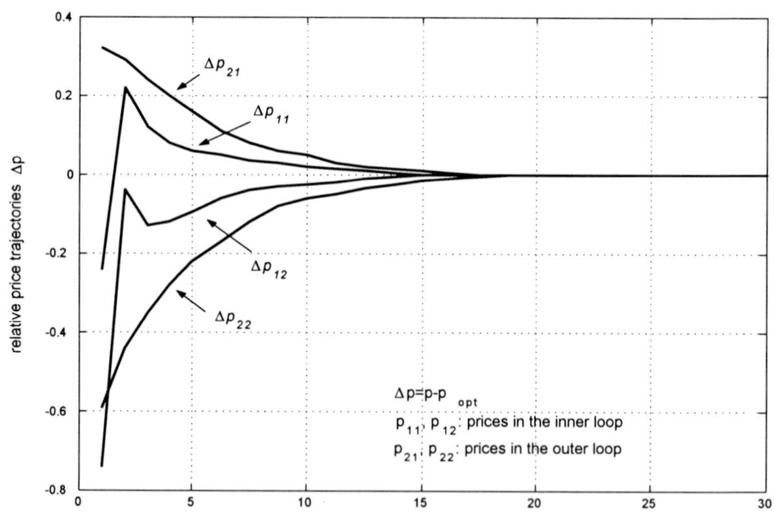

Fig. 7.32 Trajectories of prices in the inner and outer loops of the model-based algorithm in the example problem 7.5c.

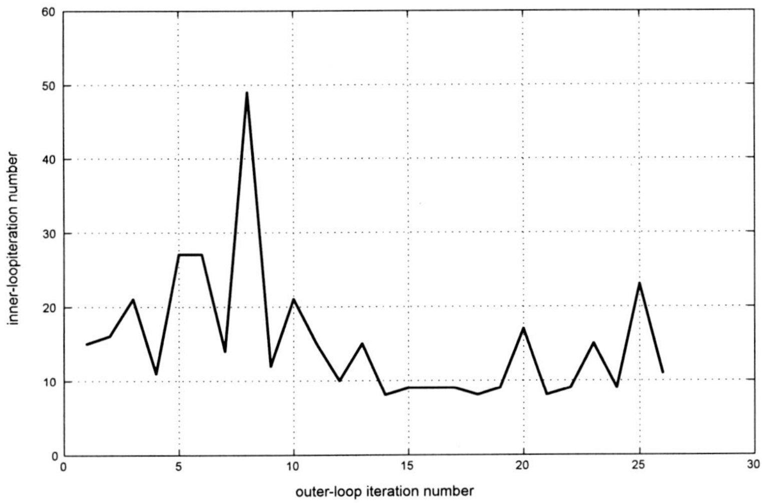

Fig. 7.33 Number of inner loop iterations required for each outer loop iteration of the model-based algorithm in the example problem 7.5c.

Appendix A

Proof of Theorem 4.1

Proof. Owing to the assumptions (i), (ii), (iv), (v) and according to the assumption (4.14) the mappings $\hat{v}_S(c, \alpha)$ and $\hat{\alpha}_S(c)$ are well defined on $C \times A$, respectively. Therefore, the mapping $\mathcal{A}(\cdot)$ is also well defined.

Owing to the assumptions (i) and (v), function $b(\cdot)$ is continuous on A which is a compact set (assumption (ii)). Hence, there exists $\tilde{\rho}_1 > 0$ such that, for every $\rho_1 > \tilde{\rho}_1$ the inequality (4.34) is satisfied. On the other hand, assumption (iv) implies that a number δ given by (4.37) is not negative and finite. Therefore, numbers τ and ε exist such that inequality (4.35) is satisfied. Moreover, because $\inf_{\alpha \in \mathcal{A}} b(\alpha) + 2\rho_1 \leq \inf_{\alpha \in \hat{\alpha}_S(c)} b(\alpha) + 2\rho_1$, for every $c \in C$, then inequality (4.35) implies that the mapping

$$\gamma(c, d) = \{c + k_c d : \tau \leq k_c \leq B(c)\} \tag{A.1}$$

with $B(c)$ defined by Eq. (4.36) is well defined. Therefore, the algorithmic mapping $\mathcal{A}(\cdot)$ defined by (4.29) is well defined on C. Moreover, assumption (i) and the fact that $k_c \leq 1$ following from definition (4.36) imply that $\mathcal{A}(c) \in C$. Proof of assertion (a) has now been completed.

To start the proof of (b) let us observe that, owing to the assumptions (i) and (iv), $q_*^{'}(\cdot)$ is Lipschitz continuous on C with constant δ (see (4.37)). This, and convexity of C, imply that the following inequality holds (Kantorovich and Akilov, 1963)

$$q_*(c) - q_*(v) \geq q_*^{'}(c)(c - v) - \frac{\delta}{2}\|c - v\|^2, \quad \forall v, c \in C. \tag{A.2}$$

Let $v \in \mathcal{A}(c)$ where c is arbitrary chosen from C. Hence, there are $\hat{v} \in$

$\hat{v}_S(c,\alpha)$, $\alpha \in \hat{\alpha}_S(c)$ and $\tau \leq k_c \leq B(c)$ such that $v = c + k_c(\hat{v} - c)$ and

$$q_*(c) - q_*(v) \geq k_c q_*'(c)(c - \hat{v}) - \frac{k_c^2 \delta}{2} \|c - \hat{v}\|^2. \tag{A.3}$$

To estimate the first term on the right hand side of inequality (A.3), we will utilize the fact that $\hat{v}_S(\cdot)$ is a set of minimizing points and that C is convex. Hence (Luenberger, 1973)

$$[q_c'(\hat{v},\alpha) - \lambda(c,\alpha)^T - 2\rho_1(c - \hat{v})](c - \hat{v}) \geq 0 \tag{A.4}$$

which, together with (see (4.40))

$$\lambda(c,\alpha)^T = q_c'(c,\alpha) - q_*'(c) \tag{A.5}$$

implies that

$$q_*'(c)(c - v) \geq [q_c'(c,\alpha) - q_c'(\hat{v},\alpha)](c - \hat{v}) + 2\rho_1 \|c - \hat{v}\|^2. \tag{A.6}$$

The assumptions (i) and (v) imply that θ exists, $0 \leq \theta \leq 1$, such that (see also (4.33))

$$\begin{aligned} [q_c'(c,\alpha) - q_c'(\hat{v},\alpha)](c - \hat{v}) &= (c - \hat{v})^T q_{cc}''(\hat{v} + \theta(c - \hat{v}),\alpha) \\ &\geq \lambda_{\min}(\hat{v} + \theta(c - \hat{v}),\alpha))\|c - \hat{v}\|^2 \\ &\geq b(\alpha)\|c - \hat{v}\|^2. \end{aligned} \tag{A.7}$$

Therefore

$$q_*'(c)(c - \hat{v}) \geq (b(\alpha) + 2\rho_1)\|c - \hat{v}\|^2. \tag{A.8}$$

Finally, combining inequalities (A.3) and (A.8), we obtain

$$q_*(c) - q_*(v) \geq \tau(b(\alpha) + 2\rho_1 - \frac{k_c \delta}{2})\|c - \hat{v}\|^2. \tag{A.9}$$

Since

$$k_c \leq B(c) \leq \frac{2b(\alpha) + 4\rho_1}{\delta + \epsilon} \tag{A.10}$$

then

$$b(\alpha) + 2\rho_1 - \frac{k_c \delta}{2} > 0. \tag{A.11}$$

Hence $q_*(v) \leq q_*(c)$ and $q_*(v) < q_*(c)$ if $c \notin \Omega$. Thus, the proof of (b) is complete.

To prove assertion (c) of the Theorem Zangwill's theorem (Zangwill, 1969) will be applied. Examining assumptions of Zangwill's theorem we can now state that, except for closeness of $\mathcal{A}(\cdot)$ outside Ω, all of them have been proved to be satisfied. The closeness of $\mathcal{A}(\cdot)$ outside Ω will now be verified by proving closeness of the mapping components and next by applying a suitable composition theorem. Let us start with mapping $\hat{\alpha}_S(\cdot)$ and note that it can be described in the following way: for a given $c \in C$

$$\hat{\alpha}_S(c) = \operatorname{Arg}\min_{\alpha \in A} \|F_*(c) - F(c, \alpha)\|. \tag{A.12}$$

As the set A is compact (assumption (ii)) and the function $\|F_*(\cdot) - F(\cdot, \cdot)\|$ is continuous on $C \times A$ (see assumption (iii)), then the mapping $\hat{\alpha}_S(\cdot)$ is closed on C (Hogan, 1973).

Now, let us consider a function

$$C \times (C \times A) \ni (v, c, \alpha) \to q(v, \alpha) - \lambda(c, \alpha)^T v + \rho_1 \|c - v\|^2 \in R^1. \tag{A.13}$$

Let c and α be fixed in C and A, respectively. The hessian of this function is equal to $q''_{cc}(v, \alpha) + 2\rho_1 I$ and

$$h^T(q''_{cc}(v, \alpha) + 2\rho_1 h \geq (2\rho_1 + \min_{\alpha \in A} b(\alpha))\|h\|^2. \tag{A.14}$$

Hence, according to definition of ρ_1 (see (4.34)), the modified model-based optimization problem (MMOP, see (4.12)) consists in minimizing a uniformly convex function on a compact and convex set. Therefore, the solution set $\hat{v}_S(c, \alpha)$ consists of a single point $\hat{v}(c, \alpha)$. Owing to continuity of the considered function on $C \times (C \times A)$ (assumptions (iv) and (v)) and according to the compactness of C, $\hat{v}(\cdot, \cdot)$ is a continuous function on $C \times A$.

Now let us consider the following mapping

$$C \ni c \to \{(c, \alpha) : \ \alpha \in \hat{\alpha}_S(c)\ \} \in 2^{C \times A}. \tag{A.15}$$

As $\hat{\alpha}_S(\cdot)$ is closed on C, then this mapping is also closed on C. Hence, because $C \times A$ is compact, then one can apply the theorem on closeness of mapping composition (Zangwill, 1969) and obtain that the mapping

$$C \ni c \to \{\hat{v}(c, \alpha) : \ \alpha \in \hat{\alpha}_S(c)\ \} \in 2^C \tag{A.16}$$

is closed on C.

Finally, applying the theorem on closeness of mapping composition to the above mapping and to an identity mapping on C, one concludes that

that the mapping $\tilde{c}(\cdot) : C \to 2^{C\times(C-C)}$,

$$\tilde{c}(c) = \{(c, v-c) :\ v \in \hat{v}_S(c,\alpha), \alpha \in \hat{\alpha}_S(c)\ \} \tag{A.17}$$

is closed on C.

Continuity of $q''_{cc}(\cdot,\cdot)$ on $C\times A$ and compactness of C (see assumptions (v) and (i), respectively) imply that $b(\cdot)$ is upper semicontinuous on A. Therefore, owing to the assumption (vi), the function

$$C \ni c \to \inf_{\alpha\in\hat{\alpha}_S(c)} b(\alpha) \in R^1 \tag{A.18}$$

is upper semicontinuous on c (Hogan, 1973). Thus, $B(\cdot)$ is upper semicontinuous on C.

Closeness of $\gamma(\cdot)$, defined by (A.1), at every point $(c,d) \in C\times(C-C)$ such that $d \neq 0$ will now be verified. Let

$$C\times(C-C) \ni (c^i, d^i) \to_{i\to\infty} (c,d) \in C\times(C-C), \quad d\neq 0 \tag{A.19}$$

and

$$\gamma(c^i, d^i) \ni y^i = c^i + k^i d^i \to_{i\to\infty} y. \tag{A.20}$$

Thus, for sufficiently large i the following holds

$$k^i = \frac{\|y^i - c^i\|}{\|d^i\|} \to_{i\to\infty} \frac{\|y-c\|}{\|d\|}. \tag{A.21}$$

Clearly, $y = c + \bar{k}d$, where $\bar{k} = \frac{\|y-c\|}{\|d\|}$. Since

$$\tau \leq k^i \leq B(c^i) \tag{A.22}$$

and because the function B is upper semicontinuous, then

$$\tau \leq \bar{k} \leq \lim_{i\to\infty}\sup B(c^i) \leq B(c). \tag{A.23}$$

Therefore, $y \in \gamma(c,d)$ and, consequently, $\gamma(\cdot,\cdot)$ is closed at (c,d).

The set C is compact, thus the sets $C-C$ and $C\times(C-C)$ are also compact. Therefore, the mapping $\mathcal{A}(\cdot)$, as a composition of the mappings $\tilde{c}(\cdot)$ and $\gamma(\cdot)$, is also closed at every point $c\in C$ such that $c \notin \hat{v}_S(c,\alpha)$ for all $\alpha \in \hat{\alpha}_S(c)$, that is at every point $c \notin \Omega$. Hence, all the assumptions of the Zangwill's theorem have now been verified, resulting in the proof of the assertion (c) of the theorem. Therefore, the proof of the theorem is complete.

Appendix B

Proofs of Theorems 7.1, 7.2 and 7.3

B.1 Proof of Theorem 7.1

Let c^n be prescribed by the outer loop. The corresponding interaction input in the real system is $u_*^n = HK_*(c^n)$. Let us denote $x_*^n \triangleq ((c^n)^T, (u_*^n)^T)^T$. The model parameter estimate α^n results by solving local model parameter estimation problems (see (7.120b)). The inner loop iterative algorithm generates price vectors $p^{l,n}$ according to (7.132) and the control $\nu^{l,n}$ and interaction inputs $u^{l,n}$ by solving local modified model-based optimization problems (7.121). Let us denote, $s^{l,n} \triangleq (\nu^{l,n}, u^{l,n})$. As $M_\rho > 0$ the $q(\cdot, \alpha^n)$ is strictly convex. Hence and because $G(\cdot, \alpha^n)$ is linear, the modified model-based optimization problem solution exists and the necessary and sufficient condition for $s^{l,n}$ to be its solution can be written as (see (7.133), (7.134), (7.135), (7.136), (7.137) and (7.138)):

$$M_\rho(s^{l,n} - x_*^n) + Mx_*^n + \widetilde{B}^T p^{l,n} - Md = 0. \tag{B.1}$$

The equation (B.1) yields

$$s^{l,n} = -M_\rho^{-1}\widetilde{B}^T p^{l,n} + (I_x - M_\rho^{-1}M)x_*^n + M_\rho^{-1}Md. \tag{B.2}$$

Hence, the inner loop iterative scheme (7.132) can be written as

$$\begin{aligned} p^{l+1,n} &= p^{l,n} + k_p B(s^{l,n} - x_*^n) \\ &= (I_u - k_p B M_\rho^{-1}\widetilde{B}^T)p^{l,n} + (I_u - M_\rho^{-1}M - k_p B)x_*^n + M_\rho^{-1}Md. \end{aligned} \tag{B.3}$$

Since the last two terms on the right-hand side of (B.3) are constant, the sequence $\{p^{l,n}\}_{l\in\mathbb{N}}$ is convergent iff the discrete time dynamical system

$$p^{l+1,n} = (I_u - k_p B M_\rho^{-1} \widetilde{B}^T) p^{l,n} \tag{B.4}$$

is asymptotically stable. Notice, that the matrix $BM_\rho^{-1}\widetilde{B}^T$ is not symmetric. Hence, its positiveness doesn't immediately guarantee an existence of a suitable value of k_p that is sufficient for the stability. We shall show the stability by employing Lyapunov approach. Let us define a candidate for the Lyapunov function as

$$T(p) \triangleq \| p \|^2 . \tag{B.5}$$

Let us denote

$$A(k_p) \triangleq (I_u - k_p B M_\rho^{-1} \widetilde{B}^T) \tag{B.6}$$

$$W \triangleq B M_\rho^{-1} \widetilde{B}^T. \tag{B.7}$$

The dynamical system (B.4) can now be written as $p^{l+1,n} = A(k_p)p^{l,n}$. Then

$$\begin{aligned} T(A(k_p)p) &= p^T p - k_p p^T W p - k_p p^T W^T p + (k_p)^2 p^T W^T W p \\ &= p^T p - k_p p^T (W^T + W) p + (k_p)^2 p^T W^T W p \\ &\leq \| p \|^2 - k_p \lambda_{\min}(W + W^T) + k_p^2 \lambda_{\max}(W^T W) \| p \|^2 \\ &= \varsigma(k_p) \| p \|^2 \end{aligned} \tag{B.8}$$

where

$$\varsigma(k_p) \triangleq 1 - k_p \lambda_{\min}(W + W^T) + k_p^2 \lambda_{\max}(W^T W). \tag{B.9}$$

Since $\varsigma(0) = 1$ and because $\lambda_{\min}(W + W^T) > 0$ and $\lambda_{\max}(W^T W) > 0$, then one can easily verify that there exists a value of $\overline{k_p}$ such that for every $k_p \in (0, \overline{k_p})$

$$0 < \varsigma(k_p) < 1. \tag{B.10}$$

Hence, if $k_p \in (0, \overline{k_p})$ then values of the function $T(\cdot)$ on the dynamical system trajectories satisfy:

$$T(p^{l+1,n}) < \| p^{l,n} \|^2 = T(p^{l,n}). \tag{B.11}$$

It has now been demonstrated that $T(\cdot)$ is Lyapunov function of the system (B.4) for every $k_p \in (0, \overline{k_p})$.

Let $k_p \in (0, \overline{k_p})$. Hence, the inner loop algorithm (B.3) generates sequence $\{p^{l,n}\}_{l\in\mathbb{N}}$ that is convergent. Let us denote the limit p^n.That is $\{p^{l,n}\}_{l\in\mathbb{N}} \to p^n$ as $l \to \infty$. It immediately follows from (B.3) that p^n satisfies jointly with the corresponding limit s^n of the sequence $\{(\nu^{l,n}, u^{l,n})\}_{l\in\mathbb{N}}$ the inner loop interaction balance condition (7.118). Hence, p^n is the inner loop solution. The assertions (a) and (b) of the Theorem 7.1 have now been proved.

We shall consider now convergence properties of the outer loop algorithm (7.127). Under control input c^n prescribed by the outer loop at its iteration n the inner loop algorithm generates (s^n, p^n) such that the following holds (see (B.1):

$$M_\rho(s^n - x_*^n) + Mx_*^n + \widetilde{B}^T p^n - Md = 0 \qquad \text{(B.12a)}$$

$$B(s^n - x_*^n) = 0. \qquad \text{(B.12b)}$$

It is straightforward to show that the $SOCP$ solution (c^{opt}, p^{opt}) satisfies

$$M\overline{x}_* + B_*^T p^{opt} - Md = 0 \qquad \text{(B.12c)}$$

$$B_*\overline{x}_* - b_* = 0. \qquad \text{(B.12d)}$$

where

$$\overline{x}_* = ((c^{opt})^T, (u_*^{opt})^T)^T \qquad \text{(B.13a)}$$

$$u_*^{opt} = HK_*(c^{opt}) \qquad \text{(B.13b)}$$

$$b_* = -(I_u - HD_{*2})^{-1} Hd_*. \qquad \text{(B.13c)}$$

The equation (B.12c) can also be written (see (7.140) and (7.137)) as

$$M\overline{x}_* + \widetilde{B}^T \overline{p} - Md = 0. \qquad \text{(B.14)}$$

The equations (B.12a), (B.12b), (B.12d) and (B.14) constitute a basis for further considerations.

Applying (7.128) and (B.12b) into (B.12a) the latter can be expressed in terms of control inputs generated by the upper loop as follows:

$$\frac{1}{k_c} M_\rho L(c^{n+1} - c^n) + Mx_*^n + \widetilde{B}^T p^n - Md = 0. \qquad \text{(B.15)}$$

Multiplying the equality (B.15) by $(x_*^n - x_*^{n+1})^T$ and utilizing that $\widetilde{B}((x_*^n -$

$x_*^{n+1}) = 0$ yields

$$\frac{-1}{k_c}(c^{n+1} - c^n)^T L_*^T M_\rho L(c^{n+1} - c^n) + (c^n - c^{n+1})^T L_*^T M x_*^n + \\ -(x_*^n - x_*^{n+1})^T M d = 0. \quad \text{(B.16)}$$

Note that owing to (B.14) the following holds:

$$\begin{aligned}(x_*^{n+1} - x_*^n)^T M d &= (x_*^{n+1} - x_*^n)^T (M\overline{x}_* + \widetilde{B}^T \overline{p}) \\ &= (x_*^{n+1} - x_*^n)^T M\overline{x}_* \\ &= (c^{n+1} - c^n)^T L_*^T M L_* (c^{opt})^T. \end{aligned} \quad \text{(B.17)}$$

Utilizing this in (B.16) and then completing the squares yields

$$\frac{-1}{k_c}(c^{n+1} - c^n)^T L_*^T M_\rho L(c^{n+1} - c^n) + (c^{opt} - c^n)^T L_*^T M L_*(c^{n+1} - c^n) = \\ \frac{-1}{k_c}(c^{n+1} - c^n)^T L_*^T M_\rho L(c^{n+1} - c^n) + \frac{1}{2}(c^n - c^{opt})^T L_*^T M L_*(c^n - c^{opt}) + \\ -\frac{1}{2}(c^{n+1} - c^{opt})^T L_*^T M L_*(c^{n+1} - c^{opt}) + \\ +\frac{1}{2}(c^{n+1} - c^n)^T L_*^T M L_*(c^{n+1} - c^n) = 0. \quad \text{(B.18)}$$

Since $B_* L_*(c^n - c^{opt}) = 0$ for any k then the above equality can be written in the form

$$\frac{1}{2k_c}(c^{n+1} - c^n)^T [k_c L_*^T M L_* - 2L_*^T M_\rho L](c^{n+1} - c^n) = \\ = \frac{1}{2}(c^{n+1} - c^{opt})^T L_*^T M_{*\gamma} L_*(c^{n+1} - c^{opt}) + \\ -\frac{1}{2}(c^n - c^{opt})^T L_*^T M_{*\gamma} L_*(c^n - c^{opt}) \quad \text{(B.19)}$$

where

$$M_{*\gamma} \triangleq M + \gamma B_*^T B_*, \quad \gamma > 0. \quad \text{(B.20)}$$

Due to the assumption (7.139) and because the matrix B_* is assumed to be of full rank, there exists such value of γ that (Luenberger, 1973)

$$M_{*\gamma} > 0. \quad \text{(B.21)}$$

Finally, let us define the function

$$T_c(c) \triangleq \frac{1}{2}(c - c^{opt})^T L_*^T M_{*\gamma} L_*(c - c^{opt}) \tag{B.22}$$

where γ is chosen to satisfy (B.20). The matrix L_*^T has full rank (see (7.143)). Hence the matrix is positively defined. According to (B.19) and (B.22) values of the function $T_c(\cdot)$ on the outer loop algorithm trajectory satisfy the following:

$$T_c(c^n) - T_c(c^{n+1}) = \frac{1}{2k_c}(c^{n+1} - c^n)^T[k_c L_*^T M L_* - 2L_*^T M_\rho L](c^{n+1} - c^n). \tag{B.23}$$

Therefore, according to assumption (*iii*) (see (7.144)), sequence $\{T_c(c^n)\}_{n\in\mathbb{N}}$ is decreasing and bounded by $T_c(c^0)$. Hence, the sequence $\{c^n\}_{n\in\mathbb{N}}$ is bounded, and consequently the sequences $\{x_*^n\}_{n\in\mathbb{N}}$ and $\{s^n\}_{n\in\mathbb{N}}$ are also bounded (see (B.12b)). The matrix B_* has full rank so that the matrix $\widetilde{B}$ also has full rank (see (7.138)). Therefore, the inverse $(\widetilde{B}M_\rho\widetilde{B}^T)^{-1}$ exists and (B.15) yields

$$p^n = (\widetilde{B}M_\rho\widetilde{B}^T)^{-1}[Md - \frac{1}{k_c}M_\rho L(c^{n+1} - c^n) + Mx_*^n] \tag{B.24}$$

which implies that the sequence $\{p^n\}_{n\in\mathbb{N}}$ is also bounded.

Thus, the sequence $\{x_*^n, s^n, p^n\}_{n\in\mathbb{N}}$ has at least one convergent subsequence. Due to (7.128), $s^n - x_*^n \to 0$ when $n \to \infty$. Hence, the equations (B.12a) and (B.12b) show that any a limit of any subsequence of the sequence $\{x_*^n, s^n, p^n\}_{n\in\mathbb{N}}$ satisfies (B.14) and (B.12d). It has been proved, however, that $\{c^{opt}, \overline{p}\}$ is the only single point to satisfy these equalities. Therefore, a proof of the assertion (*c*) and a proof of Theorem 7.1 have now been completed.

B.2 Proof of Theorem 7.2

In principle, the notation and definitions that are available in the proof of Theorem 7.1 will not be repeated here. The necessary and sufficient condition for s^n and p_1^n to be the solution of the inner loop problem, corresponding to prescribed outer loop values (x_*^n, p_2^n) of (x, p_2), can be written as

$$M_\rho(s^n - x_*^n) + Mx_*^n + B^T(p_1^n - p_2^n) + \widetilde{B}^T p_2^n - Md = 0 \tag{B.25}$$

and

$$B(s^n - x_*^n) = 0. \tag{B.26}$$

according to the assumptions that $M_\rho > 0$ and the matrix B has full rank. Therefore, there is a unique solution of the (B.25) and (B.26) with respect to (s^n, p_1^n). Hence, the iterative scheme (7.166) is well defined. Its convergence follows directly from the convergence properties properties of the IBM algorithm (Findeisen *et al.*, 1980) applied to the inner loop problem (7.149). Hence the assertions (i) and (ii) ot the Theorem (7.2) have now been proved. We shall now consider the outer loop algorithm described by (7.161) and (7.163). Conditions (B.25) and (B.26) constitute a basis for further considerations. Using (7.161) and (7.163) we can express (B.25) in terms of points generated by the outer loop (see also (7.142)) as

$$\frac{1}{k_c} M_\rho L(c^{n+1} - c^n) + M x_*^n + \frac{1}{k_{p2}} B^T(p_2^{n+1} - p_2^n) + \widetilde{B}^T p^n - Md = 0. \tag{B.27}$$

Using (B.14) and the equality $\widetilde{B}(x_*^n - \overline{x}_*) = 0$, where

$$\begin{aligned} \overline{x}_* &= ((c^{opt})^T, (u_*^{opt})^T)^T & \text{(B.28)} \\ u_*^{opt} &= HK_*(c^{opt}) & \text{(B.29)} \\ b_* &= -(I_u - HD_{*2})^{-1} H d_* & \text{(B.30)} \end{aligned}$$

we transform (B.27) to the form

$$\frac{1}{k_c} M_\rho L(c^{n+1} - c^n) + M_\gamma(x_*^n - \overline{x}_*) + \frac{1}{k_{p2}} B^T(p_2^{n+1} - p_2^n) + \widetilde{B}^T(p_2^n - \overline{p}) = 0. \tag{B.31}$$

Multiplying (B.31) by $(x_*^n - x_*^{n+1})^T$ and utilizing (7.143) we obtain

$$\begin{aligned} -\frac{1}{k_c}(c^{n+1} - c^n)^T L_*^T M_\rho L(c^{n+1} - c^n) + (c^{opt} - c^n)^T L_*^T M_\gamma L_*(c^{n+1} - c^n) + \\ +(\overline{p} - p_2^n)^T \widetilde{B} L_*(c^{n+1} - c^n) + \frac{1}{k_{p2}}(p_2^{n+1} - p_2^n)^T B L_*(c^n - c^{n+1}) = 0. \end{aligned} \tag{B.32}$$

The second term in B.32 can be expressed as

$$\begin{aligned} \frac{1}{2}(c^n - c^{opt})^T L_*^T M_\gamma L_*(c^n - c^{opt}) - \frac{1}{2}(c^{n+1} - c^{opt})^T L_*^T M_\gamma L_*(c^{n+1} - c^{opt}) + \\ + \frac{1}{2}(c^{n+1} - c^n)^T L_*^T M_\gamma L_*(c^{n+1} - c^n). \end{aligned} \tag{B.33}$$

The third term in (B.32) is equal to

$$(\overline{p}-p_2^n)^T\widetilde{B}L_*(c^{opt}-c^n)-(\overline{p}-p_2^{n+1})^T\widetilde{B}L_*(c^{opt}-c^{n+1})+$$
$$+(p_2^{n+1}-p_2^n)^T\widetilde{B}L_*(c^{n+1}-c^n)+(p_2^n-p_2^{n+1})^T\widetilde{B}L_*(c^{opt}-c^n). \qquad (B.34)$$

Let us now consider the fourth term in (B.34) starting from the term $\widetilde{B}L_*(c^{opt}-c^n)$. The equation (B.31) and the equation (B.14) from the proof of the previous theorem imply that

$$\overline{x}_*-x_*^n=\frac{1}{k_c}M_\gamma^{-1}M_\rho L(c^{n+1}-c^n)+\frac{1}{k_{p_2}}M_\gamma^{-1}B^T(p_2^{n+1}-p_2^n)+M_\gamma^{-1}\widetilde{B}^T(p_2^n-\overline{p}) \qquad (B.35)$$

and as $\overline{x}_*-x_*^n=L_*(c^{opt}-c^n)$ the following holds:

$$\widetilde{B}L_*(c^{opt}-c^n)=\frac{1}{k_c}\widetilde{B}M_\gamma^{-1}M_\rho L(c^{n+1}-c^n)+$$
$$+\frac{1}{k_{p_2}}\widetilde{B}M_\gamma^{-1}B^T(p_2^{n+1}-p_2^n)+\widetilde{B}M_\gamma^{-1}\widetilde{B}^T(p_2^n-\overline{p}). \qquad (B.36)$$

Hence

$$(p_2^n-p_2^{n+1})^T\widetilde{B}L_*(c^{opt}-c^n)=$$
$$=\frac{1}{k_c}(p_2^n-p_2^{n+1})^T\widetilde{B}M_\gamma^{-1}M_\rho L(c^{n+1}-c^n)+$$
$$+\frac{1}{k_{p_2}}(p_2^n-p_2^{n+1})^T\widetilde{B}M_\gamma^{-1}B^T(p_2^{n+1}-p_2^n)+$$
$$+(p_2^n-p_2^{n+1})^T\widetilde{B}M_\gamma^{-1}\widetilde{B}^T(p_2^n-\overline{p})$$
$$=-\frac{1}{k_c}(p_2^{n+1}-p_2^n)^T\widetilde{B}M_\gamma^{-1}M_\rho L(c^{n+1}-c^n)+$$
$$-\frac{1}{k_{p_2}}(p_2^{n+1}-p_2^n)^T\widetilde{B}M_\gamma^{-1}B^T(p_2^{n+1}-p_2^n)+$$
$$+\frac{1}{2}(p_2^n-\overline{p})^T\widetilde{B}M_\gamma^{-1}\widetilde{B}^T(p_2^n-\overline{p})+$$
$$-\frac{1}{2}(p_2^{n+1}-\overline{p})^T\widetilde{B}M_\gamma^{-1}\widetilde{B}^T(p_2^{n+1}-\overline{p})+$$
$$+\frac{1}{2}(p_2^{n+1}-p_2^n)^T\widetilde{B}M_\gamma^{-1}\widetilde{B}^T(p_2^{n+1}-p_2^n). \qquad (B.37)$$

Finally, let us define the function

$$T(c,p_2) \triangleq \frac{1}{2}[M_\gamma L_*(c-c^{opt})+\widetilde{B}^T(p_2-\overline{p})]^T M_\gamma^{-1}[M_\gamma L_*(c-c^{opt})+\widetilde{B}^T(p_2-\overline{p})]. \tag{B.38}$$

Applying (B.33), (B.34) and (B.37) to (B.32) yields:

$$\begin{aligned} T(c^n,p_2^n) - T(c^{n+1},p_2^{n+1}) = \\ (c^n - c^{n+1})^T[\frac{1}{k_c}L_*^T M_\rho L - \frac{1}{2}L_*^T M_\gamma L_*](c^n - c^{n+1}) + \\ +(p_2^n - p_2^{n+1})^T[\frac{1}{k_c}\widetilde{B}M_\gamma^{-1}M_\rho L + \frac{1}{k_p}BL_*](c^n - c^{n+1}) + \\ +(p_2^n - p_2^{n+1})^T[\frac{1}{k_p}\widetilde{B}M_\gamma^{-1}B^T - \frac{1}{2}\widetilde{B}M_\gamma^{-1}\widetilde{B}^T](p_2^n - p_2^{n+1}) \end{aligned} \tag{B.39}$$

or

$$T(c^n,p_2^n) - T(c^{n+1},p_2^{n+1}) = (c^n - c^{n+1})^T \times$$
$$\times \begin{bmatrix} \frac{1}{k_c}L_*^T M_\rho L - \frac{1}{2}L_*^T M_\gamma L_* & \frac{1}{2k_c}\widetilde{B}M_\gamma^{-1}M_\rho L + \frac{1}{2k_p}BL_* \\ \frac{1}{2k_c}\widetilde{B}M_\gamma^{-1}M_\rho L + \frac{1}{2k_p}BL_* & \frac{1}{k_p}\widetilde{B}M_\gamma^{-1}B^T - \frac{1}{2}\widetilde{B}M_\gamma^{-1}\widetilde{B}^T \end{bmatrix} (p_2^n - p_2^{n+1}). \tag{B.40}$$

Hence, due to the assumption (7.168) of Theorem 7.2 the sequence $\{T(c^n,p_2^n)\}_{n\in\mathbb{N}}$ is decreasing. Because $M_\gamma^{-1} > 0$, (B.38) implies that the sequence is bounded below. Hence, $\{T(c^n,p_2^n)\}_{n\in\mathbb{N}}$ converges and consequently $\{T(c^n,p_2^n) - T(c^{n+1},p_2^{n+1})\} \to 0$, as $k \to \infty$. Therefore, (B.40) implies that $\{(c^n - c^{n+1}, p_2^n - p_2^{n+1})\} \to 0$, as $k \to \infty$ and consequently $\{(x_*^n - x_*^{n+1}, p_2^n - p_2^{n+1})\} \to 0$, as $k \to \infty$, and (see (7.162) and (7.164))

$$\{(s^n - x_*^n, p_1^n - p_2^n)\} \to 0, \text{ as } k \to \infty. \tag{B.41}$$

Since $B_* x_*^n - b_* = 0$ then (see (7.137) and (7.138))

$$\widetilde{B}x_*^n - (I_u - HD_2)b_* = 0. \tag{B.42}$$

Equations (B.25) and (B.42) imply that the sequence $\{x_*^n, p_2^n\}$ converges to to a point $(\widehat{x}_*, \widehat{p}_*)$ satisfying

$$M\widehat{x}_* + \widetilde{B}^T\widehat{p}_* - Md = 0 \tag{B.43a}$$
$$\widetilde{B}\widehat{x}_* - (I_u - HD_2)b_* = 0 \text{ or } B_*\widehat{x}_* - b_* = 0. \tag{B.43b}$$

The conditions (B.43a) and (B.43b) determine a unique solution of *SOCP*. Hence, $\widehat{x}_* = c^{opt}$ and $\widehat{p}_* = [(I_u - HD_2)^{-1}]^T p^{opt}$, where the price vector p^{opt} is associated with the constraint $u - HK_*(c) = 0$ in the *SOCP* formulation.

A proof of Theorem 7.2 has now been completed.

B.3 Proof of Theorem 7.3

Applying (7.298a) and (7.298b) to (7.297) in order to eliminate output variable from the system performance index yields:

$$\begin{aligned} q(x,\alpha) &= \frac{1}{2}(x-d)^T M(x-d) + \frac{1}{2}(Dx-e)^T E(Dx-e) + \\ &\quad + P^T(\alpha)E(Dx-e) + P^T(\alpha)EP(\alpha) \\ q_*(x) &= \frac{1}{2}(x-d)^T M(x-d) + \frac{1}{2}(D_*x-f)^T E(D_*-f) \end{aligned} \tag{B.44}$$

where $f = e - d_*$. Also, (see (7.206), (7.298b) and (7.299))

$$\begin{aligned} G(x) &= u - HD_1c - HD_2u - HP(\alpha) \\ &= Bx - HP(\alpha). \end{aligned} \tag{B.45}$$

Similarly, combining ((7.205) and (7.298a)) yields (see (7.300))

$$G_*(x) = B_*x - b_* \tag{B.46}$$

where $b_* = (I - HD_{*2})^{-1}Hd_*$.

Let (v^n, w^n, p_1^n) be a solution of the inner loop problem under (c^n, u^n, p_2^n) prescribed by the outer loop. Let us denote $x \triangleq (c^T, u^T)^T$ and $s \triangleq (v^T, w^T)^T$. Hence, (s^n, p_1^n) is the inner loop problem solution and $(c^n, u^n, p_2^n) = (x^n, p_2^n)$. As $q(\cdot, \alpha^n)$ is convex and $G(\cdot, \alpha^n)$ is linear, then the necessary and sufficient condition for (s^n, p_1^n) to be the inner loop solution can be written as (see (7.300), (7.301) and (7.302)):

$$\widetilde{M}(s^n - x^n) + M_*x^n + B^T(p_1^n - p_2^n) + B_*^T p_2^n - Md - D_*^T Ef = 0 \tag{B.47a}$$

$$Bs^n - HP(\alpha^n) = 0. \tag{B.47b}$$

Derivation of the conditions (B.47a) and (B.47b) is straightforward but tedious. Hence, it is omitted. In the derivation it is convenient to utilize the expressions (7.208) and (7.209) to calculate the modifiers λ and t in the relaxed model based modified optimization problem (RMMOP, see (7.268)). The formulae (B.45) and (B.46) are useful in these calculations.

Notice that, since

$$G(x^n, \alpha^n) = G_*(x^n) \tag{B.48}$$

then

$$HP(\alpha^n) = Bx^n - B_* x^n + b_* \tag{B.49}$$

and condition (B.47b) can be written as

$$B(s^n - x^n) + B_* x^n - b_* = 0. \tag{B.50a}$$

According to the assumptions, $\widetilde{M} > 0$ and matrix B has full rank; therefore, since

$$\det \begin{bmatrix} \widetilde{M} & B^T \\ B & 0 \end{bmatrix} = (\det \widetilde{M}) \det(-B\widetilde{M}B^T) \tag{B.51}$$

then there is a unique solution of (B.47a) and (B.50a) with respect to (s^n, p_1^n). Hence, the generalized outer loop iterative schema (7.294a) and (7.294b) is well defined. A point $\overline{x}$ is a solution of $SOCP$ iff there is such $\overline{p}_2 \in \mathbb{R}^{n_u}$ that the following holds:

$$M_* \overline{x} + B_*^T \overline{p}_2 - Md - D_*^T Ef \;=\; 0 \tag{B.52a}$$

$$B_* \overline{x} - b_* \;=\; 0. \tag{B.52b}$$

Also, due to the assumptions, $M_* > 0$ and B_*has full rank; then similarly to the above there is exactly one pair $(\overline{x}, \overline{p}_2)$ satisfying (B.52a) and (B.52b). Now, using similar arguments as in the proof of Lemma (7.1), we complete the proof of part (a) of the Theorem 7.3. Conditions (B.47a) and (B.50a) constitute a basis in further considerations. Using (7.294a) and (7.294b) we express these conditions in terms of points generated by the outer loop as follows:

$$\widetilde{M} R_x^{-1}(x^{n+1} - x^n) + M_* x^n + B^T R_p^{-1}(p_2^{n+1} - p_2^n) + \\ + B_*^T p_2^k - Md - D_*^T Ef = 0 \tag{B.53a}$$

$$BR_x^{-1}(x^{n+1} - x^n) + B_* x^n - b_* = 0. \tag{B.53b}$$

Although the equations (B.53a) and (B.53b) are more general than those obtained by Cohen (1978) in a proof of his Theorem 5.1, they are similar in structure. We will utilize then a basic idea of Cohen's proof (Cohen, 1978) in our further considerations.

Multiplying the equality (B.53a) by $(x^k - x^{k+1})$ and the equality (B.53b) by $(p_2^{n+1} - p_2^n)^T$, adding the resulting equalities, and next employing assumption (ii), we obtain

$$\begin{aligned} -(x^{n+1} - x^n)\widetilde{M}R_x^{-1}(x^{n+1} - x^n) - (x^n)^T M_*(x^{n+1} - x^n) + \\ -(x^{n+1} - x^n)^T B_*^T p_2^n + (x^{n+1} - x^n)^T (Md - D_*^T Ef) + \\ +(x^n)^T B_*^T (p_2^{n+1} - p_2^n) - b_*^T (p_2^{n+1} - p_2^n) = 0. \end{aligned} \tag{B.54}$$

Notice that, due to (B.52a) and (B.52b), the following holds:

$$b_*^T (p_2^{n+1} - p_2^n) = \overline{x}^T B^T (p_2^{n+1} - p_2^n) \tag{B.55}$$

and

$$(x^{n+1} - x^n)^T (Md - D_*^T Ef) = (x^{n+1} - x^n)^T (M_* \overline{x} + B_*^T \overline{p}_2). \tag{B.56}$$

Applying these equalities to equality (B.54) and performing suitable ordering, we obtain

$$\begin{aligned} (\overline{x} - x^n)^T M_*(x^{n+1} - x^n) + (\overline{p}_2 - p_2^n)^T B_*(x^{n+1} - x^n) + \\ +(p_2^{n+1} - p_2^n)^T B_*(x^n - \overline{x}) - (x^{n+1} - x^n)^T \widetilde{M}R_x^{-1}(x^{n+1} - x^n) = 0. \end{aligned} \tag{B.57}$$

The first term in (B.57) can be expressed as

$$\begin{aligned} (\overline{x} - x^n)^T M_*(x^{n+1} - x^n) = \frac{1}{2}(x^n - \overline{x})^T M_*(x^n - \overline{x}) + \\ -\frac{1}{2}(x^{n+1} - \overline{x})^T M_*(x^{n+1} - \overline{x}) + \frac{1}{2}(x^{n+1} - x^n)^T M_*(x^{n+1} - x^n). \end{aligned} \tag{B.58a}$$

The second term in (B.57) may be written as

$$\begin{aligned} (\overline{p}_2 - p_2^n)^T B_*(x^{n+1} - x^n) = (\overline{p}_2 - p_2^n)^T B_*(\overline{x} - x^n) + \\ -(\overline{p}_2 - p_2^{n+1})^T B_*(\overline{x} - x^{n+1}) + (p_2^{n+1} - p_2^n)^T B_*(x^{n+1} - x^n) + \\ +(p_2^n - p_2^n)^T B_*(\overline{x} - x^n). \end{aligned} \tag{B.59}$$

Let us now compute the term $B_*(\overline{x} - x^n)$. Multiplying (B.52a) and (B.53a) through by M_*^{-1} gives

$$x^n + M_*^{-1}\widetilde{M}R_x^{-1}(x^{n+1} - x^n) + M_*^{-1}B^T R_p^{-1}(p_2^{n+1} - p_2^n) + \\ + M_*^{-1}B_*^T p^n - M_*^{-1}(Md + D_*^T Ef) = 0 \quad \text{(B.60)}$$

and

$$\overline{x} + M_*^{-1}B_*^T \overline{p}_2 - M_*^{-1}(Md + D_*^T Ef) = 0. \quad \text{(B.61)}$$

Hence, substracting the above equalities and multiplying the result by B_*, yields

$$B_*(\overline{x} - x^n) = B_* M_*^{-1}\widetilde{M}R_x^{-1}(x^{n+1} - x^n) + \\ + B_* M_*^{-1} B^T R_p^{-1}(p_2^{n+1} - p_2^n) + B_* M_*^{-1} B_*^T(p_2^n - \overline{p}_2) \quad \text{(B.62)}$$

and so

$$\begin{aligned} 2(p_2^n - p_2^{n+1})^T B_*(\overline{x} - x^n) &= 2(p_2^n - p_2^{n+1})^T B_* M_*^{-1}\widetilde{M}R_x^{-1}(x^{n+1} - x^n) + \\ &+ 2(p_2^n - p_2^{n+1})^T B_* M_*^{-1} B^T R_p^{-1}(p^{n+1} - p^n) + \\ &+ 2(p_2^n - p_2^{n+1})^T B_* M_*^{-1} B_*^T (p^n - \overline{p}_2) \\ &= -2(p_2^{n+1} - p_2^n)^T B_* M_*^{-1}\widetilde{M}R_x^{-1}(x^{n+1} - x^n) + \\ &- 2(p_2^{n+1} - p_2^n)^T B_* M_*^{-1} B^T R_p^{-1}(p^{n+1} - p^n) + \\ &+ (p_2^n - \overline{p}_2)^T B_* M_*^{-1} B_*^T (p^n - \overline{p}_2) + \\ &- (p_2^{n+1} - \overline{p}_2)^T B_* M_*^{-1} B_*^T (p^{n+1} - \overline{p}_2) + \\ &+ (p_2^{n+1} - p_2^n)^T B_* M_*^{-1} B_*^T (p^n - p_2^n). \end{aligned} \quad \text{(B.63)}$$

Finally, let us define the following function:

$$T(x, \overline{p}_2) \triangleq \frac{1}{2}(p_2 - \overline{p}_2)^T B_* M_*^{-1} B_*^T (p_2 - \overline{p}_2) + \\ + \frac{1}{2}[M_*(x - \overline{x}) + B_*^T(p_1 - \overline{p}_2)]^T M_*^{-1}[M_*(x - \overline{x}) + B_*^T(p_2 - \overline{p}_2)]. \quad \text{(B.64)}$$

Applying now (B.63), (B.59) and (B.58a) to (B.57) yields:

$$T(x^n, p_2^n) - T(x^{n+1}, p_2^{n+1}) = (x^n - x^{n+1})^T(\widetilde{M}R_x^{-1} - \frac{1}{2}M_*)(x^n - x^{n+1}) + \\ + 2(p_2^n - p_2^{n+1})(B_* M_*^{-1}\widetilde{M}R_x^{-1} - \frac{1}{2}B_*)(x^n - x^{n+1}) + \\ + (p_2^n - p_2^{n+1})^T(2B_* M_*^{-1} B^T R_p^{-1} - B_* M_*^{-1} B_*^T)(p_2^{n+1} - p_2^n). \quad \text{(B.65)}$$

Employing assumptions (*iii*) and (*iv*), and utilizing the identity

$$a^T X a + 2b^T Y a = (Xa + Y^T b)^T X^{-1}(Xa + Y^T b) - b^T Y X^{-1} Y^T b \quad \text{(B.66)}$$

where a and b are vectors while X and Y are matrices such that X^{-1} exists and $X = X^T$, we obtain from (B.65) that

$$\begin{aligned}
T(x^n, p_2^n) - T(x^{n+1}, p_2^{n+1}) = \\
& [(x^n - x^{n+1}) + M_*^{-1} B_*^T (p_2^n - p_2^{n+1})]^T \{\widetilde{M} R_x^{-1} + \\
& - \frac{1}{2} M_*\}[(x^n - x^{n+1}) + M_*^{-1} B_*^T (p_2^n - p_2^{n+1})] + \\
& + (p_2^n - p_2^{n+1})^T \{2B_* M_*^{-1} B^T R_p^{-1} + \\
& - B_* M_*^{-1} (\widetilde{M} R_x^{-1} + \frac{1}{2} M_*) M_*^{-1} B_*^T\} (p_2^n - p_2^{n+1}).
\end{aligned} \quad \text{(B.67)}$$

Let us consider an infinite sequence $\{x^n, p^n\}_{n \in \mathrm{N}}$ generated by the algorithm. It means that $(x^n, p^n) \notin \Omega$ fory every $n \in \mathrm{N}$ and, due to the assumptions $(ii), (iv)$ and (v), the sequence the sequence $\{T(x^n, p_2^n)\}_{n \in \mathrm{N}}$ is strictly decreasing (see (B.67)). Since $B_* M_*^{-1} B_*^T > 0$ and $M_*^{-1} > 0$, then (B.64) implies that the sequence is bounded below. Hence, $\{T(x^n, p_2^n)\}_{n \in \mathrm{N}}$ converges and consequently

$$T(x^n, p_2^n) - T(x^{n+1}, p_2^{n+1}) \to 0 \quad \text{as} \quad k \to \infty. \quad \text{(B.68)}$$

Therefore, (B.67) implies that $\{(x^n - x^{n+1}, p_2^n - p_2^{n+1})\}_{n \in \mathrm{N}}$ is convergent to zero. Thus, the equations (B.53a) and (B.53b) converge to the equations (B.52a) and (B.52b). As the solution $(\overline{x}, \overline{p}_2)$ of (B.52a) and (B.52b) is unique then

$$(x^n, p^n) \to (\overline{x}, \overline{p}_2) \quad \text{as} \quad k \to \infty \quad \text{(B.69)}$$

The proof of Theorem 7.3 has now been completed.

Bibliography

Arkun, Y. and Morgan, C. O. (1987). "Robustness analysis of distillation column control schemes using structured singular values", *Proc. of 1987 ACC, Minneapolis*, 319–325.

Bailey, W. N. (1981) *Hierarchical Control and Multilievel Optimization of a Reheat Furnace and Steel Reversing Mill,* PhD Thesis, City University (London), Control Engineering Centre.

Bamberger, W. and Isermann, R. (1978). "Adaptive on-line steady-state optimization of slow dynamic processes", *Automatica* **14**, 223–230.

Banyasz, C., Haber, R. and Keviczky, L. (1973). "Some estimation methods for nonlinear discrete time identification", *Proc. of 3rd IFAC Symposium on Identification and System Parameter Estimation,* 793–802.

Bertsekas, D. P. (1982) *Constrained Optimization and Lagrange Multiplier Methods,* Academic Press, New York.

Bertsekas, D. P. (1997) *Nonlinear Programming*, Athena Scientific, Belmont, Massachusetts.

Brdyś, M. and Michalak, P. (1978a). "On-line coordination with local feedback for steady-state systems", *Arch. Automat. Telemech.,* **23**, (4), 403–422.

Brdyś, M. and Ulanicki, B. (1978b) "On the completely decentralized control with local feedback in large-scale systems", *Arch. Automat. Telemech.,* **23**, (12), 21–36.

Brdyś, M. and Malinowski, K. (1979). "Problems in design of on-line optimizing control for large-scale processes", *Proc., JACC,* Denver.

Brdyś, M., Findeisen, W. and Tatjewski, P. (1980a). "Hierarchical control for systems operating in steady-state". *Large Scale Systems.* **1**, (3), 193–213.

Brdyś, M. (1980b). *Hierarchical Control of Complex Systems Operating in Time-Varying Steady-State*, Warsaw University of Technology Publications, Warsaw.

Brdyś, M., Michalak, P. and Ulanicki, B. (1982). "Optimizing control of large-scale systems under time-varying disturbances by price mechanism with local feedback", *Large Scale Systems.* **3**, (1), 123–142.

Brdyś, M. (1984) "Hierarchical optimizing control of steady-state large-scale systems under model-reality differences of mixed type - a mutually interacting approach", *Proc. 3rd IFAC Symp. on LSS, Warsaw 1983* , Pergamon Press, 49–57.

Brdyś, M. and Ruszczynski, A. (1985). *Optimization Techniques*, WNT, Warsaw (in Polish).

Brdyś, M., Chen, S. and Roberts, P. D. (1986) "An extension to the modified two-step algorithm for steady-state system ptimization and parameter estimation", *Int. J. Systems Sci.* **17**, (8), 1229–1243.

Brdyś, M. and Roberts, P. D. (1986) "Optimal structures for steady-state optimising control of large-scale industrial processes", *Int. J. Systems Sci.* **17**, (10), 1449–1474.

Brdyś, M. and Roberts, P. D. (1987) "Convergence and optimality of modified two-step algorithm for integrated system optimization and parameter estimation", *Int. J. Systems Sci.* **18**, (7), 1305–1322.

Brdyś, M., Ellis, J. E. and Roberts, P. D. (1987) "Augmented integrated system optimization and parameter estimation technique: derivation, optimality and convergence", *IEE Proceedings-D* **134**, (3), 201–209.

Brdyś, M., Roberts, P. D., Badi, M. M., Kokkinos, I. C. and Abdullah, N. (1989). "Double loop iterative strategies for hierarchical control of industrial processes", *Automatica,* **25**, (5), 743–751.

Brdyś, M., Abdullah, N. and Roberts, P. D. (1990a). "Augmented model-based ouble loop techniques for hierarchical control of complex industrial processes", *Int. J. Control,***52**, (3), (549–570).

Brdyś, M., Abdullah, N. and Roberts, P. D. (1990b). "Hierarchical adaptive techniques for optimizing control of large-scale steady-state systems: optimality, iterative strategies, and their convergence", *IMA Journal of Mathematical Control&Information,* **7**, (199–233).

Brdyś, M. and Tatjewski, P. (1994) "An algorithm for steady-state optimizing dual control of uncertain plants", *Prepr. 1st IFAC Workshop on New Trends in Design of Control Systems, Smolenice*, 249–254.

Brdyś, M. (1999). "Robust estimation of variables and parameters in dynamic networks", *Proc. of the 14th IFAC World Congress*, Beijing, P. R. China, 5–9 July.

Brdyś, M. A. and Chang, T. (2001). "Robust model predictive control under output constraints", *Proc. of the 15th IFAC World Congress*, Barcelona, July.

Brdyś, M. A. and Chang, T. (2002). "Modelling for control of quality in drinking water distribution systems", *1st Annual Environmental & Water Resources Systems Analysis (EWRSA) Symposium, A.S.C.E. Environmental & Water Resources Institute (EWRI) Annual Conference*, Roanoke, Virginia, USA, May 19–22.

Brdyś, M. A., Grochowski, M., Duzinkiewicz, K., Chotkowski, W. and Liu, Y.

(2002b). "Design of control structure for integrated wastewater treatment plant- sewer systems", *Proc., 1st International Conference on Technology, Automation and Control of Wastewater and Drinking Water Systems-TiASWiK'02*, Gdansk-Sobieszewo, Poland, June, 19–26.

Brdyś, M. A., Grochowski, M. and Konarczak,K. (2004). "Estimation of wastewater treatment plant state for model predictive control of n-p removal at medium time scale". *Prepr. 10th IFAC/IFORS/IMACS/IFIP Symposium on Large Scale Complex Systems*, Osaka, Japan, July, 283–289.

Byrski, W., Duda, J., Gajek, J. and Turnau, A. (1987) "Computer on-line control system for a rectification process", *Pomiary, Automatyka, Kontrola* **12**, 281–284 (in Polish).

Byrski, W., Duda, J. and Turnau, A. (1988) "Problems of on-line control of multistage distillation", *Proc. X National Control Conference, Lublin*, 69–77.

Byrski, W., and Duda, J. (1990) "Modelling of distillation columns for control purposes", *Int. J. Systems Sci.* **21**, (4), 21–31.

Caffin, R. N., (1972). *Design and Modelling of an Experimental Travelling Load Furnace for Computer Control,* PhD Thesis, City University (London), Control Engineering Centre.

Chang, T, Brdyś, M. A. and Duzinkiewicz, K. (2003). "Quantifying uncertainties for chlorine residual control in drinking water distribution systems", *World Water & Environmental Resources Congress - EWRI2003,* June 22–26, Philadelphia, USA.

Chang, T, Duzinkiewicz, K. and Brdyś, M. A. (2004). "Bounding approach to parameter estimation of point-parametric models", *10th IFAC/IFORS/IMACS/IFIP Symposium on Large Scale Complex Systems,* Osaka, Japan, July 26–28.

Cohen, G. (1978). "Optimization by decomposition and coordination: a unified approach", *IEEE Trans. on Automatic Control* **23**, 222–232.

Dennis, J. E. and Schanbel, R. B. (1983). *Numerical Methods for Unconstrained Optimisation and Nonlinear Equations,* Prentice Hall, Series in Computational Mathematics.

Duda, J. (1991) *Some Problems of Synthesis of the Supervisory Computer Control Algorithms for Slow-Varying Continuous Industrial Processes*, Scientific Publications of the Academy of Mining and Metallurgy, no.56 (Automatics), Cracow, (in Polish).

Duda J., Brdyś, M. A., Tatjewski, P. and Byrski, W. (1995) "Multilayer decomposition for optimizing control of technological processes", *Prepr. 7th IFAC Symposium Large Scale Systems: Theory and Applications, London*, Pergamon, 111–116.

Ellis, J. E., and Roberts, P. D. (1978). "Mathematical models of a simple chemical vaporiser", *UKSC Conf. Computer Simulation,* Chester, UK.

Feldbaum, A.A. (1965). *Optimal Control Systems*, Academic Press, New York

Findeisen, W. (1974). *Multilevel Control Systems* (in Polish), PWN, Warsaw. German translation: *Hierarchische Steuerungssyteme*, Verlag Technik, Berlin, 1997.

Findeisen, W., Brdyś, M., Malinowski, K., Tatjewski, P. and Woźniak, A. (1978). "On-line hierarchical control for steady-state systems", *IEEE Trans. on Automatic Control* **23**, 189–209.

Findeisen, W., Bailey, F. N., Brdyś, M., Malinowski, K., Tatjewski, P. and Woźniak, A. (1980) *Control and Coordination in Hierarchical Systems*, J. Wiley & Sons, Chichester.

Findeisen, W., Brdyś, M. A. and Malinowski, K. (1994). "Control problems and control systems. An overview", In: *Computer Aided Control Systems Design*, Brdyś, M. A. and Malinowski, K. (**Eds**.), World Scientific Publisher, 1–15.

Fletcher, R. (1987) *Practical Methods of Optimization*. J. Wiley & Sons, Chichester.

Gmiński, T., Grochowski, M., Brdys, M. and Drewa, M. (2004). "Model predictive controller for integrated wastewater systems", *Prepr. 10th IFAC/IFORS/IMACS/IFIP Symposium on Large Scale Complex Systems*, Osaka, Japan, July, 593–598.

Grochowski, M. (2004). *Intelligent Control of Integrated Wastewater Treatment Systems Under Full Range of Operating Conditions*, PhD Thesis, Gdansk University of Technology.

Grochowski, M., Brdyś, M. A. and Gmiński, T. (2004a). "Intelligent control structure for integrated wastewater systems", *Prepr. 10th IFAC/IFORS/IMACS/IFIP Symposium on Large Scale Complex Systems*, Osaka, Japan, July, 263–268.

Grochowski, M., Brdyś, M. A., Gmiński, T. and Deinrych, P. (2004b). "Softly switched model predictive control for control of integrated wastewater treatment systems at medium time scale", *Prepr. 10th IFAC/IFORS/IMACS/IFIP Symposium on Large Scale Complex Systems*, Osaka, Japan, July, 269–274.

Goldberg, D. E. (1989) *Genetic Algotithms in Search, Optimization and Machine Learning*, Addison-Wesley, Reading.

Goodwin, G. C., Graebe, S. F. and Salgado,. M.E. (2001) *Control System Design*, Prentice Hall, Upper Saddle River.

Hogan, W.W. (1973). "Point-to-set maps in mathematical programming", SIAM Rev, **15**, (3), 591–603.

Ingber, L. and Rosen, B. (1992) "Genetic algorithms and very fast simulated reannealing: a comparison", *Mathl. Comput. Modelling* **16**, (11), 87–100.

Kantorovich, L.and Akilov, G. (1963). *Functional Analysis in Normed Spaces*, Pergamon Press, Oxford.

Kim, H., Hao, O. J. and McAvoy, T. J. (2001). "SBR system for phosphorus removal: ASM2 and simplified linear model", *J. Env. Engng ASCE*, 127, 98–104.

Konarczak, K., Rutkowski, T., Brdyś, M. A. and Gmiński, T. (2004),

"Weighted least squares parameter estimation for model predictive control of integrated wastewater systems at medium time scale", *Prepr. 10th IFAC/IFORS/IMACS/IFIP Symposium on Large Scale Complex Systems*, Osaka, Japan, July, 290–296.

Lasdon, L. S. (1970) *Optimization Theory for Large Systems*, MacMillan, New York.

Leontaritis, I. J. and Billings, S. A. "Input-output parametric models for non-linear systems. Part I: Deterministic non-linear systems; Part II: Stochastic non-linear systems", *Int. J. Control*, **41**, 303–344.

Lewis, F. L., Syrmos, V. L. (1995). *Optimal Control,* Wiley, New York, USA.

Lin, J., Chen, S. and Roberts, P. D. (1988) "Modified algorithm for steady-state integrated system optimization and parameter estimation", *IEE Proceedings-D* **135**, (2), 119–126.

Lin, J., Han, C., Roberts, P. D. and Wan, B. (1989) "New approach to stochastic optimizing control of steady-state systems using dynamic information", *Int. J. Control* **50**, (6), 2205–2235.

Lin, J., Wang, M. and Roberts, P. D. (1990) "Improvements in the formulation and solution approach for stochastic optimizing control of steady-state industrial processes", *Int. J. Control* **52**, (3), 517–548.

Lin, J., Gionas, I. and Roberts, P. D. (1992) "Stability of steady-state optimizing control systems", *IEE Proceedings-D* **139**, (6), 481–494.

Liu, Z. Y. and Roberts, P. D. (1989) "Non-derivative algorithm for optimization and control of steady-state systems", *Int. J. Systems Sci.* **20**, (8), 1483–1479.

Ljung, L. and Soderstrom, T. (1987). *Theory and Practice of Recursive Identification,* Massachusetts Institute of Technology Press, Cambridge, Massachusetts, USA, London, England.

Luenberger, D. G. (1984) *Linear and Nonlinear Programming,* Addison-Wesley, Reading.

Maciejowski, J. M. (2002) *Predictive Control, with constraints,* Prentice Hall, Harlow, England.

Mayne, D., Rawlings, J. B., Rao, C.V, and Sckokaert P.O.M. (2000). "Constrained model predictive control: stability, and optimality", *Automatica,* **36**, 789–814.

Morari, M. and Lee, J. H. (1999). "Model predictive control: past, present and future", *Computers and Chemical Engineering,* **23**, 667–682.

Olsson, G. and Newell, R. (1999). *Wastewater Treatment Systems. Modelling, Diagnosis and Control,* IWA Publishing, London.

Ortega, J. M. and Rheinboldt, W. C. (1970) *Iterative Solution of Nonlinear Equations in Several Variables*, Academic Press, New York.

Piotrowski, R., Duzinkiewicz, K. and Brdyś, M. A. (2004). "Dissolved oxygen tracking and control of blowers at fast time scale", *Prepr. 10th IFAC/IFORS/IMACS/IFIP Symposium on Large Scale Complex Systems*, Osaka, Japan, July, 297–303.

Roberts, P. D. (1979) "An algorithm for steady-state system optimization and parameter estimation", *Int. J. System Science* **10**, 719–734.

Roberts, P. D. and Williams, T. W. C. (1981). "On an algorithm for combined system optimization and parameter estimation", *Automatica* **17**, (4), 455–472.

Rosen, J. B. (1960). "The gradient projection method for non-linear programming, Part 1, Linear Constraints", *J. Soc. Indust. Appl. Math.*, **8**, (1), 181–217.

Rutkowski, T., Brdys, M. A., Konarczak, K. and Gmiński, T. (2004). "Set-bounded joined parameter and state estimation for model predictive control of integrated wastewater treatment plant systems at medium time scale", *10th IFAC/IFORS/IMACS/IFIP Symposium on Large Scale Complex Systems,* Osaka, Japan, July 26–28.

Rutkowski, T. (2004). *Grey-Box Model Based State and Parameter Estimation for Model Predictive Control of Integrated Wastewater Systems*, PhD thesis (in Polish), Gdansk University of Technology, PhD Thesis, Gdansk University of Technology.

Sage, A. P. (1968). *Optimum Systems Control,* Prentice Hall, Electrical Engineering Series, England.

Schwefel, H. P. (1981) *Numerical Optimization of Computer Models*, J. Wiley & Sons, Chichester.

Schweppe,F. C. (1973). *Uncertain Dynamic Systems, Prentice Hall Inc., USA.*

Shao, F.Q., and Roberts, P.D. (1983) "A price correction mechanism with global feedback for hierarchical control of steady-state systems", *Large Scale Systems* **4**, 67–80.

Sheena, H. (1977). *Computer Control of a Travelling Load Furnace,* PhD Thesis, City University (London), Control Engineering Centre.

Skogestad, S., and Morari, M. (1987). "Understanding the dynamic behaviour of distillation columns", *IEC Res.* **28**, (10), 1848–1862.

Stephanopoulos, G., and Westerberg, A. W. (1975). "The use of Hestenes' method of multipliers to resolve dual gaps in engineering system optimization", *Journal of Optimization Theory and Applications*, **15**, 285–309.

Stevenson, I. A., Wadwani, D. S., Ellis, J. E., Li, C. W., and Roberts, P. D. (1984). "A distributed computer network and its application to systems control", *PPL Conf. Publ.* 22, 189–193.

Stevenson, I. A., Brdys, M. A., and Roberts, P. D. (1985). "Integrated system optimization and parameter estimation for travelling load furnace control", *7th IFAC/IFORS Symp. Identification and Parameter Estimation,* University of York, England, 3–7 July.

Stevenson, I. A. (1985). *On-line adaptive computer control of a travelling load furnace,* PhD Thesis, City University (London), Control Engineering Centre.

Tadej, W. (2001). *Analysis of the dual algorithm for set-point optimization of nonlinear processes*, PhD Thesis, Warsaw University of Technology, Faculty of Electronics and Information Technology (in Polish).

Tadej, W. and Tatjewski, P. (2001). "Analysis of an ISOPE-type dual algorithm for optimizing control and nonlinear optimization", *Int. J. of Appl. Math. and Computer Sci.* **11**, (2), 429–457.

Tatjewski, P. (1985). "On-line hierarchical control of steady-state systems using augmented interaction balance method with feedback", *Large Scale Systems,* **8,** 1–18.

Tatjewski, P. (1986). "New dual-type decomposition algorithms for nonconvex separable optimization problems", *4th IFAC Symp. Large Scale Systems Theory and Applications*, Zurich, Switzerland, 296–303.

Tatjewski, P. and Roberts, P. D. (1987). "Newton-like algorithm for integrated system optimization and parameter estimation technique", *Int. J. Control* **46**, (4), 1155–1170.

Tatjewski, P. (1988). *Hierarchical Methods for Solving Nonconvex Problems of Optimization and Steady-State Control for Complex Processes*, Warsaw University of Technology Publications, No.50 (Electronics), Warsaw.

Tatjewski, P. (1989). "New dual type decomposition algorithm for nonconvex separable optimization problems", *Automatica,* **25,** (2), 233–242.

Tatjewski, P., Brdyś, M. A. and Duda, J. (1995). "Optimizing control of uncertain plants with feedback controlled output constraints", *Prepr. 7th IFAC Symp. Large Scale Systems: Theory and Applications , London*, Pergamon, 123–128.

Tatjewski, P. and Duda, J. (1997). "Process models for optimising control of uncertain plants", *Proc. 4th Int. Symp. Methods and Models in Automation and Robotics, Międzyzdroje*, 135–140.

Tatjewski, P. (1998). "Two-phase dual-type optimising control algorithm for uncertain plants", *Proc. 5th Int. Symp. Methods and Models in Automation and Robotics, Miedzyzdroje*, 171–176.

Tatjewski, P. (1999). "Two-phase dual-type optimising control algorithm for uncertain plants with active output constraints", *European Control Conference ECC'99, Karlsruhe*, paper F0347 (CD-ROM).

Tatjewski, P., Brdyś M. A. and Duda, J. (2001). "Optimizing control of uncertain plants with constrained, feedback controlled outputs", *Int. J. Control* **74**, (15), 1510–1526.

Tatjewski, P. (2002a). "Iterative set-point optimizing control – the basic principle redesigned", *Proc. 15th IFAC World Congress, Barcelona*, paper No. 246 (CD ROM).

Tatjewski, P. (2002b). *Advanced Control of Industrial Processes, Structures and Algorithms,* Academic Publishers EXIT, Warsaw, (in Polish, English translation in preparation).

Yen, J. and Langari, R. (1999). *Fuzzy Logic: Intelligence, Control and Information,* Prentice Hall Inc., Upper Saddle River, New Jersey.

Walter, É., Pronzato, L. (1997). *Identification of Parametric Models from Experimental Data,* Springer-Verlag, Berlin Heidelberg, New York Masson, Paris,

Milan, Barcelone.
Wismer, D. A. (1971). *Optimization Methods for Large-Scale Problems,* McGraw-Hill, New York.
Zangwill, W.I. (1969). *Nonlinear Programming: a Unified Approach*, Prentice Hall, Englewood Cliffs, N.J.
Zhang, H. and Roberts, P. D. (1990). "On-line steady-state optimization of nonlinear constrained processes with slow dynamics", *Trans. Inst. MC* **12**, (5), 251–261.

Index